The Benthic
Boundary Layer

The Benthic Boundary Layer

Edited by

I. N. McCave
University of East Anglia
Norwich, England

PLENUM PRESS • NEW YORK AND LONDON

Library of Congress Cataloging in Publication Data

Nato Science Committee Conference on the Benthic Boundary Layer, Les Arcs,
France, 1974
The benthic boundary layer.

Includes bibliographies and index.
1. Ocean bottom – Congresses. 2. Benthos – Congresses. I. McCave, I. N.,
1941-   II. North Atlantic Treaty Organization. Science Committee. III. Title.
GC87.N37 1974   551.4'607   76-2641

ISBN-13 : 978-1-4615-8749-1   e-ISBN : 978-1-4615-8747-7
DOI : 10.1007/978-1-4615-8747-7

Proceedings of the NATO Science Committee Conference
on the Benthic Boundary Layer held at Les Arcs, France,
November 4-9, 1974

© 1976 Plenum Press, New York
A Division of Plenum Publishing Corporation
227 West 17th Street, New York, N.Y. 10011

United Kingdom edition published by Plenum Press, London
A Division of Plenum Publishing Company, Ltd.
Davis House (4th Floor), 8 Scrubs Lane, Harlesden, London, NW10 6SE, England

Foreword

The Conference on the Benthic Boundary Layer was held under the auspices of the NATO Science Committee as part of its continuing effort to promote the useful progress of science through international cooperation.

Science Committee Conferences are deliberately designed to focus attention on unsolved problems, with carefully selected participants invited to provide complementary expertise from a variety of relevant disciplines. Through intensive discussion in small groups they seek to reach a consensus on assessments and recommendations for future research emphasis, which it is hoped will be of value to the larger scientific community. The subjects treated over the past few years have been as varied as science itself—e.g., computer software engineering, chemical catalysis, and materials and energy research.

The present effort evolved from informal discussions between marine geologists, chemists, and biologists which underlined the desirability of improved communication among those concerned with the benthic layer. In both scientific and technological terms this is an exciting frontier, rich in promise but poorly understood at present. It is particularly striking to realize that there are at least as many definitions of the benthic layer as there are disciplines involved, and it seemed clear that there was much to be gained by a detailed exchange of views on research capabilities, trends, and priorities. The results of the meeting appear to have confirmed the hopes of the sponsors.

Each participant received in advance a collection of working papers—both published and new—which enabled plenary lectures to be limited to the minimum number necessary for an overview of the status of benthic chemistry, physics, biology, geology, and engineering. This format allowed rapid initiation of intensive interdisciplinary discussions. All participants gave generously and enthusiastically of their wisdom and knowledge during the week of the meeting—always long past normal working hours—and we extend to them our deep gratitude.

Special thanks are due to Dr. I. N. McCave, for his diligent efforts as Chairman of the meeting, to his colleagues on the Organizing Committee—Professor K. F. Bowden (UK), Dr. S. E. Calvert (UK), Professor E. D. Goldberg (USA),

Dr. E. L. Mills (Canada), Professor H. Postma (Netherlands), and Professor E. Seibold (Germany)—for their wise counsel, and to the leaders and rapporteurs of the working groups, as listed, for their indispensable dedication.

Eugene G. Kovach
Deputy Assistant Secretary General
for Scientific Affairs
NATO

Preface

This book is the offspring of the Benthic Boundary Layer Conference supported by the NATO Science Committee and held at Les Arcs in the French Alps November 4-9, 1974. The success of the organization and smooth running of the conference owed much to the efforts of Dr. E. G. Kovach, Deputy Assistant Secretary General for Scientific Affairs of NATO, and his staff.

The aim of the conference was to take stock of our knowledge of the sea-bed region and to look for those areas where critically directed research might rapidly advance the subject. During the first day and a half, five plenary lectures were given by M. Wimbush (physics), R. A. Berner (chemistry), P. de Wilde (biology), C. D. Hollister (geology), and R. B. Krone (engineering). These acquainted everyone with the basic concepts and problems in these fields, so that some common background could be assumed for all participants. I want to thank the plenary lecturers for the work they put into preparing their lectures.

Most of the work of this conference took place within the six working groups whose reports are given here. The members of those groups worked long and hard, often into the night, and I would like to thank them, and particularly their chairmen, for their efforts. They have provided a summary of views and questions concerning a region which is of considerable importance but is largely unknown.

It is unknown because the bottom of the deep ocean provides the severest test of instrumentation, a test which most pieces of apparatus initially fail. There are of course problems of pressure, but equally important is the problem of sensor size. Close to the boundary a logarithmic sampling grid is appropriate, but designing sensors to measure some physical or chemical parameter at 1, 2, 4, 8, and 16 cm above or below the sediment–water interface, at a depth of 5000 m and for a period of weeks to months, has proved very difficult. Most oceanographic measurements have been made from the reference plane of the sea surface. Boundary layer work requires the deployment of free vehicles designed to sit on the sea bed and provided with sufficient power to operate sensors and data-loggers over a long period of time, to take measurements at positions defined by reference to the sediment–water interface. The engineering problems of abyssal boundary layer research will have to be solved before we can get down to many of the scientific questions.

Two particular facets of our ignorance recur in the working group reports. First, there has not been a single good measurement of the physical properties of the flow near the bed of the deep sea. In particular, velocity (including turbulence) and other gradient measurements should be made at one or two sites over a long period of time. Until we have this information, precise questions about physical processes in the abyssal environment cannot be formulated. This information is necessary not only for physical but also for all other aspects of boundary layer research. Second, the chemistry and biology of organic matter enters so many realms of boundary layer work—metal–organic complexes, nutrient regeneration, available calorific value, sediment binding by mucus (all imperfectly understood)—that investigation of the role of organic matter in benthic boundary layer processes should also have high priority.

I hope the reports and papers contained here will prove both timely and useful to workers and intended workers in the field.

Norwich   Nicholas McCave

Contents

Chapter 8
Marine Geotechnology: Average Sediment Properties, Selected Literature
   and Review of Consolidation, Stability, and Bioturbation-
   Geotechnical Interactions in the Benthic Layer. 157
   Adrian F. Richards and James M. Parks

Chapter 9
Flow Phenomena in the Benthic Boundary Layer and Bed Forms
   beneath Deep-Current Systems. 183
   C. D. Hollister, J. B. Southard, R. D. Flood, and P. F. Lonsdale

Part II ● WORKING GROUP REPORTS

I. PAPERS

1
The Physics of the Benthic Boundary Layer

MARK WIMBUSH

Physical laws governing profiles of velocity, temperature, and dissolved substance in the benthic boundary layer are discussed (some measurements of highly unstable temperature profiles near the deep-sea floor remain unexplained). Deep-sea and shelf boundary layers are compared. The effects of fluctuations with a wide range of time scales are described.

This survey begins with a discussion of the mean temperature structure of the benthic boundary layer. This may seem perverse to the nonphysicist (for whom the survey is intended), since temperature variation in the boundary layer is only a minute fraction of a degree. The reason for starting with this topic is partly historical and partly because temperature measurements have been the most puzzling; also the initial temperature discussion provides a peg on which to hang the rest of the physics.

Around the middle of the nineteenth century, Georges Aimée, a Frenchman, lowered thermometers into the sea. He observed that temperature varied seasonally and decreased with depth down to 300–400 m (the depth limit of his instruments). Below this, he expected temperature to increase with depth because of the upward geothermal heat flux H from the solid earth beneath.

We now know that, independent of H, the temperature does eventually (at several kilometers) increase with depth because of the adiabatic compression of water. The temperature gradient associated with this effect (analogous to the "lapse rate" in the atmosphere) is $\Gamma \approx 10^{-6}$ °C cm^{-1}. But as the sea floor is approached, the upward geothermal heat flux H should increase this temperature

MARK WIMBUSH ● Physical Oceanographic Laboratory, Nova University, Dania, Florida, U.S.A.

gradient (as Aimée predicted). To what elevation z above bottom is the effect of H appreciable?

A century after Georges Aimée's measurements, Sir Edward Bullard in England developed a probe to measure temperature gradient and conductivity (and hence heat flux) in submarine sediments. Temperature gradients measured with this device in deep-sea sediments are typically $dT/dz \approx 10^3 \Gamma$. Multiplying this by typical sediment conductivity $c \approx 10^{-3}$ cal cm^{-1} s^{-1} (°C)$^{-1}$ gives heat flux $H \approx 10^{-6}$ cal cm^{-2} s^{-1}.

Molecular conductivity of sea water and sediment are of the same order ($c_{\text{water}} \approx \frac{1}{2} c_{\text{sediment}}$), so the conductive temperature gradient in the water is approximately the same as in the sediment, $10^3 \Gamma$. However, a gradient of this magnitude will exist only in a thin layer just above the bottom. Critical Rayleigh number considerations indicate that this conductive layer extends, in the absence of shear, to an elevation $z \approx$ a few centimeters. Above this, the gradient is actively reduced by free convection. The presence of shear further attenuates the thickness of this conductive layer: typical tidal currents ($U \approx$ a few centimeters per second) produce shear stress ($\tau \approx 10^{-2}$ dyn cm^{-2}) that will allow a conductive layer only up to $z_\gamma \approx 1$ cm. Above this, the effects of shear turbulence dominate the heat transfer process. (Reynolds number considerations for the entire dynamical boundary layer indicate that the layer is indeed turbulent.)

In the lower part of the benthic boundary layer ($z <$ a meter or two), usually:

1. Shear forces $\gg$ buoyancy forces (due to temperature gradients); $\therefore$ temperature field is dynamically passive.
2. Shear forces $\gg$ Coriolis forces (due to earth's rotation); $\therefore$ structure is similar to the "wall region" of a nonrotating boundary layer.
3. Longest time scale $\approx z/u_* < 1$ h $\ll \frac{1}{2}$ day (friction velocity $u_* \equiv \sqrt{\tau/\rho}$ is a convenient kinematic representation of the bottom shear stress τ; $\frac{1}{2}$ day is the shortest tidal period); $\therefore$ steady-state boundary layer theory applies.

So by analogy with the homogeneous, nonrotating, steady, turbulent boundary layer (i.e., the simplest laboratory turbulent boundary layer on a flat plate), we may expect logarithmic profiles in both current velocity and temperature near the ocean floor (see Fig. 1).

To use the logarithmic relation (1') in Fig. 1 to predict the temperature gradient, u_* and z_l must be determined. This may be done as follows: measure $\bar{U}$ at a number of levels z, then assume these levels to be within the logarithmic layer and use equation (1) to calculate u_* in each case. For pairs of levels that are both truly within the logarithmic layer, these calculated u_* values should agree, but u_* calculated in this way from a level $z > z_l$ should be too large. Figure 2 shows plots of four pairs of u_* values computed in this way from 4-min

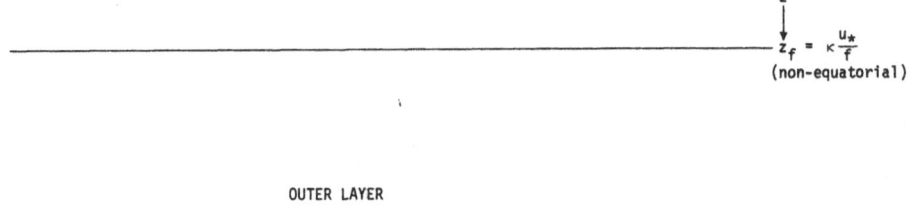

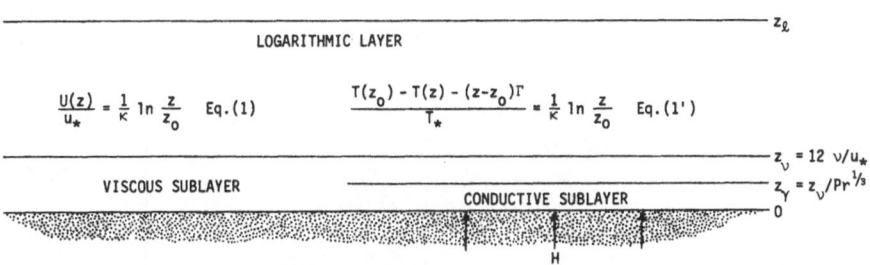

Fig. 1. Dynamical and thermal structure of the benthic boundary layer. f, Coriolis parameter; $\kappa = 0.4$, von Kármán constant; $T_* = H/c_p \rho u_*$, scaling temperature; c_p, specific heat; ρ, density; ν, molecular kinematic viscosity; γ, molecular thermal diffusivity; $P_r = \nu/\gamma$, Prandtl number; and $z_0 = 0.11 \, \nu/u_*$ (for smooth bottom). (In the case of a rough bottom with characteristic roughness size d of order z_γ, z_ν, or larger, the sublayers are broken up. For $d \gg z_\nu$, $z_0 \approx d/30$; for intermediate d, see appropriate formulas in Weatherly [1972].)

averages of current speed measured at levels above the deep-sea floor in the range $z = 7.5$–234 cm. It appears from this figure that z_l lies between 100 cm and 160 cm, and u_* fluctuates from 0.03 to 0.3 cm/s. Using these results, we find that at $z = 100$ cm $(1')$ applies and (on differentiation) gives $|d\tau/dz| < 2\Gamma$. This simple theory predicts that the magnitude of the temperature gradient should decrease from 1000Γ at $z = 0$ to less than 2Γ at $z = 100$ cm (and ultimately to Γ for large z). The influence of H on the temperature gradient should therefore be appreciable only in the lowest meter of the water column.

Many records [including a carefully conducted measurement by the author (Wimbush, 1970)] are consistent with the theory outlined above. However, since 1960, several investigators have reported "hyperadiabatic" gradients (10Γ to 1000Γ) extending to elevations of meters or even tens of meters above the deep-sea floor. If such gradients do indeed exist, how can we account for them? Since $-dT/dz = H_e/c_e$, a large value on the left of this equation will clearly result from either (1) a small value of the effective (eddy plus molecular) conductivity c_e, or (2) a large value of effective upward heat flux H_e.

Mark Wimbush

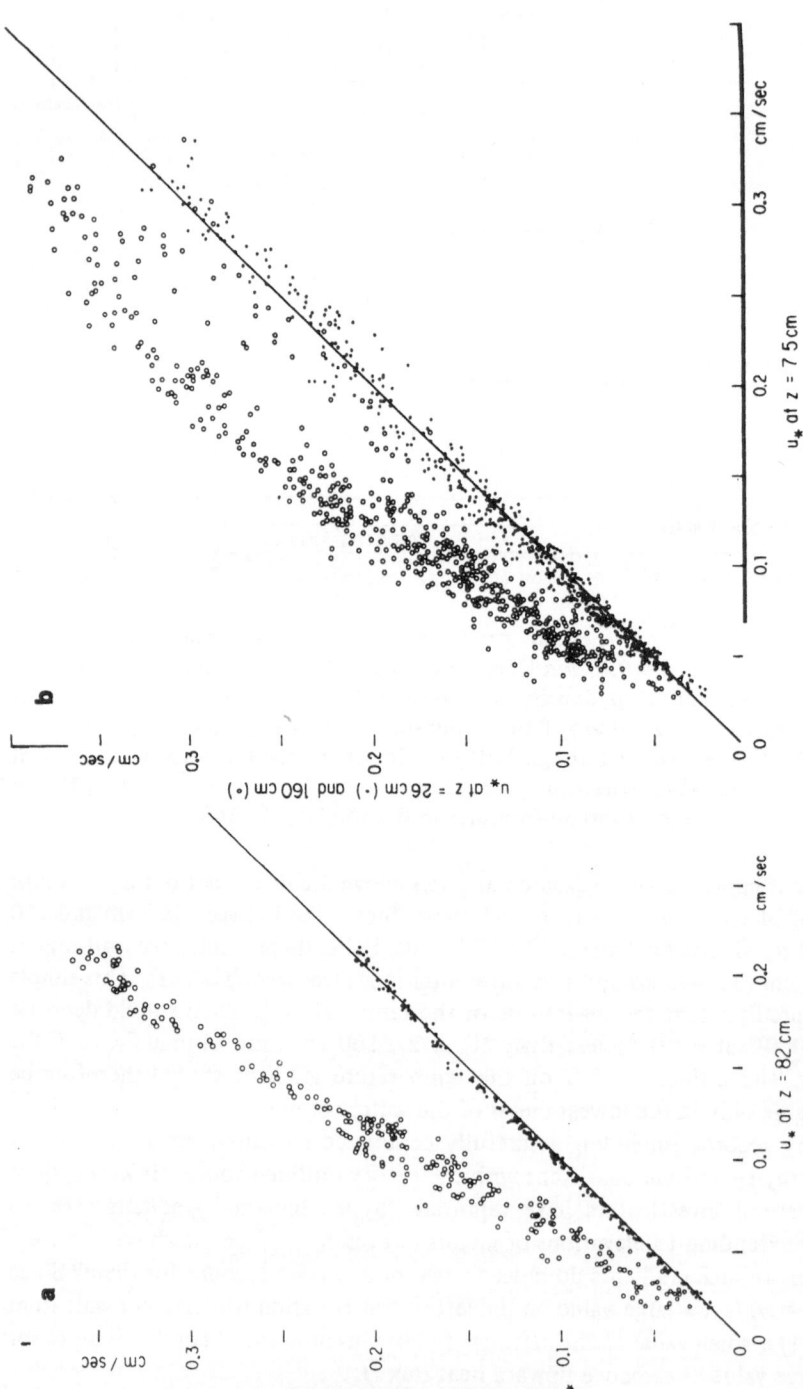

Fig. 2. Regression plots of u_* estimates, computed (via the logarithmic formula for a smooth bottom) from simultaneous measurements of mean current speed at pairs of elevations. In (a) the solid dots being near the 45° line show that measurements at 82 cm and 100 cm elevations were consistent. The open circles lie above the line, indicating that the measurements at 234 cm were taken above the logarithmic layer. Similarly, (b) indicates that the 7.5 cm and 26 cm levels were within the

Consider these possibilities in turn:

1. c_e would be reduced if the vertical component of turbulent motion were suppressed by a stable density gradient. Such a gradient in density might conceivably result from a gradient of substance (either dissolved or particulate) derived from the underlying sediment. But probable ion fluxes out of the sediment and competencies of deep-sea currents to suspend sediment particles are both inadequate.

2. H_e would be larger in regions where the geothermal flux H is larger. Maximum values for H (near mid-ocean ridges) are one order of magnitude larger than normal. Alternatively, H_e will be very different from H in regions where the temperature of the water overlying the sediment fluctuates significantly with time. To get an estimate of the possible importance of this effect, I computed $H_e - H$ as a function of time from a $6\frac{1}{4}$-year record of ocean bottom temperature at 2 km depth on the Blake Escarpment. This record is described by Broek (1969) and was kindly made available to me by Dr. Broek. Figure 3 shows histograms of the record itself and of the resulting excess heat flux $H_e - H$. At this location (which appears more variable thermally than a typical deep-sea site) the instantaneous flux H_e is almost as likely to be down (negative) as up (positive), and H_e is often one order of magnitude larger than a typical H (10^{-6} cal cm^{-2} s^{-1}). For $H_e = 20 \times 10^{-6}$ cal cm^{-2} s^{-1}, the temperature gradient at $z = 1$ meter would be less than 20Γ. This can account for only the weakest of the reported hyperadiabatic gradients.

One might wonder if frictional dissipation in the boundary layer could be a significant source of heat flux, but this is not so, for the dissipative flux $H_d = \rho u_*^2 U \approx 10^{-3} H$. Moreover, integrated over the entire deep ocean, this is

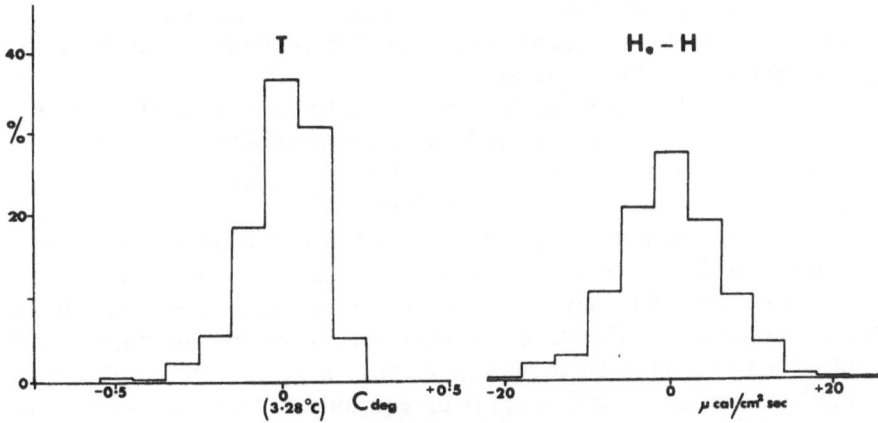

Fig. 3. Histograms of recorded sea-bottom temperature and resulting sediment–water heat flux.

Table 1

Parameter	Deep sea	Shelf
U (cm/s)	3	30
u_* (cm/s)	0.1	1
ν_e (cm^2/s)	2	200
z_f (cm)	500	5000
z_l (cm)	100	1000
z_ν (cm)	2	——
z_γ (cm)	1	——
z_δ (cm)	0.2	——

only 1% of the total tidal dissipation. Most of the tidal energy in the oceans is dissipated on the continental shelves.

Shelf conditions vary greatly from place to place, but typically u_* and U are an order of magnitude larger than in the deep sea. As seen by this relatively swift current, the bottom on the shelf will almost always appear rough, so there is no simple conductive or viscous sublayer. Table 1 compares "typical" (midlatitude) deep-sea and shelf benthic boundary layer properties. ν_e in Table 1 is a characteristic eddy viscosity for the layer, obtained by inverting the classical relation for Ekman layer thickness, $z_f = \pi(\nu_e/f)^{1/2}$. The values of z_δ represent thickness of the diffusive sublayer for a typical electrolyte. $z_\delta = z_\nu/\mathrm{Sc}^{1/3}$, where the Schmidt number $\mathrm{Sc} = \nu/\delta$ (δ = molecular diffusivity). The theory for transport (upward or downward) of dissolved substance through the overall boundary layer is the same as that for transport of heat (except that there is no concentration gradient analogous to the thermal adiabatic gradient). However, since $\delta \ll \gamma$, the diffusive sublayer (in which the transport of substance is principally by molecular processes) is considerably thinner than the corresponding thermal conductive sublayer.

On the shelf, the geothermal heat flux is a negligible component of the total heat flux across the sediment–water interface. Seasonal effects are likely to be important all the way down to the sea floor (Pingree, 1976).

Fluctuations in deeper benthic boundary layers are also of interest. The thickness of the bottom mixed layer observed in temperature profiles sometimes seems to change considerably, with a decrease in water temperature tending to accompany a thickening of the mixed layer (Weatherly and Niiler, 1974). On a somewhat shorter time scale ($\frac{1}{2}$–1 day), tidal fluctuation introduces special problems: tidal periods being of the same order as the response time scale for the overall boundary layer ($2\pi/f$, except in the equatorial zone), steady-state theory is inapplicable in the outer layer ($z_l < z < z_f$). Mathematical models of time-varying boundary layers have been constructed in which $\nu_e(z, t)$ is a prescribed function [Ekman (1905) did this for $\nu_e(z, t)$ equal to a constant in his classic paper on rotating boundary layers]. But recently a numerical model has been constructed in which $\nu_e(z, t)$ is an output rather than an input (Weatherly, 1975). Closure is achieved by using the turbulence energy equation with similarity arguments relating this energy to a mixing length and to ν_e. The model allows for the buoyancy effects of a density gradient and is even generalized to include a sloping bottom.

At very short time scales ($\ll 1$ h) one is looking at turbulent fluctuations with horizontal wavelengths $\lambda \ll 1$ km. Turbulence spectra of horizontal velocity and temperature near the deep-sea floor seem well behaved, inasmuch as they give estimates for u_* and H that agree with values obtained by other means (Wimbush and Sclater, 1971). Spectra of turbulent shear stress ($\tau = -\rho \overline{uw}$), measured at a given level z above a shallow sea bed, have been observed to peak at $\lambda \approx 20z$ (Bowden, 1962). Studies (Gordon, 1974; Heathershaw, 1974) of the stress record $\tau(t)$ together with the individual horizontal $u(t)$ and vertical $w(t)$ velocity records indicate that events occupying 10% of the time contribute 60% of the total shear stress, with somewhat more stress generated by upward "ejections" ($u < 0$, $w > 0$) than by downward "sweeps" ($u > 0$, $w < 0$). A similar intermittent character has been observed in laboratory boundary layers.

Much of this review article is a summary of Wimbush and Munk (1970). Further detail may be sought there and in the other references.

References

Broek, H. W., 1969, Fluctuations in bottom temperature at 2000-meter depth off the Blake Plateau, *Journal of Geophysical Research 74:* 5449–5452.

Bowden, K. F., 1962, Measurements of turbulence near the sea bed in a tidal current, *Journal of Geophysical Research 67:* 3181–3186.

Ekman, V. W., 1905, On the influence of the earth's rotation on ocean-currents, *Arkiv för Matematik, Astronomi och Fysik 2:* 11.

Gordon, C. M., 1974, Intermittent momentum transport in a geophysical boundary layer, *Nature 248:* 392–394.

Heathershaw, A. D., 1974, "Bursting" phenomena in the sea, *Nature 248:* 394–395.

Pingree, R. D., 1976, The thermal boundary layer of the continental shelf, *Deep-Sea Research* (in press).

Weatherly, G., 1972, A study of the bottom boundary layer of the Florida Current, *Journal of Physical Oceanography 2:* 54–72.

Weatherly, G. L., 1975, A numerical study of time-dependent turbulent Ekman layers over horizontal and sloping bottoms, *Journal of Physical Oceanography 5:* 288–299.

Weatherly, G. L., and Niiler, P. P., 1974, Bottom homogeneous layers in the Florida Current, *Geophysical Research Letters 1:* 316–319.

Wimbush, M., 1970, Temperature gradient above the deep-sea floor, *Nature 227:* 1041–1043.
Wimbush, M., and Munk, W., 1970, The benthic boundary layer, in: *The Sea*, (A. E. Maxwell, ed.), Vol. 4, (1), Chapter 19, pp. 731–758, Wiley-Interscience, New York.
Wimbush, M., and Sclater, J. G., 1971, Geothermal heat flux evaluated from turbulent fluctuations above the sea floor, *Journal of Geophysical Research 76:* 529–536.

2
Measurements of Turbulence in the Irish Sea Benthic Boundary Layer

A. D. HEATHERSHAW

A description is given of the instruments and techniques that have been developed to observe the turbulent structure and related hydrodynamic parameters (e.g., drag coefficient, roughness length, friction velocity) of the benthic boundary layer in the sea. This work has involved the use of vertically separated electromagnetic current meters mounted on a probe which was lowered to the sea bed from a vessel at the surface. The horizontal and vertical turbulent velocity fluctuations u and w, have been recorded simultaneously at two heights in the lower 2 m of the boundary layer together with velocity profile measurements between the surface and the sea bed. These observations are an extension of those of other workers in estuarine and coastal waters and have been made under more or less open sea conditions. Measurements have been made over a large area of the Irish Sea and in the Menai Strait, at depths of 10–60 m and maximum surface currents of the order of 1 m s^{-1}. These have corresponded to a range of sediment types, which, in addition to their known distributions, have been observed from grab samples and with an underwater camera mounted on the probe. Time series of the turbulent velocity fluctuations have been analyzed using standard digital methods and for each record the spectra, cross-spectra, phase, and coherency of u and w have been computed. These data are being used to determine those scales of motion contributing to the Reynolds stress (−ρuw, where ρ is the fluid density). The distributions of u, w, and uw have been examined in order to determine those mechanisms involved in the generation and transfer of turbulent energy and, in particular, the intermittent nature of the Reynolds stress.

A. D. HEATHERSHAW • Department of Physical Oceanography, Marine Science Laboratories, Menai Bridge, Gwynedd, U.K. (Present address: Institute of Oceanographic Sciences, Taunton, Somerset TA1 2DW, U.K.)

Introduction

In contrast with the atmospheric boundary layer, the bottom boundary layer in the sea has not been extensively studied. Measurements have generally been confined to shallow coastal waters and estuaries and therefore typified by length and time scales not necessarily characteristic of circulation in the sea. The greatest interest has been directed toward measuring those readily observed parameters related to sediment transport, particularly the drag coefficient, roughness length, and bottom stress. There has been little emphasis on studying the turbulent structure of this important example of a geophysical boundary layer.

During the period 1971–73, the Department of Physical Oceanography, University College of North Wales, completed a program of observations of near-bottom turbulence in the Irish Sea and adjoining waters. These are believed to be some of the first large-scale measurements of high-frequency turbulence to be made on the continental shelf. This chapter briefly reviews the instruments and techniques that have been developed to study the bottom boundary layer and discusses some initial results and their implications.

Background and Objectives

The primary objective of our study has been to examine the frictional interactions between a tidal current and the sea bed and to quantify these by conventional means in terms of a quadratic friction law of the form

$$\tau_0 = \rho C_D |\overline{U}| \overline{U} \tag{1}$$

where τ_0 is the horizontal shear stress exerted at the sea bed, C_D is a drag coefficient, $\overline{U}$ is the mean current at a specified height above the sea bed (usually 1 m), and ρ is the fluid density. Additionally, we have attempted to extend the measurements to a large and, we hope, representative area of the Irish Sea.

The bottom stress has been determined by equating it to measured values of the Reynolds stress (eddy correlation method), given by $\tau = -\rho\overline{uw}$, where u and w are the horizontal and vertical turbulent velocity fluctuations and the overbar denotes averaging in time; consequently we have invoked the constant-stress-layer hypothesis. The work has thus involved observations of the turbulent velocity field in the boundary layer and as a further consideration has examined those processes responsible for the generation of turbulence and the maintenance of the Reynolds stress.

In addition to the preceding type of measurement, it is possible to determine the bottom stress from the velocity gradient close to the sea bed. This is known as the profile method. Suitably fitted logarithmic profiles of the form

$$\overline{U}_z = (U_*/\kappa) \ln (z/z_0) \tag{2}$$

where κ is von Kármán's constant ($\kappa = 0.4$ in a flow free of suspended sediment), are used to calculate the friction velocity, $U_* = (\tau_0/\rho)^{1/2}$, and the roughness length z_0. In this work, relation (2) has been used to calculate a roughness length from known values of $\overline{U}$ and τ_0. It has been implicitly assumed in these derivations that a logarithmic velocity profile existed at all times during which measurements were made. Sternberg (1968) has reported a mean occurrence of the logarithmic profile of some 85% based upon a ±10% fit to a linear relationship. This was estimated as being equivalent to the accuracy of the current meters used in that study. In addition, the work of Charnock (1959) and Bowden et al. (1959) in the Irish Sea suggests that this may not be an unreasonable assumption.

Our present knowledge of typical values of C_D and z_0 in the marine environment is based largely upon profile measurements. Recent observations have included those of Sternberg (1968), Sternberg and Creager (1966), Dyer (1970), Channon and Hamilton (1971), and McCave (1973). These have in general indicated considerable scatter in individual values of C_D and z_0 at any one location (of the order 10^{-4}-10^{-2} and 10^{-6}-10^{1} cm respectively) for flows ranging from hydrodynamically transitional to rough. The summary given by Kagan (1971) of published values of the drag coefficient since 1918 indicates a range for C_D of 6×10^{-4} to 2×10^{-2} (deduced by direct and indirect methods), these values being representative of widely differing sedimentary deposits and bed forms.

Direct observations of turbulent velocity fluctuations, particularly those contributing to the Reynolds stress, are less common. Previous work has included the early studies of Bowden (1962) and Bowden and Fairbairn (1956) in the Irish Sea, and Seitz (1973) and Gordon and Dohne (1973) in tidal passages adjacent to Chesapeake Bay. Recent observations in the deep sea boundary layer have been made by Wimbush and Munk (1970) and Thorpe et al. (1973). However, these have not included measurements of the Reynolds stress and thus give little further insight into the turbulent structure of the boundary layer.

In the bottom boundary layer, no comparison of profile and eddy correlation methods is available. However, in the atmosphere Miyake et al. (1970) have shown that for the boundary layer over water, estimates of the friction velocity obtained by both methods agree to within 4%. This evidence suggests that, technical limitations apart, a determination of the bottom stress comparable with values from profile observations should be possible from a single-point measurement of the Reynolds stress, provided that this is within the constant stress layer and consequently that the flow is both uniform and steady. In addition, this type of measurement is to be favored as it represents a fundamental parameter of a turbulent shear flow and thus contains much useful information relating to its structure, in particular the length and time scales and types of motion that occur within it.

Fig. 1. Bottom turbulence probe: A—swivel link, B—underwater camera units, C—fin, D—inclinometer, E—electronics, F—5-cm-diameter electromagnetic current meter sensors. The height of the probe is approximately 2 m.

Experimental Methods

Electromagnetic current meters (ECMs) were chosen for this study as they represented a comparatively robust and sensitive velocity sensor measuring more than one component of flow, and they were readily available from commercial sources. Development of the ECM by the Institute of Oceanographic Sciences, England, bringing about improvements in mechanical and electrical design, has resulted in its wider application in oceanographic research (e.g., Thorpe *et al.*, 1973; Huntley and Bowen, 1973). The principles of their operation are well described in the literature (Tucker *et al.*, 1970), and are not discussed here.

Initial observations were made using a single 10-cm-diameter sensor energized by a 11.2-Hz square wave, and later work has involved the use of two 5-cm-diameter sensors coupled in a series-parallel configuration and energized by a 40-Hz square wave from a single drive circuit. The time constants of these systems were 1.0 s and 0.1 s, respectively.

The probe that has been developed for this work, shown in Fig. 1, consists of a tubular alloy frame attached to a heavy steel triangular base. Two ECM sensors were mounted on the frame so as to record simultaneously at two heights the vertical and horizontal turbulent velocity fluctuations u and w, and the mean component $\bar{U}$. The probe, weighing some 300 kg when fully ballasted, was lowered to the sea bed on a single 200-m length of 35-core electromechanical cable having a breaking strain of the order of 3000 kg; a swivel link permitted alignment of the frame with the mean current, as it was lowered to the sea bed, through the action of a large rectangular fin.

All analog signals and power were transmitted between the sensors and the ship via the multicore cable.

The data-logging system consisted of a Racal (T3000) four-channel FM tape recorder and Dynamco Microscan data logger with paper tape output device, recording simultaneously the voltage analogs of the turbulent velocity fluctuations and mean components. Both magnetic and paper tape records were returned to the laboratory for subsequent computer analysis.

Analysis of Data and Results

Observations of near-bottom turbulence have been made at stations in the Irish Sea (Fig. 2) and in the Menai Strait. These were chosen on the basis that they provided a wide range of depths, current speeds, and bottom sediments. Depths at these stations varied between 10 and 60 m, with maximum surface currents of the order of 100 cm s^{-1}. Bottom currents ranged from 20 to 80 cm s⁻¹. Approximately 150 records have been obtained, which include measurements of u and w at heights of 50, 100, 150, and 200 cm above the sea bed. Typical records are shown in Figs. 3 and 4.

Magnetic tape analogs of u and w were replayed at half their recorded speed, through second-order low-pass filters, into the data logger, and processed into digital format on paper tape. The digitizing interval for records from the Irish Sea was 0.35 s. For each record the mean, standard deviation, skewness, kurtosis, spectra, cospectra, quadrature spectra, phase, and coherence were computed using standard digital methods. During spectral analyses corrections were made

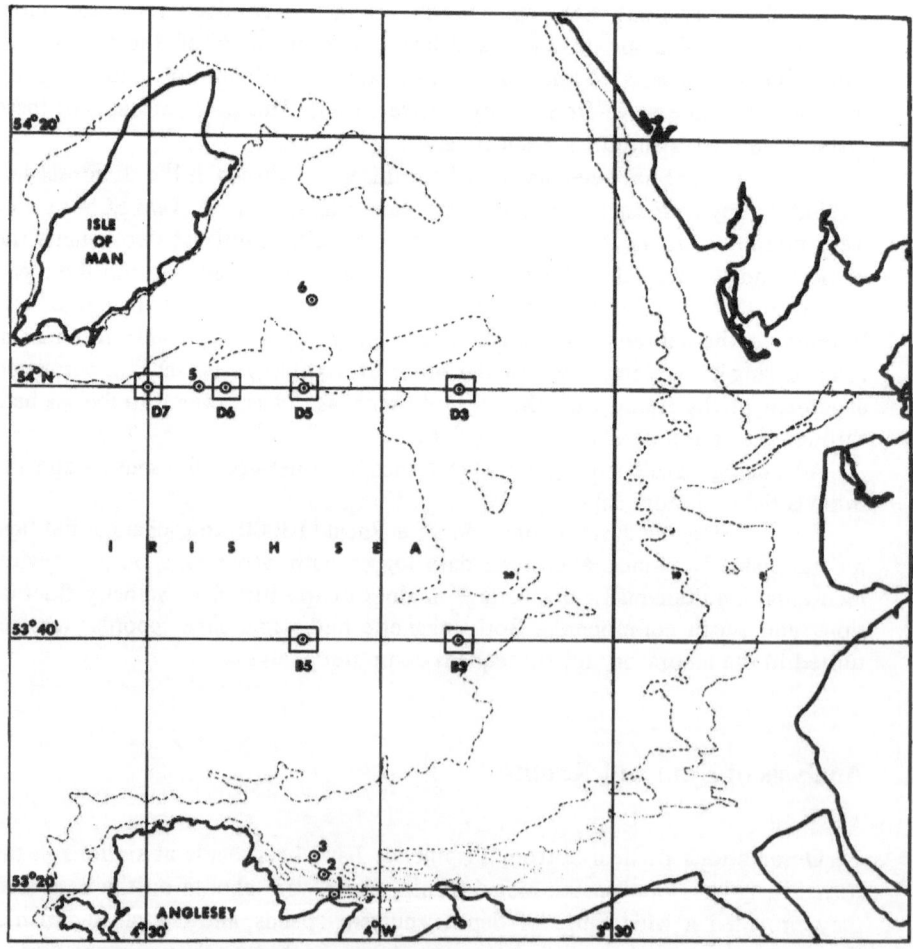

Fig. 2. Irish Sea stations at which measurements of turbulence were made between 1971 and 1973.

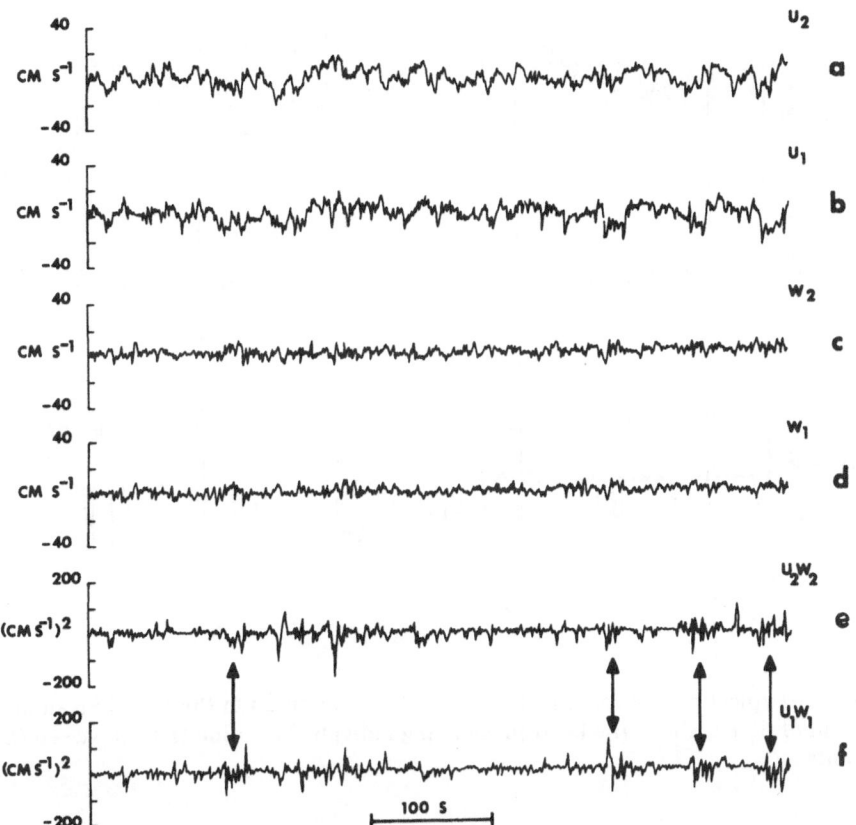

Fig. 3. Typical analog record of u, w, and uw measured simultaneously at 100 cm (b, d, f) and 150 cm (a, c, e) above the sea bed at Station B5; $\overline{U}_{100} \simeq 45$ cm s^{-1}, $\overline{U}_{150} \simeq 57$ cm s^{-1}, $\tau_{100} \simeq 3.54$ dyn cm^{-2}, $\tau_{150} \simeq 4.30$ dyn cm^{-2}. Intermittent events are marked by arrows.

for nonsimultaneity of sampling and the overall transfer function of the measuring system (both phase and amplitude); transients or "spikes" were removed from records on a statistical basis prior to spectral analyses.

Records of u and w

A summary of the results from stations B3–D7 is given in Table 1. These data were calculated, using equations (1) and (2), from 10-min time series of u and w measured at a height of 100 cm above the sea beds and recorded at or near times of maximum current speed. The results of measurements at 150 cm

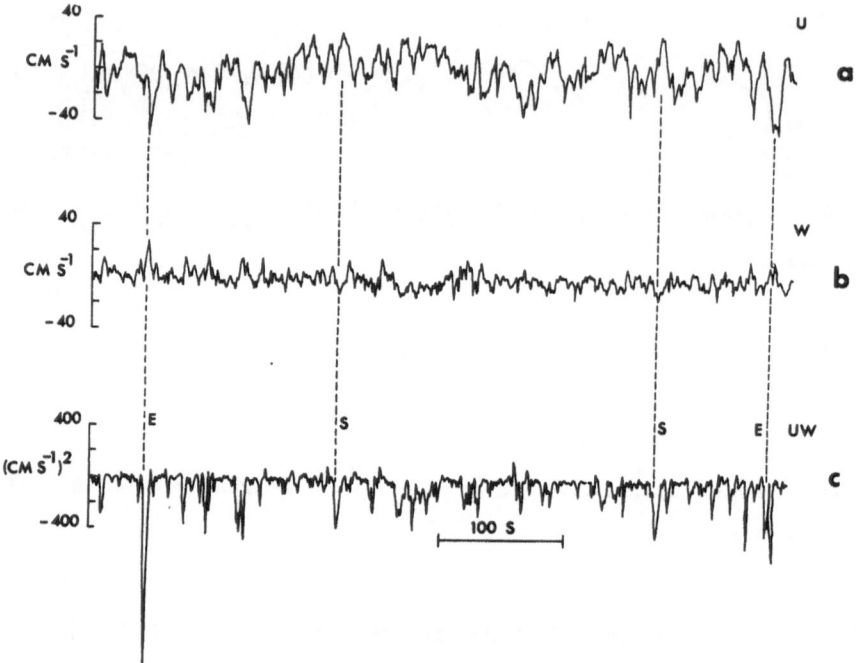

Fig. 4. Typical analog record of u, w, and uw measured in the Menai Strait at a height of 50 cm above the bottom, showing individual ejection (E) and sweep (S) events.

are not included here. Values of the Reynolds stress have been calculated from integrated cospectra up to a limit of 1.0 Hz (Nyquist frequency—$f_N \simeq 1.4$ Hz) at which contributions from aliased energy up to $4f_N$ are of the order of 2% for an assumed high-frequency slope of f^{-2}. Cumulative cospectra from a number of records indicate that contributions to the Reynolds stress are complete within ±5% at a frequency of approximately 0.5 Hz. However, instruments with a lower cutoff than this may seriously overestimate or underestimate the stress, depending on cospectral forms (Fig. 6).

Values of the drag coefficient have been calculated corresponding to measurements of u and w at 100 cm above the sea bed. Also shown in Table 1 are mean values of the correlation coefficient r_{uw} and the ratios of rms turbulent velocities to the mean convection speeds $(\overline{u^2})^{1/2}/\overline{U}$ and $(\overline{w^2})^{1/2}/\overline{U}$.

Typical spectra from records collected at station B5 are shown in Fig. 5. These have been plotted as wave number spectra, where a wave number has been defined by $k = f/\overline{U}$. Taylor's hypothesis of a frozen pattern of turbulence is normally assumed on the basis of $(\overline{u^2})^{1/2}/\overline{U} \ll 1$, and from this work a mean value for this ratio from all records can be taken as 0.15. However, observed

Table 1. Details of Records from Irish Sea Stations Measurements Only at $z = 100$ cm[a]

Station	Number of records	Mean speed[b] $\overline{U}$, cm s^{-1}	Drag coefficient $C_{100} \times 10^3$	Roughness length $z_0 \times 10^3$, cm	Reynolds stress,[b] dyn cm^{-2}	Correlation coefficient r_{uw}	$(\overline{u^2})^{1/2}/\overline{U}$	$(\overline{w^2})^{1/2}/\overline{U}$
B3	6	33–39	1.89 ± 0.38	26.73 ± 14.73	1.08–4.10	-0.21 ± 0.037	0.14 ± 0.010	0.067 ± 0.0034
B5	7	38–49	1.58 ± 0.34	17.32 ± 12.16	0.19–6.74	-0.15 ± 0.031	0.15 ± 0.0068	0.073 ± 0.0018
D3	2	18–21	1.81 ± 0.56	20.11 ± 13.97	0.48–0.85	-0.15 ± 0.035	0.17 ± 0.0062	0.072 ± 0.0028
D5	5	32–47	2.58 ± 0.50	72.21 ± 43.31	1.49–3.06	-0.27 ± 0.045	0.14 ± 0.0043	0.072 ± 0.0074
D6	4	41–56	0.89 ± 0.14	0.46 ± 0.30	0.59–1.29	-0.087 ± 0.018	0.18 ± 0.0060	0.068 ± 0.00050
D7	4	48–55	1.49 ± 0.22	6.02 ± 3.42	0.95–2.14	-0.20 ± 0.029	0.14 ± 0.0096	0.060 ± 0.0024
All records	28	18–56	1.73 ± 0.18	25.31 ± 10.0	0.19–6.74	-0.18 ± 0.018	0.15 ± 0.0039	0.069 ± 0.0018

[a]Mean values are shown with their standard errors.
[b]For $\overline{U}_{100}$ and τ, ranges only are indicated.

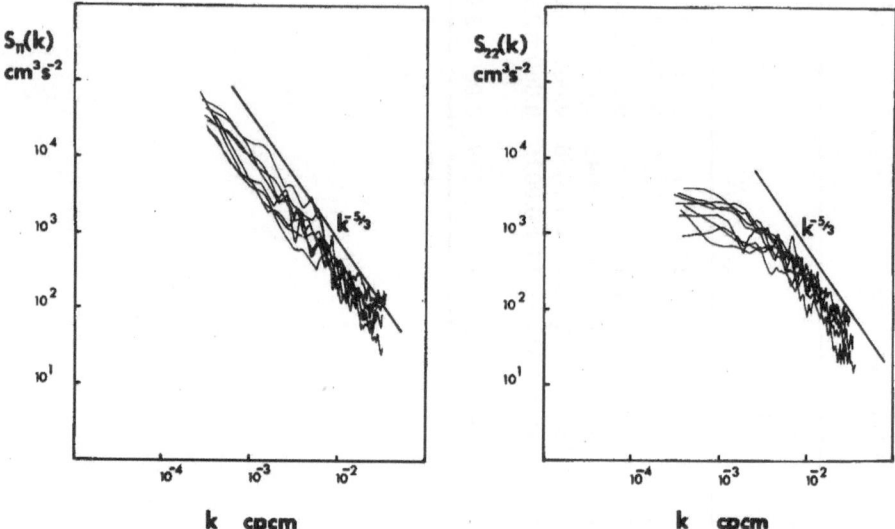

Fig. 5. Spectra of the horizontal and vertical turbulent velocity fluctuations, $S_{11}(k)$ and $S_{22}(k)$ respectively, measured at Station B5 at a height of 100 cm above the sea bed. $\overline{U}_{100} \simeq 38\text{--}49$ cm s^{-1}, $\tau_{100} \simeq 0.19\text{--}6.74$ dyn cm^{-2}.

spectra may well extend over several decades and ideally we should satisfy a criterion having a wave number dependence. Seitz (1973) has used a modified Taylor's hypothesis, first proposed by Lumley (1965), in which the ratio $u'/\overline{U}$ is examined and where u is determined from $u'^2 = kS(k)$. From Fig. 5, for a range of k from $10^{-4}\text{--}10^{-2}$ cm^{-1} and of 20 cm s^{-1} and 50 cm s^{-1}, we obtain ratios of $u'/\overline{U}$ of the order 0.25–0.05 and 0.10–0.02, respectively. These are significantly higher than those of Seitz (0.007 for $\overline{U} = 25$ cm s^{-1} and $k = 0.01$ cm^{-1}), but we conclude that for length scales of the order of 10 m and less, the spectra correspond reasonably well to wave number spectra.

Typical cospectra are shown in Fig. 6, plotted as spectral densities weighted by wave numbers to give an "equal area–equal energy" representation. These correspond to Reynolds stress values of 0.19, 2.59, and 6.74 dyn cm^{-2} (Figs. 6a, 6b, and 6c, respectively), which are fairly typical of boundary layer flows on the continental shelf.

Figure 7 shows the frequency distributions of u, w, and uw for the analog records shown in Fig. 3. Values of the standard deviation, skewness, and kurtosis were calculated on 1650 data points and distributions plotted to include those out to ±4 standard deviations from the mean (removed and set to zero). These figures illustrate the essentially non-Gaussian behavior of contributions to the covariance, the resultant distributions of which are negatively skewed and highly

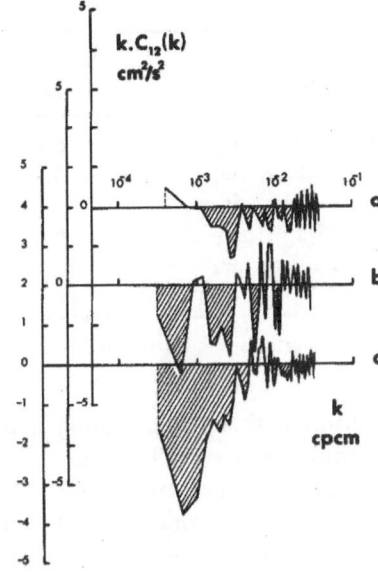

Fig. 6. Typical Reynolds stress cospectra measured at Station B5, $z = 100$ cm: (a) $\overline{U}_{100} \simeq 38$ cm s^{-1}, $\tau_{100} \simeq 0.19$ dyn cm^{-2}; (b) $\overline{U}_{100} \simeq 48$ cm s^{-1}, $\tau_{100} \simeq 2.59$ dyn cm^{-2}; (c) $\overline{U}_{100} \simeq 46$ cm s^{-1}, $\tau_{100} \simeq 6.74$ dyn cm^{-2}.

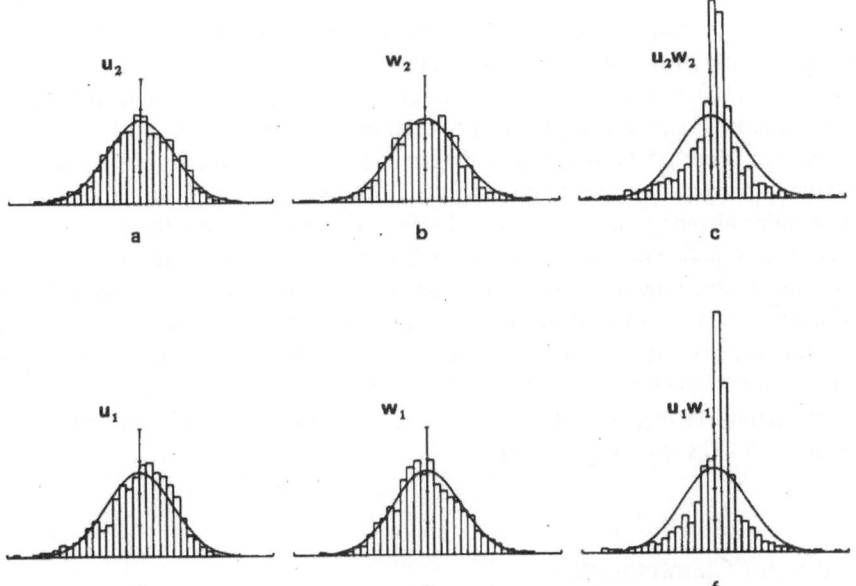

Fig. 7. Frequency distribution of u, w, and uw analog shown in Fig. 3, illustrating the intermittent nature of contributions to the Reynolds stress, $-\rho\overline{uw}$ (7c and 7f). Details of these distributions are given in Table 2.

Table 2

Record		Standard deviation	Skewness	Kurtosis[a]	Figure
11/35/2	u_2	6.48	−0.13	2.96	7a
$z = 150$ cm	w_2	3.17	−0.04	3.50	7b
	$u_2 w_2$	21.23	−0.99	9.88	7c
11/35/1	u_1	6.67	−0.56	3.35	7d
$z = 100$ cm	w_1	2.92	0.18	3.24	7e
	$u_1 w_1$	19.48	−0.05	9.65	7f

[a]Here Kurtosis is defined as $\overline{u^4}/(\overline{u^2})^2 = 3$ for a Gaussian distribution.

kurtosed, thereby indicating a possibly intermittent process. Details of these records are given in Table 2.

Intermittency manifests itself in our records as "patches" of high energy (marked by arrows) in the uw analog, which occur more or less simultaneously at both sensors. This is shown in Fig. 3. In laboratory flow visualization experiments (Corino and Brodkey, 1969; Kim et al., 1971) a sequence of well-ordered motions has been identified as the mechanism responsible for generating what are known as "bursts." These have been shown to consist of two dominant motions—"ejections" and "sweeps." Ejections correspond to an outward movement of fluid from the boundary accompanied by a longitudinal deceleration ($u < 0$, $w > 0$), whereas a sweep consists of an inward movement of fluid toward the boundary and longitudinal acceleration ($u > 0$, $w < 0$). Both events are major contributors to the Reynolds stress (Lu and Willmarth, 1973) and may account for the intermittency observed in our work. Figure 4 shows large amplitude intermittency occurring at a height of 50 cm above the sea bed. This record was collected from the Menai Strait in an intense and complex flow. Individual ejections and sweeps can be seen in the record with typical ratios of $uw/\overline{uw}$ of the order of 10:1 and extreme values of $uw = 30\,\overline{uw}$. In the open sea this ratio would appear to remain approximately the same and agrees well with ratios of peak to mean values of 30:1, obtained in laboratory experiments (Eckelmann, 1974). The duration of events in the bottom boundary layer is typically 5–20 s from Figs. 3 and 4.

Sediment Characteristics

In addition to their known distribution, sediment characteristics have been studied from grab samples and with an underwater camera system mounted on the probe (Fig. 1). At certain stations echo sounding surveys were also carried out to determine the presence of any large-scale topographic features. Sediment samples have been analyzed in some detail but only general conclusions are presented here. Figure 8 illustrates a typical selection of bottom deposits and small-scale bed forms. Figure 8a shows long, crested ripples, indicated by ac-cretions of shell fragments in troughs running from left to right, recorded at station 2 in a depth of water of approximately 16 m. These ripples would ap-pear to have a height of 1-2 cm and a crest-to-crest separation of 15-20 cm. Figures 8b and 8c from station 5 show the variations in bottom deposits that can occur within comparatively short distances at any one location (400-500 m), ranging from a sandy and irregularly rippled bottom to a bottom comprising small pebbles and shell fragments (1-2 cm). These photographs il-lustrate that fairly dense coverages of benthic organisms can occur and also that considerable amounts of material may be in suspension, as indicated by the haziness in Fig. 8b.

Discussion of Results

Drag Coefficients and Roughness Lengths

This work has shown that there is comparatively good agreement between values of the drag coefficient derived from eddy correlation measurements and those values obtained by other workers using the profile method (e.g., Sternberg, 1968; McCave, 1973). Table 1 indicates a range of C_{100} from 0.89-2.58×10^{-3} and a mean value for all records and all stations in the Irish Sea of $1.73 \pm 0.18 \times 10^{-3}$. This is lower than the mean value of 3.0×10^{-3} suggested by Sternberg (1972) for estimating the bottom stress on naturally sorted sand and gravel de-posits. Sternberg's measurements were made in hydrodynamically transitional to rough flows in the tidal passages of Puget Sound. Our work in the Menai Strait, a similar tidal channel, has indicated a mean C_{100} value of 4.56×10^{-3} (5 records only), thereby suggesting that values of C_D measured in the open sea may be lower than those measured in estuaries or rivers with comparable bottom deposits, bed forms, and mean flows. However, the scatter in most published values of C_D obscures such a trend.

Bowden and Fairbairn (1956) and Charnock (1959), working close to sta-tions 1-3 in Fig. 2, found values of C_{100} of 2.0-2.5×10^{-3} and 11.6×10^{-3}, respectively. McCave (1973), working in the North Sea and English Channel, obtained C_{100} values of 3.0-8.7×10^{-3} for flows over sand and gravel beds. While some of this variability might be explained in terms of differing bed forms, and unsteady and nonuniform flows (McCave, 1973), much may be due to the

Fig. 8. Photographs of bottom sediments, taken using the underwater camera system mounted on the probe: (a) long crested ripples at Station 2; (b) and (c) variation within short distances at Station 5 from a sandy irregularly rippled bottom to small pebbles and shell fragments.

essentially probabilistic description of the boundary layer and our attempts to parameterize the bottom stress in terms of a mean flow variable.

This distinction between turbulent and mean flow characteristics has been pointed out by Bradshaw *et al.* (1967), who proposed "that there is a much closer connection between the shear-stress profile and the other parameters describing the turbulence structure than between the shear-stress profile and the mean velocity profile." In particular they demonstrated that the shear stress was closely related to the turbulent kinetic energy. Gordon and Dohne (1973) found that in an estuarine boundary layer, the Reynolds stress varied linearly

Fig. 8b

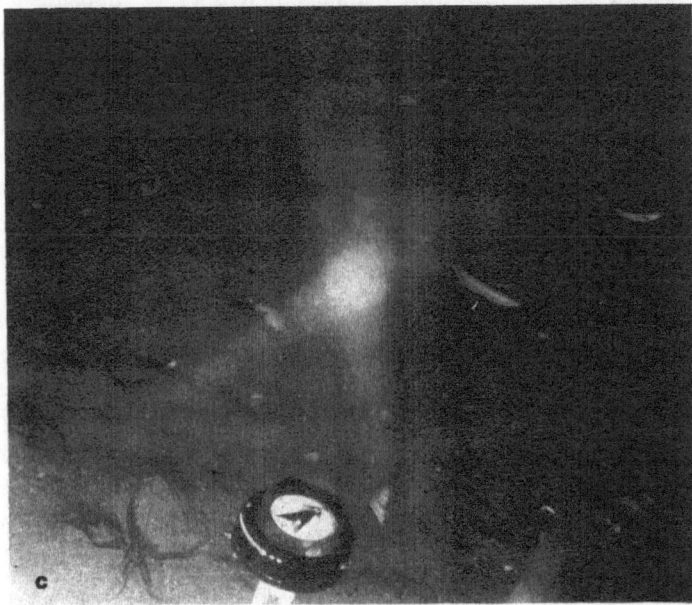

Fig. 8c

with the turbulent kinetic energy $\frac{1}{2}\rho(\overline{u^2} + \overline{v^2} + \overline{w^2})$. The evidence from our work is that a similar relationship may exist in the open sea but that it is complicated by the effects of pressure gradients. In view of these findings, predictions of the bed shear stress should be based upon direct observation of the Reynolds stress close to the sea bed.

This work has shown no clear correlation between measured values of the roughness length and observed grain size characteristics; the latter have usually indicated bimodal distributions. Assuming that all the observed flows were hydrodynamically rough, then a typical grain size can be defined by $d \simeq 30\, z_0$. Thus, for the measured values of z_0 from 0.46-72.21 $\times$ 10^{-3} cm (Table 1), we would anticipate a range of grain sizes from 0.014-2.16 cm. While this adequately covers the observed sediment sizes, individual distributions at any one location show no obvious relation to z_0, and we conclude that this may be due to small variations in bed form and the nature of sedimentary deposits, and the variability in flow conditions.

Spectral and Statistical Measurements

The spectra of u and w shown in Fig. 5 are similar to those of Seitz (1973) for an estuarine flow. At high wave numbers there is a $k^{-5/3}$ dependence indicating an inertial subrange and locally isotropic conditions. Little variation is found between spectra corresponding to different $\overline{U}$ except in those of the vertical fluctuations where there is a tendency toward increasing energy at low wave numbers with increasing $\overline{U}$. The effect of boundary induced anisotropy on the vertical motions is shown by the break from a $k^{-5/3}$ slope at low wave numbers in the spectra of w. This occurs at a wave number, which would also appear to be the upper limit to the energy-containing scales of motion in the cospectra (shaded areas Fig. 6) and is equivalent to a length scale of the order of the height of the sensor (100 cm).

The spectra of horizontal turbulent velocity fluctuations indicate a $k^{-5/3}$ dependence over the entire range of wave numbers considered. This suggests that the inertial subrange extends in a continuum down to wave numbers associated with large-scale, advective tidal motions, even though the Kolmogorov hypothesis fails to predict its existence here. Such behavior has been observed by Cannon (1971) for large-scale, horizontal velocity fluctuations in an estuary.

In the inertial subrange, dimensional analysis yields a one-dimensional spectrum for the horizontal component of the form

$$S_{11}(k') = (18/55)\,\alpha\epsilon^{2/3}k'^{-5/3} \tag{3}$$

where $k' = 2\pi k$ is the radian wave number, α is a constant and equal approximately to 1.44, and ϵ is the rate of energy dissipation per unit mass. Applying this relation to data from station B5, with u and w measured at 100 cm, $\overline{U} \simeq$ 38–49 cm s^{-1}, $\tau = 0.19$–6.74 dyn cm^{-2}, and values of $S_{11}(k)$ evaluated at $k \simeq 0.01$ cm^{-1}, we obtain a mean dissipation rate of 1.24 cm^2 s^{-3}. Seitz (1973), from an equivalent measurement in an estuarine boundary layer, found $\epsilon \simeq$ 0.0179 cm^2 s^{-3}. Most of the energy dissipation occurs at a wave number corresponding to the Kolmogorov fine scale, $\eta = (\nu^3/\epsilon)^{1/4}$, which for this work gives $\eta \simeq 0.05$ cm or a wave number $k \simeq 20$ cm^{-1}. This is about two decades above the upper limit of the present system and suggests that attempts to examine the spectra at higher wave numbers would be greatly restricted by the size and response time of the velocity sensor.

In the constant stress layer, turbulent energy production should equal dissipation, that is,

$$\rho\epsilon = \tau \, d\overline{U}/dz \qquad (4)$$

We can estimate an order of magnitude for the right-hand term. For the data from station B5, a mean value of the Reynolds stress is 3.32 dyn cm^{-2}, and assuming $\tau \simeq \tau_0$ at $z = 100$ cm, we obtain $(\tau/\rho) \, d\overline{U}/dz \simeq 0.15$ cm^2 s^{-3}. Thus, from these few estimates, we find that energy dissipation exceeds energy production by nearly an order of magnitude. This is in disagreement with the work of Seitz (1973). Similarly the mean dissipation rate calculated in this work exceeds that of Seitz by nearly two orders of magnitude. The reasons for these discrepancies are not clear, although they may be due in part to our assumptions regarding the constant stress layer and the validity of (4) in describing the energy balance.

These measurements were made in an area of the Irish Sea where the water column was well mixed by strong tidal currents and thus our spectra do not show any effects due to density stratification.

The cospectra (Fig. 6) indicate that nearly all energy production in the boundary layer occurs at wave numbers $3 \times 10^{-4} < k < 3 \times 10^{-2}$ cm^{-1}, and that with increasing shear there is increased production at low wave numbers with a subsequent shift in the cospectral peak. This range of wave numbers is well within the resolution of the present electromagnetic current meter system, although possibly there is a minor instrumental restriction at low wave numbers due to poor zero stability.

An important aspect of this work has been the confirmation that turbulent energy production occurs intermittently in a bounded shear flow of geophysical proportions (Heathershaw, 1974). This has also been observed by Gordon (1974) in an estuarine boundary layer. The role of a possibly intermittent bed shear stress in sediment transport does not appear to have previously been investigated in the sea. From the typical covariance frequency distribution shown in Figs. 7c and 7d, it is possible to calculate that as much as 70% of the Reynolds stress is contributed in about 5% of the total time of the record. Intermittency has previously been studied in terms of an increase in the kurtosis of probability distributions of successively higher order derivatives of the turbulent velocity fluctuations (Batchelor and Townsend, 1949). Additional techniques have included the examination of electronically filtered, high-frequency portions of turbulent velocity analogs, revealing their "burst"-like quality (Sandborn, 1959; Rao *et al.*, 1971). In particular, the former method has been applied to grid-generated isotropic fields of turbulence. However, in a bounded turbulent shear flow, where energy production occurs at anisotropic scales, it is possible to study intermittency in terms of contributions to the Reynolds stress. The instantaneous values of uw plotted in Fig. 3 show the same behavior that has been observed in laboratory experiments (Wallace *et al.*, 1972). The distributions of uw shown in Fig. 7 indicate that intermittency is nearly always associated with a high kurtosis and a negative skewness (see Table 2), whereas the distributions for u and w are more nearly Gaussian in form.

In the light of recent laboratory experiments (Corino and Brodkey, 1969), it seems likely that such intense motions penetrate the boundary layer to the sediment–water interface. In their experiments, the viscous sublayer was observed to be periodically disrupted by ejections and sweeps. Figure 3 shows intermittency occurring simultaneously at two sensors separated by 50 cm, and Fig. 4 shows the same types of motion occurring only 50 cm above the sea bed. Thus, we might conclude that intermittency in the bottom boundary layer is comprised of large-scale coherent movements of water that may well be capable of extending their influence to the sea bed, thereby playing an important role in sediment transport and entrainment.

One of the most frequently overlooked sources of error in stress determinations by the eddy correlation methods is that of misalignment of sensors in the plane of the measured components. Estimates of this error in atmospheric boundary layer observations vary from 8% deg^{-1} to 100% deg^{-1}.

The covariance of u and w can be shown to vary with θ as

$$\overline{uw}' \simeq \overline{uw} \left\{ 1 + \frac{\theta\,[\overline{u^2} - \overline{w^2}]}{\overline{uw}} \right\} \tag{5}$$

to the first order in θ, where θ is the misalignment between sensor axes and a preferred orientation in radians, and $\overline{uw}'$ is the covariance in the rotated frame of reference. Substituting typical values of $\overline{uw}$, $(\overline{u^2})^{1/2}$, and $(\overline{w^2})^{1/2}$ obtained in this work yields a possible error of the order 10%/deg for small angles ($<5°$). In this work, sensor orientation has been measured with respect to the gravitational vertical using an inclinometer (Fig. 1) and misalignments have been typically $\pm 2°$. However, the correct choice of a suitable frame of reference for our measurements is not entirely clear and might in fact be defined in terms of the directions given by the mean flow streamline, the slope of the sea bed, or the gravitational vertical. Due to this uncertainty we have not corrected any measurements for tilt, and if equation (5) holds, then stress estimates may be in error by as much as ±20%. This suggests that sensor misalignments may be a seriously limiting factor in obtaining accurate measurements of the Reynolds stress; if a preferred orientation were known, it would be necessary to locate a sensor to within ±0.5° of this in order to achieve an accuracy of ±5%. The alignment problem is a continuing aspect of this study.

Conclusions

A full analysis of our data remains to be completed, in particular those results relating to observed roughness lengths and sediment characteristics. However, several important features have emerged and these are summarized as follows:

1. This work has demonstrated the suitability of electromagnetic current meters for measurements of turbulence in the bottom boundary layer.

2. Measured values of the drag coefficient have agreed comparatively well with those of other workers using the profile method. For the Irish Sea and for sediments of a sand–gravel texture, we have obtained a mean value for the drag coefficient of

$$C_{100} = 1.73 \pm 0.18 \times 10^{-3}$$

3. Roughness length measurements have given only poor agreement with observed sediment grain size. However, in accordance with the definitions (1) and (2), a well-defined trend of C_D with increasing z_0 has been found, higher values being associated with rippled sandy bottoms. Bottom photographs have revealed the small-scale variability of bed forms and bottom deposits and at some locations have indicated considerable coverages of biological material.

4. Peak Reynolds stress values of the order 10 dyn cm^{-2} have been observed in the Irish Sea from measurements of u and w at $z = 150$ cm (not shown in Table 1).

5. Well-defined peaks in the Reynolds stress cospectra have indicated that nearly all turbulent energy production occurs at wave numbers

$$3 \times 10^{-4} < k < 3 \times 10^{-2} \text{ cm}^{-1}$$

6. Energy production and dissipation scales would appear to be well separated and to correspond to wave numbers of the orders 10^{-3} cm^{-1} and 10^1 cm^{-1} respectively. Thus, conditions of isotropy and independence of detail of the mean motion extend to a range of wave numbers well below the dissipating scales of motion; this has enabled estimates of the energy dissipation rate (ϵ) to be made with an instrument having only a moderate frequency response (time

constant $\simeq 0.10$ s). A typical value for ϵ in the bottom boundary layer has been found to be of the order 1.0 cm^2 s^{-3}.

7. In common with other geophysical boundary layer flows, the spectra of turbulent velocity fluctuations at high wave numbers agree well with the Kolmogorov hypothesis, showing well-defined inertial subranges and a $k^{-5/3}$ dependence.

8. Intermittency of turbulent energy production would seem to be a universally common feature of this example of a geophysical boundary layer. Peak values of uw 10–30 times $\overline{uw}$ have been observed.

9. This work has shown that sensor misalignment may introduce serious errors into measurements of the Reynolds stress and that these may be of the order $\pm10\%$/deg.

Acknowledgment

I wish to thank my colleagues in the Marine Science Laboratories for their assistance during this study. The work was carried out under NERC Research Studentship No. GT4/70/0F/36.

References

Batchelor, G. K., and Townsend, A. A., 1949, The nature of turbulent motion at large wave-numbers, *Proceedings of the Royal Society of London*, Series A, *199:* 238–255.

Bowden, K. F., 1962, Measurements of turbulence near the sea bed in a tidal current, *Journal of Geophysical Research 67:* 3181–3186.

Bowden, K. F., and Fairbairn, L. A., 1956, Measurements of turbulent fluctuations and Reynolds stresses in a tidal current, *Proceedings of the Royal Society of London*, Series A, *237:* 422–438.

Bowden, K. F., Fairbairn, L. A., and Hughes, P., 1959, The distribution of shearing stresses in a tidal current, *Geophysical Journal of the Royal Astronomical Society 2:* 288–305.

Bradshaw, P., Ferriss, D. H., and Atwell, N. P., 1967, Calculation of boundary layer development using the turbulent energy equation, *Journal of Fluid Mechanics 28:* 593–616.

Cannon, G. A., 1971, Statistical characteristics of velocity fluctuations at intermediate scales in a coastal plain estuary, *Journal of Geophysical Research 76:* 5852–5858.

Channon, R. D., and Hamilton, D., 1971, Sea bottom velocity profiles on the continental shelf south-west of England, *Nature 231:* 383–385.

Charnock, H., 1959, Tidal friction from currents near the sea bed, *Geophysical Journal of the Royal Astronomical Society 2:* 215–221.

Corino, E. R., and Brodkey, R. S., 1969, A visual investigation of the wall region in turbulent flow, *Journal of Fluid Mechanics 37:* 1–30.

Dyer, K. R., 1970, Current velocity profiles in a tidal channel, *Geophysical Journal of the Royal Astronomical Society 22:* 153–161.

Eckelmann, H., 1974, The strucutre of the viscous sublayer and the adjacent wall region in a turbulent channel flow, *Journal of Fluid Mechanics 65:* 439–459.

Gordon, C. M., 1974, Intermittent momentum transport in a geophysical boundary layer, *Nature 248:* 392–394.

Gordon, C. M., and Dohne, C. F., 1973, Some observations of turbulent flow in a tidal estuary, *Journal of Geophysical Research 78:* 1971–1978.

Heathershaw, A. D., 1974, "Bursting" phenomena in the sea, *Nature 248:* 394–395.

Huntley, D., and Bowen, A., 1973, Field observations of edge waves, *Nature 243:* 160–162.

Kagan, B. A., 1971, Sea bed friction in a one-dimensional tidal current, *Izvestiya, Atmospheric and Oceanic Physics 8:* 780–785.

Kim, H. T., Kline, S. J., and Reynolds, W. C., 1971, The production of turbulence near a smooth wall in a turbulent boundary layer, *Journal of Fluid Mechanics 50:* 133–160.

Lu, S. S., and Willmarth, W. W., 1973, Measurements of the strucutre of the Reynolds stress in a turbulent boundary layer, *Journal of Fluid Mechanics 60:* 481–511.

Lumley, J. L., 1965, Interpretation of time spectra measured in high intensity shear flows, *The Physics of Fluids 8:* 1056–1062.

McCave, I. N., 1973, Some boundary-layer characteristics of tidal currents bearing sand in suspension, *Memoires Société Royale des Sciences de Liège*, 6th series, *VI:* 107–126.

Miyake, M., Stewart, R. W., and Burling, R. W., 1970, Comparison of turbulent fluxes over water determined by profile and eddy correlation techniques, *Quarterly Journal of the Royal Meteorological Society 96:* 132–137.

Rao, N. K., Narasimha, R., and Narayanan, Badri, M. A., 1971, The "bursting" phenomenon in a turbulent boundary layer, *Journal of Fluid Mechanics 48:* 339–352.

Sandborn, V. A., 1959, Measurements of intermittency of turbulent motion in a boundary layer, *Journal of Fluid Mechanics 6:* 221–240.

Seitz, R. C., 1973, Observations of intermediate and small scale turbulent water motion in a stratified estuary (Parts I and II), Chesapeake Bay Institute, Technical Report No. 79, Ref. 73-2.

Sternberg, R. W., 1968, Friction factors in tidal channels with differing bed roughness, *Marine Geology 6:* 243–260.

Sternberg, R. W., 1972, Predicting initial motion and bedload transport of sediment particles in the shallow marine environment, in: *Shelf Sediment Transport* (D. J. P. Swift, D. B. Duane and O. H. Pilkey, eds.), p. 61–82, Dowden, Hutchinson and Ross, Inc., Stroudsburg, Pa.

Sternberg, R. W., and Creager, J. S., 1966, An instrument system to measure boundary-layer conditions at the sea floor, *Marine Geology 3:* 475–482.

Thorpe, S. A., Collins, E. P., and Gaunt, D. I., 1973, An electromagnetic current meter to measure turbulent fluctuations near the ocean floor, *Deep-Sea Research 20:* 933–938.

Tucker, M. J., Smith, N. D., Pierce, F. E., and Collins, E. P., 1970, A two-component electromagnetic ship's log, *Journal of the Institute of Navigation 23:* 302–316.

Wallace, J. M., Eckelmann, H., and Brodkey, R. S., 1972, The wall region in turbulent shear flow, *Journal of Fluid Mechanics 54:* 39–48.

Wimbush, M., and Munk, W., 1970, The benthic boundary layer, in: *The Sea*, Vol. 4 (A. E. Maxwell, ed.), p. 731–758, Wiley-Interscience, New York.

3
The Benthic Boundary Layer from the Viewpoint of a Geochemist

ROBERT A. BERNER

The nonequilibrium assemblage of minerals, organic matter, and sea water deposited together in marine sediments brings about chemical reactions which appreciably change the composition of near-surface interstitial waters from the typical sea water values present at the time of deposition. Many of the most important reactions are the result of the microbiological decomposition of organic matter. Large changes in the concentrations of dissolved O_2, NO_3^-, SO_4^{2-}, HCO_3^-, Ca^{2+}, NH_4^+, CO_2, CH_4, H_2S, Fe^{2+}, Mn^{2+}, and orthophosphate have been shown by previous studies to result directly or indirectly from microbiological activity. The rate at which sedimentary chemical reactions occur is not well known but can be determined, in principle, by laboratory studies combined with kinetic modeling of concentration–depth data. Mathematical models are presented here which express in outline form the processes of organic matter decomposition, dissolution and precipitation of minerals, rapid (equilibrium) adsorption and ion exchange, ionic diffusion, bioturbation, flow of water due to compaction, and "flow" of water plus enclosing solids away from the sediment–water interface due to depositional burial. Many of these processes are complex and are treated in the literature in an incorrect or oversimplified manner. Because of gradients in chemical composition, fluxes of dissolved constituents between sediment pore waters and overlying bottom waters must occur. Calculations of fluxes are fraught with difficulties and are often incorrect due to: incorrect formulation and estimation of gradients and of diffusion coefficients; lack of an evaluation of the role of turbulent mixing at the sediment–water interface due to waves, currents, and bioturbation; lack of correction for depositional burial of pore waters; and lack of consideration of diffusion within the viscous–conductive sublayer of the bottom water. In shallow-water sediments where deposition and bioturbation are important, it may be preferable to directly measure fluxes to and from sediments rather than to calculate them from pore-water concentrations. In pelagic sediments, by con-

ROBERT A. BERNER • Department of Geology and Geophysics, Yale University, New Haven, Connecticut 06520, U.S.A.

trast, bioturbation, current stirring, and deposition may be sufficiently unimportant that simple Fick's first law calculations can give reasonable estimates of fluxes, once limits are placed on values for ionic diffusion coefficients and the thickness of the viscous–conductive sublayer. Such calculations indicate that sediment–bottom water exchange is a major process for controlling the chemical composition of the oceans.

Introduction

The chemistry of the benthic boundary layer is intimately intertwined with the physics, geology, and biology of the same region. It is my hope here to demonstate this. The distribution of chemicals in the watery portion of the layer is a function of the chemistry of the underlying sediments, factors which disturb the sediment–water interface, and the physics of transport within the bottom water. Discussion here will be confined almost entirely to the sediments. It is in the sediments that enhanced chemical and biochemical reactions take place because of the bringing together into close proximity of constituents (minerals, organic matter, sea water) which are not at chemical equilibrium. As a result of a large ratio of solid surface area to interstitial water volume (especially in fine-grained muds), concentrations of species in pore waters may change appreciably and give rise to large concentration gradients between sediments and overlying sea water. This in turn results in fluxes of dissolved constituents to and from the sediments. In the present paper some of the more important reactions occurring in the upper few meters of sediment will be discussed and an attempt will be made to show how such reactions can be kinetically modeled and how transport between sediment and overlying sea water is brought about. As will be seen, this poses many unsolved problems.

No attempt will be made to review all or even most of the pertinent literature in the field of interstitial water chemistry. Up-to-date summaries are presented by Glasby (1973) and Manheim (1976) and, for deeper portions of sediments sampled during the Deep Sea Drilling Project, by Manheim and Sayles (1974). The interested reader is referred to these publications for details and topics not discussed in the present paper.

Diagenetic Reactions

Diagenetic chemical reactions, those occurring during and after burial, can be divided into two categories: biogenic and abiogenic. The criterion for classification is whether or not the reactions are mediated by bacteria and other microorganisms. It can be readily shown (e.g., Berner, 1971) that the common state of chemical disequilibrium in sediments is often not relieved over the time intervals represented by the benthic boundary layer (10^2–10^6 years) because of kinetic barriers. Bacteria, however, contain enzymes or catalysts which enable some of the barriers to be surmounted. This is why bacterial reactions often dominate the chemistry of pore waters (Manheim and Sayles, 1974).

Biogenic Reactions

Many of the geochemically important bacteria in sediments require the pre-existance of organic compounds for their metabolism (in other words, they are heterotrophic). Thus, biogenic reactions are intimately tied to the deposition of organic matter. This illustrates the dependence of benthic boundary layer chemical processes upon biological, geological, physical, and chemical processes occurring in the overlying water. High rates of deposition of organic matter are favored by (1) high planktonic productivity in the overlying water (which is in turn controlled by nutrient transport), and (2) quick settling and burial to avoid decomposition in the water column (which is a function of water depth, turbulence at the sediment–water interface, dissolved oxygen content, and rate of burial of enclosing sediment particles). Organic-rich sediments, if deposited rapidly enough, need not be overlain by anoxic or even nearly anoxic water as is often supposed. Good examples of this situation are provided by many near-shore shallow-water muds.

Some of the most important diagenetic reactions directly resulting from the bacterial decomposition of organic matter are the removal of dissolved oxygen, the production of carbon dioxide, the reduction of nitrate, the reduction of sulfate, and the production of ammonia, phosphate, hydrogen sulfide, and methane. (Strictly organic geochemical reactions will not be discussed here.) Representative reactions are shown in Table 1. Many studies have demonstrated

Table 1. Some Representative Overall Biogenic Chemical Reactions in Sediments[a]

Oxygen utilization; CO_2 production
$$CH_2O + O_2 \longrightarrow CO_2 + H_2O$$
Nitrate reduction
$$5CH_2O + 4NO_3^- \longrightarrow 2N_2 + 4HCO_3^- + CO_2 + 3H_2O$$
Sulfate reduction
$$2CH_2O + SO_4^{2-} \longrightarrow H_2S + 2HCO_3^-$$
Ammonia formation
$$CH_2NH_2COOH + 2(H) \longrightarrow CH_3COOH + NH_3$$
Methane formation
$$CO_2 + 8(H) \longrightarrow CH_4 + 2H_2O$$

[a]Organic compounds represented are for general illustrative purposes only.

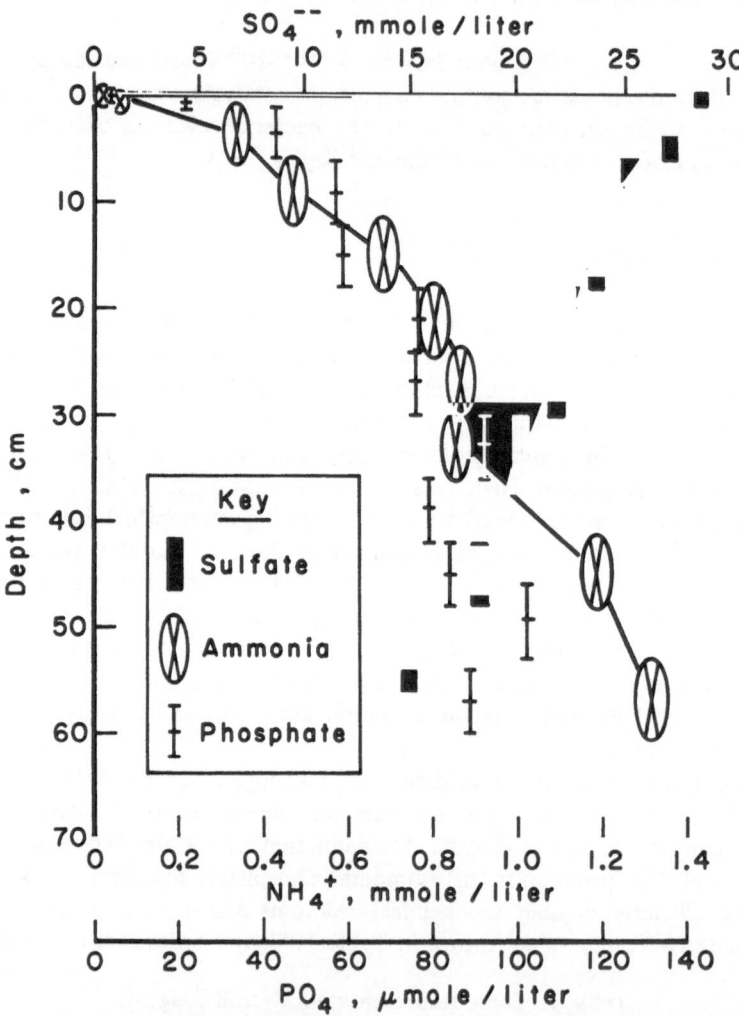

Fig. 1. Sulfate, phosphate, and ammonia vs. depth in pore waters of sediment from the Santa Barbara Basin, Calif. (Data from Sholkovitz, 1973.)

these reactions in surficial sediments (e.g., Murray and Irvine, 1895; Emery and Rittenberg, 1952; Shishkina, 1959; Rittenberg *et al.*, 1955; Kaplan *et al.*, 1963; Brafield, 1964; Bruevich, 1966 (summary of Russian work); Presley and Kaplan, 1968; Hartmann and Nielsen, 1969; Reeburgh, 1969; Friedman and Gavish, 1970; Berner *et al.*; 1970; Shishkina, 1972 (summary of Russian work); Nissenbaum *et al.*, 1972; Thorstenson and Mackenzie, 1974; Sholkovitz, 1973; Hartmann *et al.*, 1973; Bray *et al.*, 1973; Manheim and Sayles, 1974; Martens and

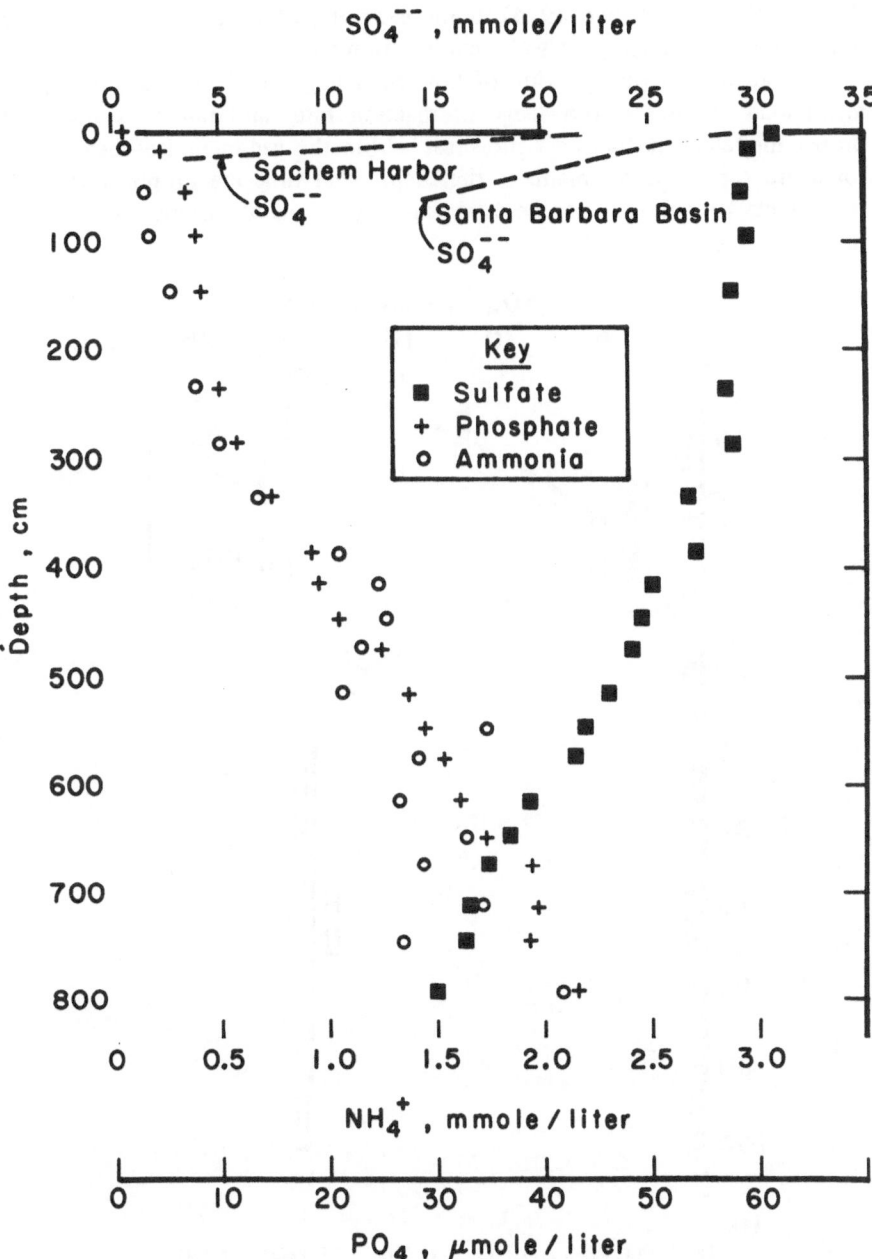

Fig. 2. Sulfate, phosphate, and ammonia vs. depth in pore waters of sediment from the West African continental borderland. Data for sulfate from Figs. 1 and 3 shown for comparison (West African data from Hartmann *et al.*, 1973.)

Berner, 1974; Goldhaber, 1974). Some examples of large differences between pore water and overlying sea water composition are shown in Figs. 1–4.

The answer to the question of how such large gradients are built up and maintained requires a knowledge of reaction rates and rate laws along with proper modeling of transport processes within the sediment (see below). One fortunate aspect of sediments is that depth and time are proportional and a time frame of reference can be established by means of various (usually radio-

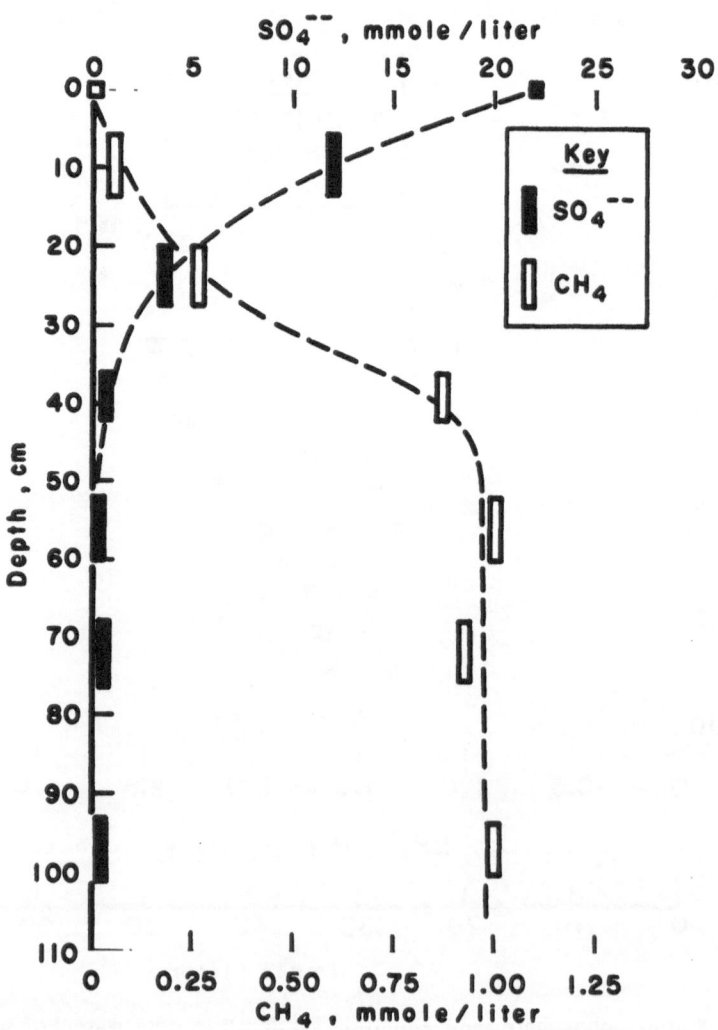

Fig. 3. Sulfate and methane vs. depth in pore waters of sediment from Sachem Harbor, Guilford, Conn. (After Martens and Berner, 1974.)

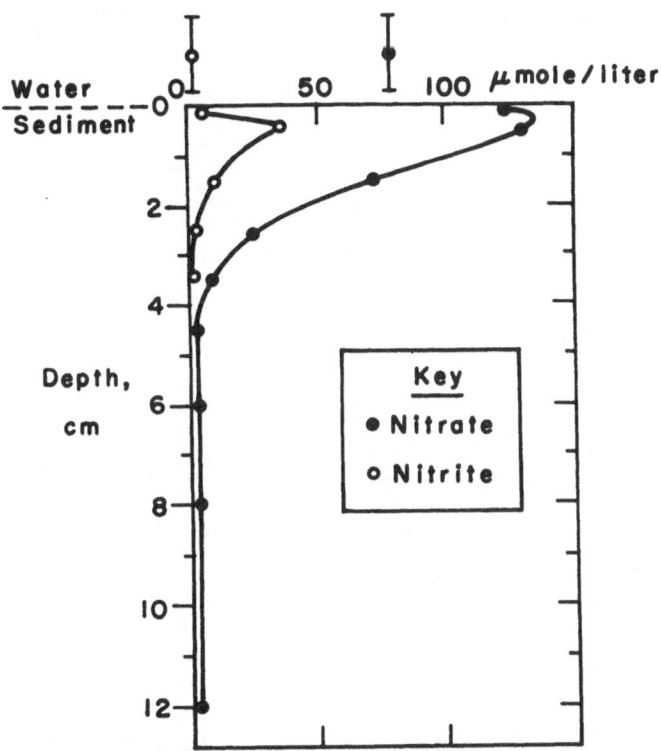

Fig. 4. Nitrate and nitrite vs. depth in pore waters of sediment from the Scheldt Estuary, Belgium (After Billen, 1975.)

metric) dating techniques (for a recent summary consult Goldberg and Bruland, 1974). This enables determination, under ideal circumstances, of the rate and the rate law for the bacterial reaction (e.g., Berner, 1971, 1974a; Goldhaber and Kaplan, 1974; Goldhaber, 1974). Otherwise, direct rate measurements using whole mud (Sorokin, 1962, 1970; Nakai and Jensen, 1964; Stuiver, 1967; Ivanov, 1968) are useful in trying to deduce the appropriate rate and/or rate law. An example is shown in Fig. 5. Rate studies in the laboratory using pure cultures and exactly defined organic substrates (e.g., Kaplan and Rittenberg, 1964) are important for elucidating mechanisms, but cannot be directly used to evaluate a rate for any given sediment because of the microbiological and organic chemical complexity of natural situations. Ideally, a combination of pore-water kinetic modeling with both rigidly controlled laboratory studies and whole-mud measurements would be the best overall approach, especially when transitional studies are also used, such as those by Ramm and Bella (1974), where organic extracts from organisms and sediments, and raw sediment cultures are employed to elucidate rate laws.

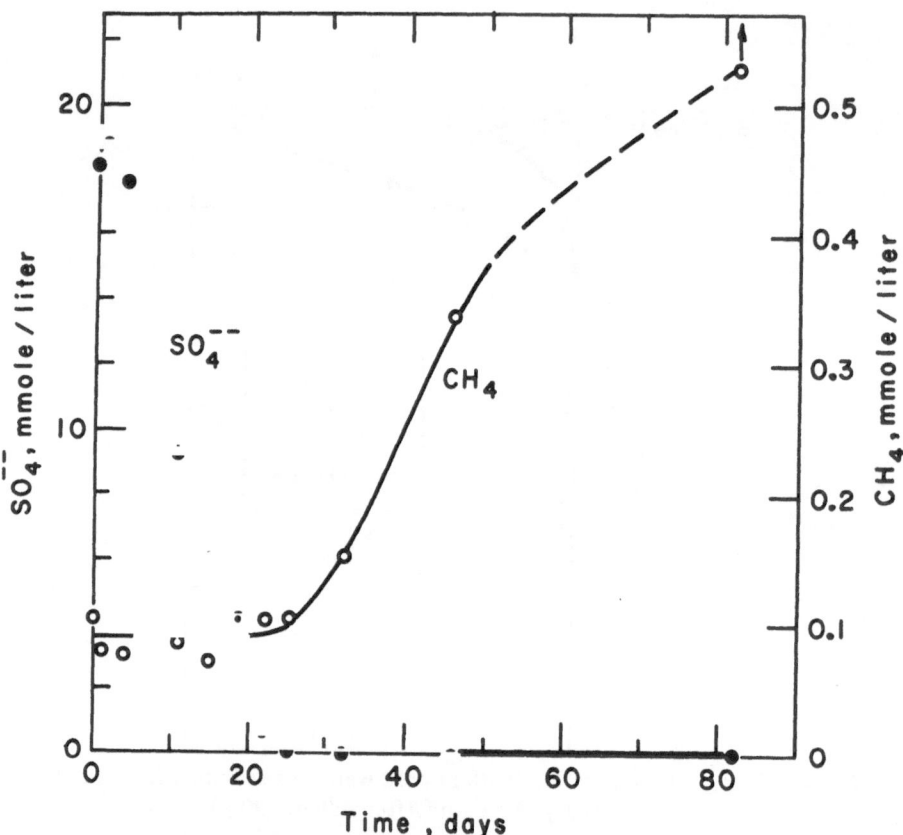

Fig. 5. Sulfate and methane in pore waters of an organic-rich salt marsh panne sediment as a function of time. Results of a serial laboratory experiment where each sample was kept in a sealed jar until analysis. The arrow on the 80-day methane point signifies a minimum value due to experimental difficulties (After Martens and Berner, 1974.)

Products of bacterial reactions can, in turn, react to bring about further changes in sediment chemistry as shown in Figs. 6 and 7. Removal of dissolved oxygen enables the solubilization of iron and manganese (Bruevich, 1938; Debyser and Rouge, 1956; Lynn and Bonatti, 1965; Hartmann, 1964; Li *et al.*, 1969; Brooks *et al.*, 1968; Bischoff and Ku, 1971; Calvert and Price, 1972; Michard, 1971). Hydrogen sulfide produced by bacterial sulfate reduction reacts with detrital iron minerals to form various iron sulfides (e.g., Goldhaber and Kaplan, 1974; Rickard, 1975). Excess bicarbonate ion is produced by sulfate reduction and ammonia formation, and this may lead to the precipitation of

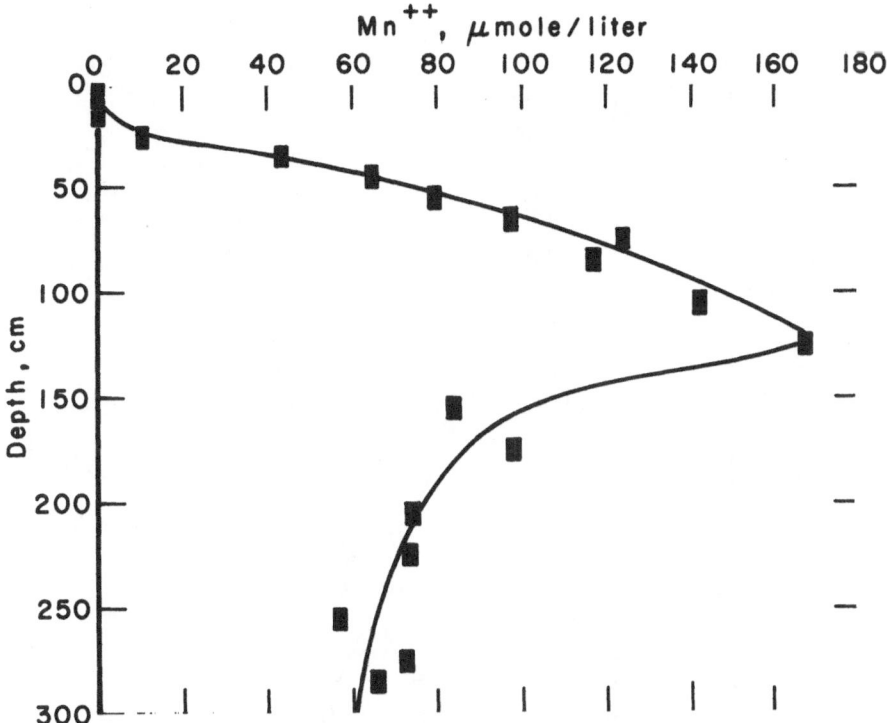

Fig. 6. Manganese vs. depth in pore waters of sediment from the Arctic Basin at 82°N; 156°W. (Data from Li *et al.*, 1969.)

$CaCO_3$ (e.g., Presley and Kaplan, 1968; Berner, 1966; Sholkovitz, 1973; Hartmann *et al.*, 1973). Dissolved phosphate buildup may bring about the precipitation of calcium phosphate as the mineral apatite (e.g., Baturin, 1971; Berner, 1974a). Magnesium may be precipitated as a result of the removal from clay minerals of iron which reacts with H_2S to form iron sulfides (Drever, 1974; Sholkovitz, 1973). Sodium, magnesium, potassium, and calcium ions may undergo enhanced ion exchange or authigenic silicate formation as a result of bacterial alteration of pore water chemistry (Manheim and Chan, 1974; Manheim and Sayles, 1974). It is even possible that the microbiological removal of organic coatings might promote the abiogenic dissolution of biogenic opaline silica (D. Hurd, personal communication). Thus, the concentrations of Na, Ca, Mg, K, Fe, Mn, and possibly SiO_2 in pore waters are affected by the microbiological decomposition of organic matter as well as by the concentrations of O_2, HCO_3^-, SO_4^{2-}, NO_3^-, N_2, H_2S, PO_4^{3-}, CO_2, and CH_4. Interstitial water chemistry in sediments would be a lot simpler if organic matter decomposition did not occur!

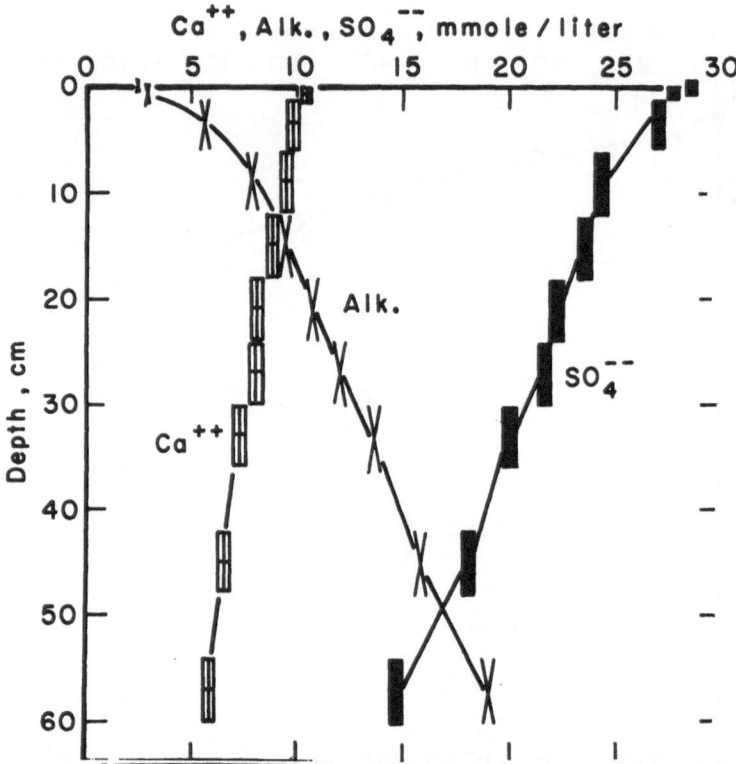

Fig. 7. Calcium, titration alkalinity (Alk.), and sulfate in pore waters of sediment from the Santa Barbara Basin, Calif. (Data from Sholkovitz, 1973.)

Abiogenic Reactions

Reactions which are not biogenically controlled, either directly or indirectly, are less numerous. Examples (see Fig. 8) include the dissolution of opaline silica (Rittenberg *et al.*, 1955; Siever *et al.*, 1965; Harris and Pilkey, 1966; Bischoff and Ku, 1970; Hurd, 1973; Sholkovitz, 1973; Fanning and Pilson, 1974; Wollast, 1974; Schink *et al.*, 1974), the dissolution of $CaCO_3$ at the sediment-water interface (e.g., Berger, 1968; Gieskes, 1974; Berner, 1974b), the recrystallization of carbonate minerals with the consequent uptake of Mg and release of Sr (Milliman and Müller, 1973; Manheim and Sayles, 1974), and various documented or suggested silicate-sea water reactions (references below) such as cation exchange between pore waters and clay minerals, the formation of authigenic clay minerals, the dissolution of feldspar, and the reaction of volcanic glass and basaltic minerals with pore waters to produce new silicate phases. Cation exchange is

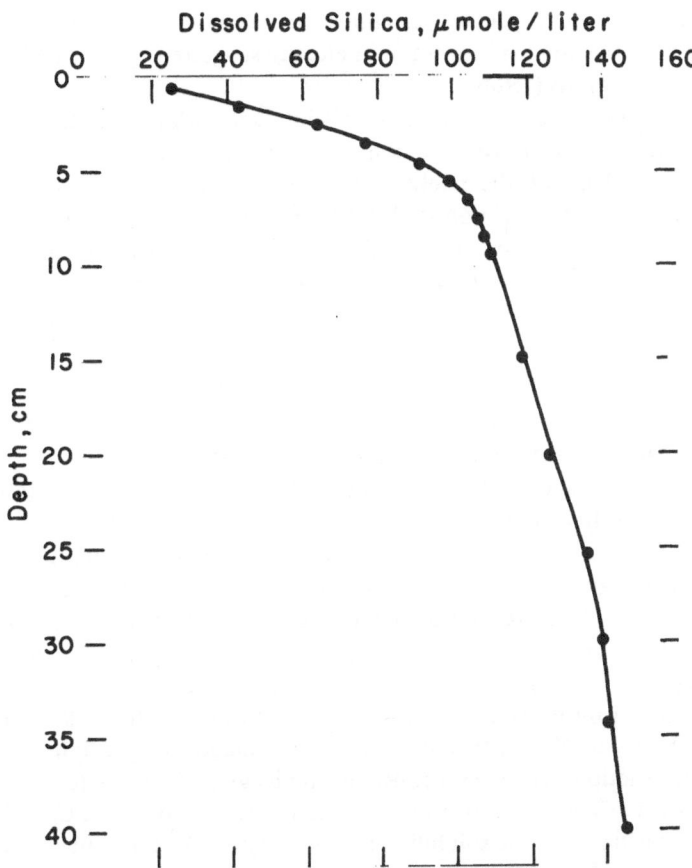

Fig. 8. Dissolved silica vs. depth in pore waters of sediment from the Bermuda
Rise, N. Atlantic (After Fanning and Pilson, 1974.)

probably the best documented silicate–pore water reaction (e.g., Shishkina, 1972;
Manheim and Chan, 1974). Nevertheless, the other silicate reactions may prove
to be more important quantitatively once they are more fully studied (e.g.,
Wollast, 1974). For example, Manheim and Sayles (1974) have shown that interstitial waters sampled just above basalts, as a part of the Deep Sea Drilling Project, often show very exotic compositions indicating extensive silicate–water
reactions. Perhaps similar purely abiogenic reactions occur near the sediment–
water interface in areas of volcanism and/or volcanogenic sedimentation. Also,
recent unpublished studies (F. Sayles and P. Mangelsdorf, personal communication) indicate that very slight but distinct gradients in HCO_3^-, Ca^{2+}, Mg^{2+}, and
K^+ occur as a result of silicate reactions near the sediment–water interface
within the benthic boundary layer. These gradients appear to be a general

phenomenon, and if so, provide an important mechanism for the control of the
major ion concentration of sea water as already suggested by Sillén (1961) and
Mackenzie and Garrels (1966).

The actual processes whereby minerals interact with pore waters via dissolution, precipitation, adsorption, or ion exchange are abiogenic whether or not
they are brought about ultimately by biogenic reactions. The kinetics of these
reactions, unfortunately, is not at all well understood. Assumptions of first and
null order mineral reactions make diagenetic equations mathematically simpler
but may not be correct. A general expression for mineral precipitation and dissolution in stagnant (or very slowly flowing) pore water (Berner, 1974a) is

$$\frac{dC}{dt} = \frac{\bar{A}D(C_s - C)}{r}$$

(1)

where C is the concentration in mass per unit volume of pore water, C_s is the
concentration in a layer of solution immediately adjacent to the surface of the
solid, D is the diffusion coefficient in the solution, $\bar{A}$ is the surface area of solid
per unit volume of pore water, r is the average radius of solid particles, and t is
the time. Although, in principle, it is possible to determine all the critical parameters in equation (1), in general there are severe practical difficulties. First of all,
the equation is strictly valid only for equidimensional, equisized particles separated by at least ten particle diameters in stagnant water. These conditions can
be approximated by many situations, but more serious is the inability to determine C_s. The value of C_s depends upon the mechanism of precipitation or dissolution. If reaction rate is controlled by molecular diffusion to or from the
particle surface, C_s is equal to the equilibrium or saturation value C_{eq} which can
be readily calculated from solubility measurements. On the other hand, if the
rate is partly or wholly controlled by the rate of reactions occurring at the particle surface, then the value of C_s is intermediate between C_{eq} and C and not, at
the present state of knowledge, generally calculable. Furthermore, during reaction C_s for surface reaction controlled kinetics is not constant during reaction, as
is C_{eq}.

It would seem appropriate at this point to stress more experimental studies
to determine mineral–solution kinetics. However, while necessary for elucidating
rate-controlling mechanisms, experimental studies in most cases cannot be
applied directly to natural sediments in order to determine absolute rates. Only
limits can be placed. For instance, if poisoning (inhibition of reaction rate due to
adsorption of specific trace inhibitors at active surface sites) and consequent
surface reaction controlled dissolution are demonstrated in the laboratory, it is
probable that dissolution in sediments is also surface controlled and inhibited
by the same or similar species. Similarly, calculation of the rate of diffusion-
controlled dissolution places a maximum limit on the rate of dissolution in a
given stagnant pore water. Laboratory rates cannot be extrapolated directly to

sediments because trace poisons may exist in natural pore waters which were not measured or studied in the lab and because different samples of the same mineral will react at different rates due both to differences in specific surface area and to differences in the concentration of reactive sites (e.g., dislocations) on the surface. From the above, it appears that the best approach to the kinetics of mineral–water interactions is by direct modeling of pore water data (see next section) combined with laboratory studies that enable intelligent extrapolation of results obtained from modeling to other areas.

Transport Mechanisms

Diagenetic Modeling

At a given depth in a sediment, chemical reaction is not the only process causing changes to occur in the chemical composition of the pore water. Due to sharp vertical gradients in concentration brought about by reaction, diffusion and advection can also bring about changes. This is expressed mathematically (Berner, 1975) in terms of a one-dimensional depth model as follows:

$$\frac{\partial(\phi C)}{\partial t} = \frac{\partial[\phi D(\partial C/\partial x)]}{\partial x} - \frac{\partial(\phi v C)}{\partial x} + \phi R(C, C_j) \qquad (2)$$

where C is the concentration of a dissolved species in mass per unit volume of pore water, x is the depth in the sediment measured positively downward from the sediment–water interface, ϕ is the porosity, i.e., volume of pore water per unit volume of sediment, D is the diffusion coefficient for the sediment including the effects of tortuosity but not adsorption or ion exchange, v is the vertical velocity of pore water relative to the sediment–water interface, and R is the rate of all chemical reactions (or radioactive decay) affecting the value of C. C_j refers to concentration(s) of other species involved in the reactions. The three terms on the right reflect the processes, respectively, of diffusion, advection, and chemical reaction. The appearance of the porosity ϕ in the expression is due to the fact that fine-grained sediments undergo compaction (decrease of ϕ) with depth, especially in fine-grained sediments within the benthic boundary layer. This gives rise to the burial of pore water more slowly than the burial of originally enclosing sediment particles. In other words, the water moves upward relative to the surrounding particles as a response to compaction. Equations used heretofore to correct for compaction (e.g., Anikouchine, 1967; Tzur, 1971) are, strictly speaking, only special, simplified cases of the more general expression, equation (2).

Compaction is often ignored in diagenetic modeling and this is justified if chemical gradients far exceed porosity gradients, as would be the case, for example, in sands. In this case, ϕ is assumed to be constant both with depth and with time. If ϕ is a constant, then $\partial\phi/\partial t = 0$, $\partial\phi/\partial x = 0$, $\partial v/\partial x = 0$ (water is assumed to be incompressible), $v = \omega$ (the rate of burial of sediment particles or "rate of deposition"), and, in general, $\partial D/\partial x = 0$, so that equation (2) greatly simplifies to

$$\frac{\partial C}{\partial t} = D \frac{\partial^2 C}{\partial x^2} - \omega \frac{\partial C}{\partial x} + R(C, C_j) \tag{3}$$

Here the diffusion term represents simply Fick's second law of diffusion, and the advection term is more readily grasped since flow is not divergent. Equations of this type have been used by many workers (e.g., Goldberg and Koide, 1963; Berner, 1964; Lynn and Bonatti, 1965; Li et al., 1969; Lerman, 1971; Lerman and Weiler, 1970; Michard, 1971; Hurd, 1973).

A digression is necessary at this point because it bears on the relative importance of advection. As stated above, the chemical rate term expresses all chemical reactions affecting the value of C. Included are adsorption and ion exchange reactions. However, adsorption equilibrium is often attained rapidly, and in this case, a different formalism is generally adopted (Duursma and Hoede, 1967; Berner, 1974a, 1976). It can be readily shown that if equilibrium adsorption or ion exchange is assumed, then equation (3), for example, is modified to

$$\frac{\partial C}{\partial t} = I(C)D \frac{\partial^2 C}{\partial x^2} - \omega \frac{\partial C}{\partial x} + I(C)R(C, C_j) \tag{4}$$

where $I(C) = (1 + d\bar{C}/dC)^{-1}$ and $\bar{C}$ is the concentration (in mass per unit volume of pore water) of an adsorbed species in equilibrium with its concentration in solution, C. Since $\bar{C}$ increases with increasing C, $d\bar{C}/dC > 0$ and $I(C) < 1$. Thus, if $I(C)$ is sufficiently low, or, in other words, adsorption is especially strong, the relative importance of advection, as compared with diffusion, may be increased. This is often ignored. Lowered diffusion due to adsorption and ion exchange is often expressed in terms of a migration or pseudodiffusion coefficient $D' = I(C)D$. However, this is confusing because $I(C)$ also appears explicit in the chemical rate term as a coefficient and this fact is generally overlooked. For strongly adsorbing species it is important that the correct formulation of ion exchange or adsorption in diagenetic equations be made. Otherwise, incorrect conclusions may arise.

A useful concept in modeling transport and chemical reactions in sediments which is very commonly employed, is that of steady-state diagenesis. This means that processes so adjust themselves that the change of concentration at a given depth due to chemical reaction is exactly balanced by diffusion and advection. In this case $\partial C/\partial t = 0$. This concept is illustrated diagrammatically in Fig. 9. (For sediments where compaction is important, the steady state concept can be expanded to include compaction so that both $\partial C/\partial t = 0$ and $\partial\phi/\partial t = 0$ and, thus,

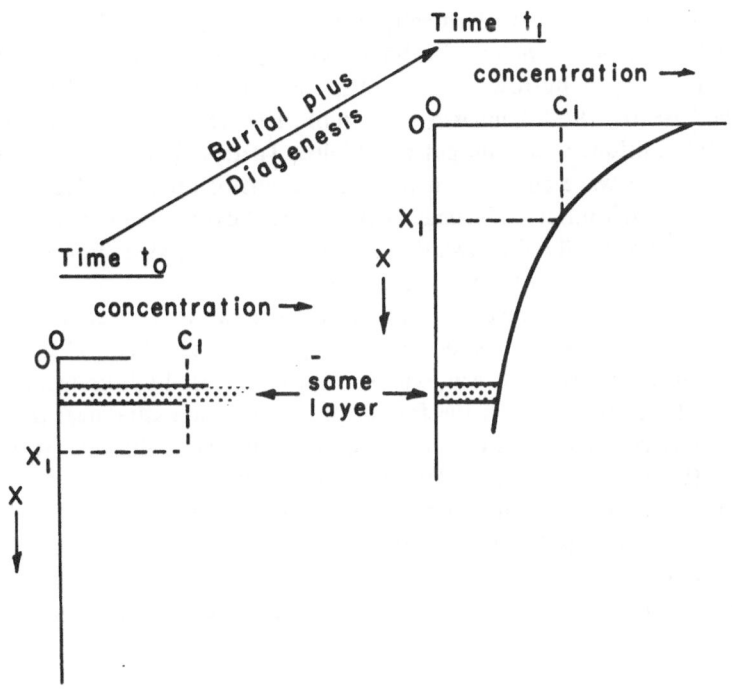

Fig. 9. Steady-state diagenesis represented diagrammatically. Note that at a fixed depth (x_1) concentration remains constant with time, but that within a given layer it changes during burial.

$\partial(\phi C)/\partial t$ in equation (2) is equal to zero.) Steady-state diagenesis means that all changes with depth are due to diagenesis and not to changes at the time of deposition.

Transport of dissolved constituents in a sediment is not affected only by advection, from burial and compaction, and by simple molecular diffusion. Upward flow of ground water driven by hydrostatic heads on nearby landmasses (e.g., Manheim and Chan, 1974) can be present and must therefore be included within the velocity v in equation (2). More importantly, within the benthic boundary layer bioturbation (stirring of sediment by the activity of burrowing benthonic organisms), stirring and resuspension by waves and currents, and tidal pumping (Morse, 1974a; Wimbush and Munk, 1970) may be significant processes for transporting dissolved constituents. The most studied of these processes is bioturbation (e.g., Rhoads, 1974). Bioturbation is especially important when calculating fluxes between sediments and overlying water, in that it is most important at the sediment–water interface. Mathematical modeling of bioturbation is in its infancy, but progress made so far suggests that mixing due to bioturbation can be treated stochastically in terms of a "diffusion" or mixing coefficient, provided that samples of sediment are large enough to average out local irregularities due to burrows, worm tubes, etc. (Goldberg and Koide, 1962; Berger and Heath, 1968; Guinasso and Schink, 1973; Hanor, personal communication). In addition, Vanderborght and Wollast (personal communication) have similarly treated wave and current mixing in terms of diffusion. Questions still arise whether such mixing coefficients are constant over a fixed depth of mixing or decrease continually downward, and whether the mixing process is better treated as an advective, rather than as a diffusive, process.

If bioturbation and wave and current stirring can be treated as a diffusive process, then equation (2) need not be modified to include them. This is because D, the diffusion coefficient, can express both biological–hydrodynamic mixing and molecular diffusion, with the former processes dominating near the surface and the latter dominating at depth. For deep-sea pelagic sediments, it has been suggested (Guinasso and Schink, 1973) that molecular diffusion is quantitatively more important than bioturbation at all depths in the sediment, but in nearshore, shallow-water sediments this situation is most unlikely, especially in the top 10 cm of sediment (Rhoads, 1974). Obviously, much more research on sediment–water mixing processes is needed to check these suggestions.

Fluxes between Sediments and the Ocean

One of the most important aspects of pore water transport is the determination of the flux of dissolved constituents between sediments and the overlying water. This can be accomplished by material balance calculations for small enclosed water bodies such as lakes (Lee, 1970), by direct measurement (Fanning and Pilson, 1974; Hallberg et al., 1972; M. Bender, personal communication), or by calculation from pore water concentration gradients. The latter technique is the usual one. From the studies summarized earlier in this paper a qualitative summary of the direction of fluxes to be expected is shown in Fig. 10.

The appropriate expression for calculating fluxes is

$$J_i = \frac{C_0 J_s}{\rho_s} \left(\frac{\phi_X}{1 - \phi_X} \right) - \phi_0 D_0 \left. \frac{\partial C}{\partial x} \right|_{x=0} \tag{5}$$

where J_i is the flux of dissolved constituent i between sediment and overlying water, measured positively downward, in terms of mass per unit area of sediment per unit time; J_s is the flux of solid particles to the sediment due to deposition; ρ_s is the mean density of solids; C_0 is the concentration at the sediment–water interface in terms of mass per unit volume of water; D_0 is the diffusion coefficient at the sediment–water interface; ϕ_0 is the porosity at the sediment–water interface; and ϕ_X is the porosity at a depth X below which either porosity re-

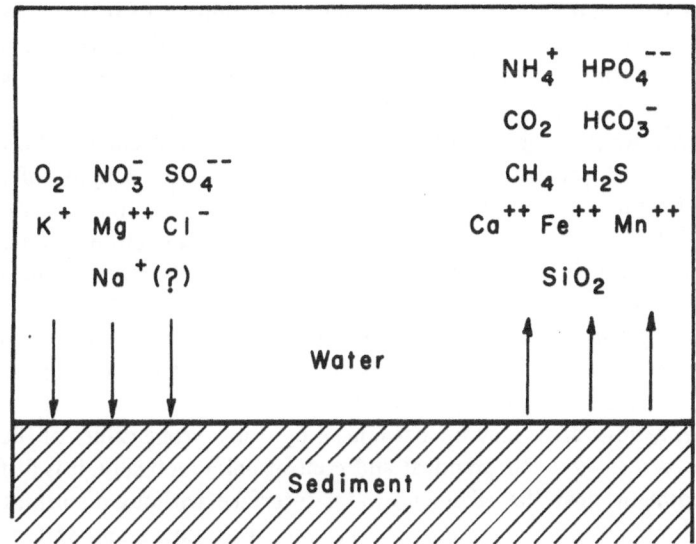

Fig. 10. Direction of fluxes (shown by arrows) expected for dissolved constituents between sea water and sediment pore waters.

mains constant or the continuous upward flow of water due to compaction is interrupted by the presence of basement, permeable sand layers, etc. The first term on the right represents burial of pore water by deposition, while the latter term represents diffusion between sediment and overlying water. Note that ϕ_X, and not ϕ_0, is used to describe burial. This is because compaction forces pore water up into surficial layers at the same time that the overlying water is being buried in the same layers, and only the water being buried is of interest. Equation (5), to be useful, implies steady-state porosity and concentration gradients near the sediment surface. This enables measurements made at one instant of time to apply over much longer periods. Averaged over the entire ocean, this assumption is probably justified but checks of steady state are always needed.

Calculation of sediment–water fluxes is often done erroneously. First, depositional burial is ignored which is probably correct for pelagic sediments where values of J_s are very low, but would be incorrect for rapidly deposited near-shore sediments, such as those within a delta complex. Second, the effect of rapid-equilibrium adsorption or ion exchange is often included within the diffusion coefficient, which is incorrect when applied to Fick's first law calculations where only true diffusion is involved. Third, turbulent mixing by bioturbation, waves, and currents is generally ignored. This is justifiable only in those cases where there is negligible chemical reaction within the zone of turbulent mixing. In that case, flux into the zone from below via molecular diffusion must, at steady state, equal flux via mixing to the overlying water. In general this is not true, and thus equation (5) must be modified to include the effects of turbulent mixing within the benthic boundary layer. From the previous discussion it is possible that mixing of this sort may be described by equation (5) if its effects are included within the diffusion coefficient D.

A fourth error is the use of incorrect values for molecular diffusion coefficients which at present is unavoidable because of the paucity of data for various sediment types and various dissolved species in sea water. Recent studies, however, are starting to make some headway in this direction (Duursma and Bosch, 1970; Manheim, 1970; Ben-Yaakov, 1972; Li and Gregory, 1974; Fanning and Pilson, 1974).

A fifth error in calculating fluxes is that ϕ_0 is often omitted in the diffusive term. This is practically justified in most cases, in that other errors are much larger, but it is not conceptually correct.

Finally, serious error may be introduced by using wrong values of $(\partial C/\partial x)$ at $x = 0$. The errors are both practical and conceptual. Practically, it is difficult to sample sediment near the sediment–water interface without disturbing it and bringing about anthropoturbation, the mixing of sediment and water by sampling. Undisturbed, accurately defined concentration gradients within the top few centimeters are necessary to obtain reliable diffusive fluxes. Often, the top few centimeters are not sampled, and concentration profiles are drawn connecting points at depths of ten or more centimeters below the surface with values for the overlying oceanic bottom water. This procedure completely misses sharp gradients in the top few centimeters which may well be present. The conceptual error is that the laminar-conductive sublayer (Wimbush and Munk, 1970) in the overlying water is ignored. There must be a molecular diffusion gradient in this layer as well as in the sediment. Thus, the concentration at the sediment–water interface, where $x = 0$, is not the value for average bottom water which is normally sampled many meters above this interface. This problem has been pointed out recently (Morse, 1974b) and is illustrated in Fig. 11. Limits on the value of C_0 (Morse, 1974b) can be made by considerations of the thickness of the laminar-conductive sublayer (at most a few millimeters) and the relative values of diffusion coefficients in the sublayer and in the immediately underlying sediment.

Perhaps the best way to determine fluxes between sediments and overlying waters is to avoid all of the above problems by measuring the fluxes directly, either *in situ* (Hallberg *et al.*, 1972) or in the laboratory using carefully obtained cores overlain by sea water of controlled composition and hydrodynamic state (Fanning and Pilson, 1974; M. Bender, personal communication). The problem here is that by introducing a measuring device at the sea bottom or by taking samples into the laboratory the system to be measured undergoes unavoidable perturbation. For instance, a device placed over the bottom to collect outfluxing ions must be closed to flow from surrounding sea water, and, thus, flow condi-

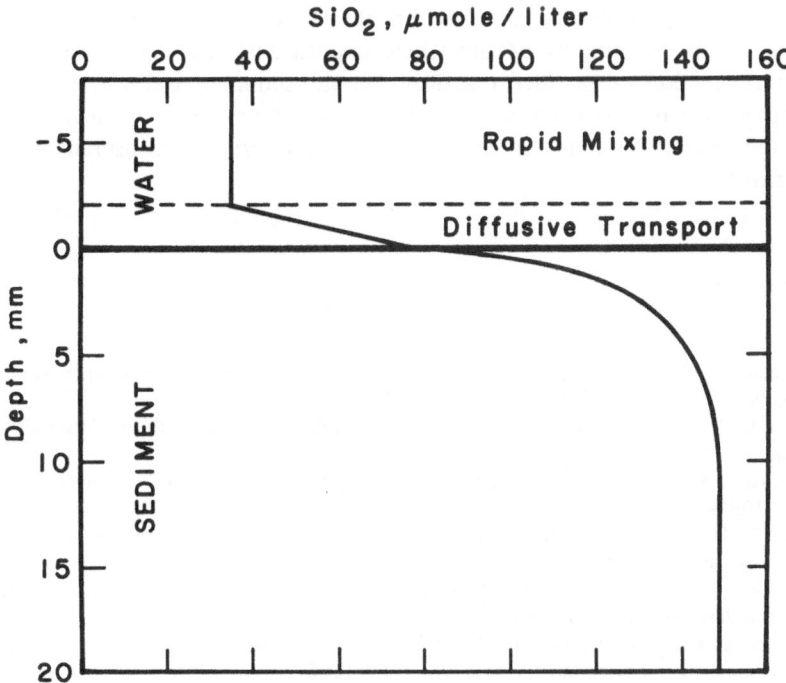

Fig. 11. Concentration profile for dissolved silica in sediment and overlying pore water according to model calculations which include consideration of the laminar-conductive sublayer (the zone labeled "diffusive transport"). (Modified from Morse, 1974b.)

tions at the bottom and chemical conditions sensitive to flow (e.g., oxygenation) are necessarily altered. Probably the best way out of the problem is to determine fluxes both by direct methods and by calculation from pore water profiles at the same site and to compare results (and hope for agreement). Obviously, much more work in this area is needed.

In the deep sea, pelagic sediments are deposited very slowly, and bioturbation plus current mixing may, in some areas at least, be slow relative to molecular diffusion. Diffusion coefficients for several species can be estimated to within a factor of 3. Also, for some species concentration profiles have been measured which, when the laminar-conductive sublayer is considered, are at least good to one order of magnitude (Manheim, 1976; Fanning and Pilson, 1974; Sayles and Mangelsdorf, personal communication). Thus, fluxes can be calculated with some assurance, and results (Manheim, 1976) show that molecular diffusive fluxes between deep sea sediments and the ocean are of the same order of magnitude as river inputs. This indicates the importance of sediment flux as a factor which controls the composition of sea water.

By contrast, in shallow water sediments where bioturbation and wave plus current mixing are important and where depositional rates are high and much chemical reaction takes place practically *at* the sediment-water interface, it is perhaps better not to make any conclusions based on simple diffusive calculations until better data are obtained and/or until *in situ* flux measurements have been made.

References

Anikouchine, W. A., 1967, Dissolved chemical substances in compacting marine sediments, *Journal of Geophysical Research*, *72:* 505–509.

Baturin, G. N., 1971, Stages of phosphorite formation on the ocean floor, *Nature Physical Science*, *232:* 61–62.

Ben-Yaakov, S., 1972, Diffusion of sea water ions, 1, *Geochimica et Cosmochimica Acta*, *36:* 1395–1406.

Berger, W. H., 1968, Planktonic foraminifera: selective solution and the lysocline, *Marine Geology*, *8:* 111–138.

Berger, W. H., and Heath, G. R., 1968, Vertical mixing in pelagic sediments, *Journal of Marine Research*, *26:* 134–143.

Berner, R. A., 1964, An idealized model of dissolved sulfate distribution in recent sediments, *Geochimica et Cosmochimica, Acta, 28:* 1497–1503.

Berner, R. A., 1966, Chemical diagenesis of some modern carbonate sediments, *American Journal of Science, 264:* 1–36.

Berner, R. A., 1971, *Principles of Chemical Sedimentology*, McGraw-Hill, N.Y., 240 pp.

Berner, R. A., 1974a, Kinetic models for the early diagenesis of nitrogen, sulfur, phosphorus, and silicon in anoxic marine sediments, *in: The Sea, 5* (E. D. Goldberg, ed.), Wiley-Interscience, pp. 427–450.

Berner, R. A., 1974b, Physical chemistry of carbonates in the oceans *in: Studies in Paleo-Oceanography*, Society of Economic Paleontologists and Mineralogists, Memoir 20, pp. 37–43.

Berner, R. A., 1975, Diagenetic models of dissolved species in the interstitial waters of compacting sediments, *American Journal of Science, 275:* 88–96.

Berner, R. A., 1976, Inclusion of adsorption in the modeling of early diagenesis, *Earth and Planetary Science Letters*, (in press).

Berner, R. A., Scott, M. R., and Thomlinson, C., 1970, Carbonate alkalinity in the pore waters of anoxic marine sediments, *Limnology and Oceanography, 15:* 544–549.

Billen, G., 1975, Nitrification in the Scheldt Estuary (Belgium and the Netherlands), *Estuarine and Coastal Marine Science, 3:* 79–89.

Bischoff, J. L., and Ku, T., 1970, Pore fluids of recent marine sediments: I. Oxidising sediments of 20°N. Continental Rise to Mid-Atlantic Ridge, *Journal of Sedimentary Petrology, 40:* 960–972.

Bischoff, J. L., and Ku, T., 1971, Pore fluids of recent marine sediments II. Anoxic sediments of 35° to 45°N. Gibraltar to mid-Atlantic Ridge, *Journal of Sedimentary Petrology, 41:* 1008–1017.

Brafield, A. E., 1964, The oxygen content of interstitial water in sandy shores, *Journal of Animal Ecology, 33:* 97–116.

Bray, J. T., Bricker, O. P., and Troup, B. N., 1973, Phosphate in interstitial waters of anoxic sediments: Oxidation effects during sampling procedure, *Science, 180:* 1362-1364.

Brooks, R. P., Presley, B. J., and Kaplan, I. R., 1968, Trace elements in the interstitial waters of marine sediments, *Geochimica et Cosmochimica Acta, 32:* 397-414.

Bruevich, S. V., 1938, Oxidation-reduction potential and the pH of sediments of the Barents and Kara Seas, *Doklady Akademiya Nauka SSSR, 19:* 637-648.

Bruevich, S. V., 1966, Chemistry of interstitial waters in sediments of the Pacific Ocean, *in: Khimiya Tikhogo Okeana, 2,* Izdatel'stvo Academiya Nauka, Moscow, pp. 263-358.

Calvert, S. E., and Price, N. B., 1972, Diffusion and reaction profiles of dissolved manganese in the pore waters of marine sediments, *Earth and Planetary Science Letters, 16:* 245-249.

Debyser, J., and Rouge, P. I., 1956, Sur l'origine du fer dans les eaux interstitielles des sediments marins actuels, *Comptes Rendues de l'Academie des Sciences, 243:* 2111-2113.

Drever, J. I., 1974, The magnesium problem, *in: The Sea, 5* (E. D. Goldberg, ed.), Wiley-Interscience, N.Y., pp. 337-357.

Duursma, E. K., and Bosch, C. J., 1970, Theoretical, experimental, and field studies concerning diffusion of radioisotopes in sediments and suspended solid particles of the sea, Part B, *Netherlands Journal of Sea Research, 4:* 395-469.

Duursma, E. K., and Hoede, C., 1967, Theoretical, experimental, and field studies concerning molecular diffusion of radioisotopes in sediments and suspended particles of the sea, Part A, *Netherlands Journal of Sea Research, 3:* 423-457.

Emery, K. O., and Rittenberg, S. C., 1952, Early diagenesis of California Basin sediments in relation to origin of oil, *American Association of Petroleum Geologists Bulletin, 36:* 735-806.

Fanning, K. A., and Pilson, M. E. Q., 1974, The diffusion of dissolved silica out of deep sea sediments, *Journal of Geophysical Research, 79:* 1293-1297.

Freidman, G. M., and Gavish, E., 1970, Chemical changes in interstitial waters from sediments of lagoonal, deltaic, river, estuarine, and salt water marsh and cove environments, *Journal of Sedimentary Petrology, 40:* 930-953.

Gieskes, J. M., 1974, The alkalinity–total carbon dioxide system in seawater, *in: The Sea, 5,* (E. D. Goldberg, ed.), Wiley-Interscience, N.Y., pp. 123-151.

Glasby, G. P., 1973, Interstitial waters in marine and lacustrine sediments: A review, *Journal of the Royal Society of New Zealand, 3:* 43-59.

Goldberg, E. D., and Bruland, K., 1974, Radioactive geochronologies, *in: The Sea, 5,* (E. D. Goldberg, ed.), Wiley-Interscience, N.Y., pp. 451-490.

Goldberg, E. D., and Koide, M., 1962, Geochronological studies of deep-sea sediments by the Io/Th method. *Geochimica et Cosmochimica Acta, 26:* 417-450.

Goldberg, E. D., and Koide, M., 1963, Rates of sediment accumulation in the Indian Ocean, *in: Earth Science and Meteoritics* (J. Geiss and E. D. Goldberg, eds.), North Holland, Amsterdam, pp. 90-102.

Goldhaber, M., 1974, Equilibrium dynamic aspects of marine geochemistry of sulfur, Ph.D. Dissertation, University of California, Los Angeles.

Goldhaber, M., and Kaplan, I. R., 1974, The sulfur cycle, *in: The Sea, 5,* (E. D. Goldberg, ed.), Wiley-Interscience, N.Y., pp. 599-655.

Guinasso, N. L., and Schink, D. R., 1973, Quantatitive evaluation of bioturbation rates in deep sea sediments, *EOS, Transactions of the American Geophysical Union, 54,* p. 337.

Hallberg, R. O., Bagnander, L. E., Engvail, A. G., and Schippel, 1972, Method for studying geochemistry of sediment–water interface, *Ambio, 1:* 71-72.

Harris, R. C., and Pilkey, O. H., 1966, Interstitial water of some deep marine carbonate sediments, *Deep Sea Research, 13:* 967-969.

Hartmann, M., 1964, Zur geochemie von Mangan und Eisen in der Ostsee, *Meyniana, 14:* 3-20.

Hartmann, M., and Nielsen, H., 1969, Delta-34 S-Werte in recenten Meeressedimenten und ihre Deutung am Beispiel einiger Sediment-profile aus der westlichen Ostsee, *Geologische Rundschau, 58:* 621-655.

Hartmann, M., Muller, P., Suess, E., and van der Weijden, C. H., 1973, Oxidation of organic matter in recent marine sediments, *Meteor Forschung Ergebnisse, Reihe C, 12:* 74-86.

Hurd, D. C., 1973, Interactions of biogenic opal, sediments, and sea water in the central equatorial Pacific, *Geochimica et Cosmochimica Acta, 37:* 2257-2282.

Ivanov, M. V., 1968, Microbiological processes in the formation of sulfur deposits, Israel Program for Scientific Translation, Jerusalem.

Kaplan, I. R., and Rittenberg, S. C., 1964, Microbiological fractionation of sulfur isotopes, *Journal of General Microbiology, 34:* 195-212.

Kaplan, I. R., Emery, K. O., and Rittenberg, S. C., 1963, The distribution of isotopic abundance of sulfur in recent marine sediments off southern California, *Geochimica et Cosmochimica Acta, 27:* 297-331.

Lee, G. F., 1970, *Factors affecting the transfer of materials between water and sediments,* Literature Review No. 1, Eutrophication Information Program, University of Wisconsin, 35 pp.

Lerman, A., 1971. Time to chemical steady-state in lakes and oceans, *in: Non-equilibrium Concepts in Natural Water Systems,* American Chemical Society Series, *106:* 30-76.

Lerman, A., and Weiler, R. R., 1970, Diffusion and accumulation of chloride and sodium in Lake Ontario sediment, *Earth and Planetary Science Letters, 10:* 150-156.

Li, Y-H., and Gregory, S., 1974, Diffusion of ions in sea water and in deep sea sediments, *Geochimica et Cosmochimica Acta, 38:* 703-714.

Li, Y-H., Bischoff, J., and Mathieu, G., 1969, The migration of manganese in the Arctic Basin sediments, *Earth and Planetary Science Letters, 7:* 265-270.

Lynn, D. C., and Bonatti, E., 1965, Mobility of mangenese in the diagenesis of deep-sea sediments, *Marine Geology, 3:* 457-474.

Mackenzie, F. T., and Garrels, R. M., 1966, Chemical mass balance between rivers and oceans, *American Journal of Science, 264:* 507-525.

Manheim, F. T., 1970, The diffusion of ions in unconsolidated sediments, *Earth and Planetary Science Letters, 9:* 307-309.

Manheim, F. T., 1976, Interstitial waters of marine sediments, *in: Chemical Oceanography, 3,* (J. P. Riley and G. Skirrow, eds.) (in press).

Manheim, F. T., and Chan, K. M. 1974, Interstitial waters of Black Sea sediments: New data and review, *in: The Black Sea—Geology, Chemistry, and Biology,* (E. T. Degens and D. A. Ross, eds.), American Association of Petroleum Geologists Memoir, *20:* 155-182.

Manheim, F. T., and Sayles, F. L., 1974, Composition and origin of interstitial waters of . marine sediments based on deep sea drill cores, *in: The Sea, 5,* (E. D. Goldberg, ed.), Wiley-Interscience, N.Y., pp. 527-568.

Martens, C. S., and Berner, R. A., 1974, Methane production in the interstitial waters of sulfate depleted marine sediments, *Science, 85:* 1167-1169.

Michard, G., 1971, Theoretical model for manganese distribution in calcareous sediment cores, *Journal of Geophysical Research, 76:* 2179-2186.

Milliman, J. D., and Müller, J., 1973, Precipitation and lithification of magnesian calcite in the deep-sea sediments of the eastern Mediterranean Sea, *Sedimentology 20:* 29-45.

Morse, J. W., 1974a, Tidal pumping as a possible major transport mechanism of surface active dissolved constituents of interstitial waters (submitted to *Journal of Geophysical Research*).

Morse, J. W., 1974b, Calculation of diffusive fluxes across the sediment–water interface, *Journal of Geophysical Research, 33:* 5045-5048.

Murray, J., and Irvine, R., 1895, On the chemical changes which take place in the composition of the sea water associated with blue muds on the floor of the ocean, *Transactions of the Royal Society of Edinburgh, 37:* 481-507.

Nakai, N., and Jensen, M. L., 1964, The kinetic isotope effect in the bacterial reduction and oxidation of sulfur, *Geochimica et Cosmochimica Acta, 28:* 1893-1912.

Nissenbaum, A., Presley, B. J., and Kaplan, I. R., 1972, Early diagenesis in a reducing fjord, Saanich Inlet, British Columbia-I. Chemical and isotopic changes in major components of interstitial water, *Geochimica et Cosmochimica Acta, 36:* 1007-1027.

Presley, B. J., and Kaplan, I. R., 1968, Changes in dissolved sulfate, calcium and carbonate from interstitial water of near-shore sediments, *Geochimica et Cosmochimica Acta, 32:* 1037-1048.

Ramm, A. E., and Bella, D. A., 1974, Sulfide production in anaerobic microcosms, *Limnology and Oceanography, 19:* 110-118.

Reeburgh, W. S., 1969, Observations of gases in Chesapeake Bay sediments, *Limnology and Oceanography, 14:* 368-375.

Rickard, D. T., 1975, Kinetics and mechanism of pyrite formation at low temperatures, *American Journal of Science, 275:* 636-652.

Rittenberg, S. C., Emery, K. O., and Orr, W. L., 1955, Regeneration of nutrients in sediments of marine basins, *Deep-Sea Research, 3:* 23-45.

Rhoads, D. C., 1974, Organism–sediment relations on the muddy seafloor, *in: Oceanography and Marine Biology Annual Review,* (H. Barnes, ed.), Allen and Unwin, London, pp. 263-300.

Schink, D. R., Fanning, K. A., and Pilson, M. E. Q., 1974, Dissolved silica in the upper pore water of the Atlantic Ocean floor, *Journal of Geophysical Research, 79:* 2243-2250.

Shishkina, O. V., 1959, Sulfate in the pore waters of Black Sea sediments, *Trudy Instituta Okeanologiya Akademiya Nauka SSSR, 33:* 178-193.

Shishkina, O. V., 1972, *Geochemistry of marine and oceanic interstitial waters,* Izdatel'stvo Nauka, Moscow, 227 pp.

Sholkovitz, E., 1973, Interstitial water chemistry of the Santa Barbara Basin sediments, *Geochimica et Cosmochimica Acta, 37:* 2043-2073.

Siever, R., Beck, K. C., and Berner, R. A., 1965, Composition of interstitial waters of modern sediments, *Journal of Geology, 73:* 39-73.

Sillén, L. G., 1961, The physical chemistry of sea water, *in: Oceanography,* (M. Sears, ed.), American Association for the Advancement of Science, Washington, D.C., pp. 549-581.

Sorokin, Y. I., 1962, Experimental investigation of bacterial sulfate reduction in the Black Sea using S^{35}, *Mikrobiologiya, 31:* 402-410.

Sorokin, Y. I., 1970, Interrelations between sulfur and carbon turnover in meromictic lakes, *Archives für Hydrobiologie, 66:* 391-446.

Stuiver, M., 1967, The sulfur cycle in lake waters during thermal stratification, *Geochimica et Cosmochimica Acta, 31:* 2151-2157.

Thorstenson, D., and Mackenzie, F. T., 1974, The variability of pore water chemistry in recent carbonate sediments, Devil's Hole, Harrington Sound, Bermuda, *Geochimica et Cosmochimica Acta, 38:* 1-19.

Tzur, Y., 1971, Interstitial diffusion and advection of solute in accumulating sediments, *Journal of Geophysical Research, 76:* 4208-4211.

Wimbush, M., and Munk, W., 1970, The benthic boundary layer, *in: The Sea, 4(i),* (A. E. Maxwell, ed.), Wiley-Interscience, N.Y., pp. 731-758.

Wollast, R., 1974, The silica problem, *in: The Sea, 5,* (E. D. Goldberg, ed.), Wiley-Interscience, N.Y., pp. 359-392.

4
Nutrients Near the Depositional Interface

ERWIN SUESS

The regeneration of nutrients in the benthic boundary layer is discussed with examples from a stagnant basin of the Baltic Sea, from the continental slope off West Africa, and from deep oceanic sediments of the Central Pacific. Under stagnant conditions in fjord waters and sediments more phosphate and less ammonia are regenerated than would be predicted from chemical decomposition models, but total carbon dioxide and hydrogen sulfide are found in the expected amounts. The excess of dissolved phosphate is related to dissolution of inor-·ganic iron–manganese–phosphate precipitates and the deficit in ammonia is thought to re-flect decomposition of nitrogen-deficient organic substances rather than the relatively nitrogen-rich substances used in the chemical decomposition model. At the continental margin off West Africa, horizontal and vertical concentration profiles of nutrient constit-uents in the interstitial water indicate an increased flux of nutrients from the organic-rich sediments to the waters of the lower continental slope. The bottom water there contains increased amounts of dissolved phosphate and is depleted in oxygen. It is suggested that this is due to the combined effect of increased nutrient regeneration from within the sediments and from the sediment surface. Sediments from the deep Pacific Ocean characteristically show decreasing organic carbon and nitrogen depth-profiles near the depositional interface. This could be attributed to organic decomposition processes, i.e., to regeneration of nutri-ents. The loss of organic carbon and nitrogen from the sediment to the water equals possible molecular diffusion transport in the form of nitrate and dissolved organic constituents.

Introduction

Regeneration of nutrients from marine sediments is an important process for those studying primary productivity in the oceans. Chemical oceanographers

ERWIN SUESS • Geologisch-Paläontologisches Institut der Universität, 23 Kiel, Ols-haustentrasse 40/60, Germany.

have accumulated enough evidence to show that in the open ocean most of the nutrients are regenerated within the water column by *in situ* decomposition of organic matter before it reaches the bottom (Riley, 1951; Redfield *et al.*, 1963; and others). On the shelf and in other shallow water areas, however, regeneration involves the sediment surface because the settling time for organic detritus is relatively short and the supply of organic matter is large (Riley, 1956; v. Bodungen *et al.*, 1974). In stagnant basins where restricted circulation lowers the turnover rate, nutrient regeneration builds up high concentrations in the free water through both an increased supply of organic matter and an increased release from within the sediment (Richards, 1965; Rittenberg *et al.*, 1955).

In all three of these oceanographic situations, some regeneration of nutrients takes place within the benthic boundary layer through various processes. It is in the benthic boundary layer that organic decomposition often changes from oxygen consumption to sulfate reduction. These processes can be described by reactions based on chemical models first suggested by Redfield *et al.* (1963) and Richards (1965) and recently modified by Almgren *et al.* (1974) using a slightly different model substance. This substance contains nitrogen in peptides and phosphorus in phosphate esters, which means that less sulfate is consumed in the reaction than in the classic Redfield-Ketchum-Richards reaction, as shown in equations (1) and (2); the decomposition of organic matter by sulfate reduction according to Richards (1965) is

$$(CH_2O)_{106} (NH_3)_{16} (H_3PO_4) + 53\ SO_4^{2-} \longrightarrow$$

$$106\ CO_2 + 53\ S^{2-} + 16\ NH_3 + H_3PO_4 + 106\ H_2O \quad (1)$$

and according to Almgren *et al.* (1974) is

$$(CH_2O)_{89} (NHCO)_{16} C (H_2PO_4)^- + 45\ SO_4^{2-} \longrightarrow$$

$$106\ CO_2 + 45\ S^{2-} + 106\ H_2O + 16\ NH_4^+ + H_2PO_4^- \quad (2)$$

Using examples from the Baltic, the continental slope off West Africa, and the deep Central Pacific, this paper seeks to show the relationships involving dissolved nutrients and other constituents released to bottom and interstitial waters from sedimentary organic and inorganic components near the sediment-water interface.

Stagnant Basins

In a recent study on organic decomposition under stagnant conditions, Almgren *et al.* (1974) showed that decomposition in the water column of a fjord on the Swedish west coast releases the predicted amount of carbon dioxide according to the above model. The expected amount of hydrogen sulfide is also produced, but there is less ammonia and more phosphate found in the water than would be predicted from equation (2).

The excess phosphate is thought to escape from the sediment surface, and this is illustrated in Fig. 1A by the deviation of the concentration curve of phosphate from the curves of total carbon dioxide and of hydrogen sulfide. One unit of the scale for the abscissa equals 0.025 mmole/liter of phosphate, 0.4 mmole/liter of ammonia, 1.125 mmole/liter of hydrogen sulfide, and 2.65 mmole/liter of total carbon dioxide, each corresponding to the molar ratio of the respective constituent in reaction (2), i.e., $CO_2 : H_2S : NH_4^+ : PO_4^{3-} = 106 : 45 : 16 : 1$. The negative deviation of ammonia, according to these authors, is probably due to oxidation of ammonia to molecular nitrogen, which cannot be measured against the high atmospheric nitrogen background of the fjord waters (Nehring, 1974).

Above the halocline, organic matter is oxidized by the consumption of free oxygen instead of by sulfate reduction and the changes in concentration of the inorganic constituents are not included here, Figure 1B shows a continuation of the concentration profiles from the stagnant basin water into the upper 30 cm of reducing sediments of the central Bornholm Basin. The scaling of the abscissa here is 0.025 mmole/liter of phosphate, 0.4 mmole/liter of ammonia, 1.325 mmole/liter sulfate, and 2.65 mmole/liter of total carbon dioxide per unit, each corresponding to the molar ratios in reaction (1) of $CO_2 : SO_4^{2-} : NH_4^+ : PO_4^{3-} = 106 : 53 : 16 : 1$, i.e., to the Redfield-Ketchum-Richards model.

Not only are the constituents about seven times more concentrated in the interstitial water than in the free water, but also the ammonia and phosphate concentrations continue to deviate from the concentrations of total carbon dioxide and reduced sulfate (= produced hydrogen sulfide) already shown in the free water. This raises the question of whether or not reactions (1) and (2) properly describe the decomposition process. If they were correct, it would mean that inorganic phosphate is still supplied from a source other than organic matter and that the oxidation of ammonia to nitrogen is continuing in the interstitial environment at the same rate as in the free water, which is impossible in the absence of nitrite or nitrate according to Rittenberg et al. (1955).

It would be more plausible to assume that the deficit in ammonia was due to a composition of organic matter with lower nitrogen content or, alternatively, to a fractionation between organic nitrogen and carbon during decomposition; both possibilities were suggested by Hartmann et al. (1973a). To explain the excess of dissolved phosphate, an inorganic source was suggested by Almgren et al. (1974), Hallberg et al. (1972), Sen Gupta (1973) and others. Both explanations, i.e., the decomposition of nitrogen-deficient sedimentary organic matter and the additional release of phosphate from an inorganic source in the sediment during decomposition, may be tested by determining the organic nitrogen and inorganic phosphate content of the sediments from which the interstitial waters were obtained.

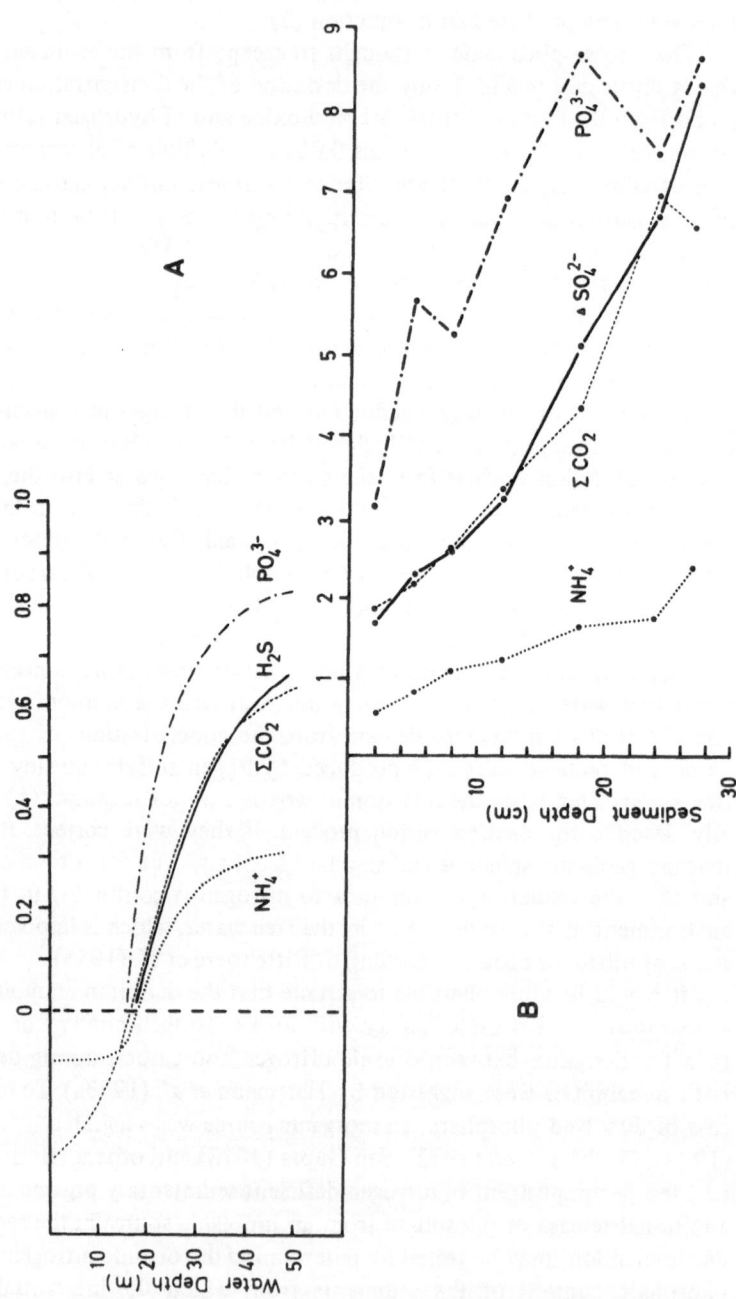

Fig. 1. Changes in dissolved ammonia, total carbonate, phosphate, and hydrogen sulfide with depth in the water column of Byfjorden, western Sweden (A), according to Almgren *et al.* (1974); and in the interstitial waters of the first 30 cm of sediment from the Bornholm Basin, Baltic Sea (B). The scaling of the dissolved constituents for the Byfjorden waters is

$$CO_2 : H_2S : NH_4^+ : PO_4^{3-} = 106 : 45 : 16 : 1$$

and for the Bornholm Basin interstitial waters is

$$CO_2 : SO_4^{2-} : NH_4^+ : PO_4^{3-} = 106 : 53 : 15 : 1$$

Zero marks the first occurrence of hydrogen sulfide. The deviations of NH_4^+ and PO_4^{3-} from the model reaction equations (2) and (1) in both free and interstitial waters indicate that excess phosphate is regenerated and that either some ammonia is lost due to oxidation to nitrogen or there is less organic nitrogen decomposed than would be expected with the reaction model. Planet Cruise 05/1974. Stations: 10193–2, 10191–1, and 10179–1.

In Table 1 are listed the organic carbon, total nitrogen, inorganic (acid-soluble) phosphate, acid-soluble iron, and acid-soluble manganese contents for three near-surface sediment cores from the Bornholm Basin. These results indicate that the molar ratio of organic carbon to organic nitrogen $(= N_{total} - NH_{4fix}^+ - NH_{4exch}^+)$ in the sediments is about 106:10, which is significantly less than that used in equations (1) and (2) but is reasonably close to that observed in the interstitial waters near the sediment surface (106:8; see Fig. 2). Consequently, scaling of the nitrogen axis in Figs. 1A and B to 8 moles of ammonia and 53 moles of reduced sulfate suggests that the composition of the

Table 1. Composition of Bornholm Basin Sediments

Depth, cm	Organic carbon, %	Total nitrogen,[a] %	HCl soluble		
			Iron, %	Phosphorus, ppm	Manganese, ppm
Station 10179-1					
0–3	5.67	0.629	1.210	819	455
5–7	5.25	0.620	1.237	685	265
15–20	5.33	0.553	1.374	732	283
25–30	4.94	0.563	0.909	692	313
30–35	5.35	0.579	1.459	724	195
35–40	4.43	0.496	1.402	764	243
Station 10191-1					
0–3	6.26	0.714	2.321	2260	2000
3–6	5.62	0.634	0.957	841	340
6–10	5.13	0.615	1.402	712	318
10–13	5.16	0.613	1.576	656	275
16–20	5.13	0.626	1.606	602	295
22–25	5.13	0.616	1.517	570	428
25–28	5.23	0.645	1.080	578	223
Station 10193-2					
0–1	5.09	0.692	1.920	1756	1475
1–2	5.09	0.653	1.904	1810	1588
2–3	4.96	0.690	1.596	986	295
3–4	4.97	0.648	1.286	718	181
4–5	4.98	0.591	1.211	730	198
5–6	5.00	0.594	1.533	958	696
6–7	4.72	0.606	1.805	1600	1250
7–8	4.58	0.572	1.241	720	210

[a]About 0.06% may be considered inorganic as fixed and exchangeable ammonia.

model substance should be changed, for equation (1), to

$$(CH_2O)_{106} (NH_3)_8 H_3PO_4 \qquad\qquad (3)$$

and for equation (2), to

$$(CH_2O)_{89} (NHCO)_8 C (H_2PO_4)^- \qquad\qquad (4)$$

in order to provide a better fit to the observed nitrogen concentrations. This also suggests that oxidation of ammonia to molecular nitrogen is necessary to ex-

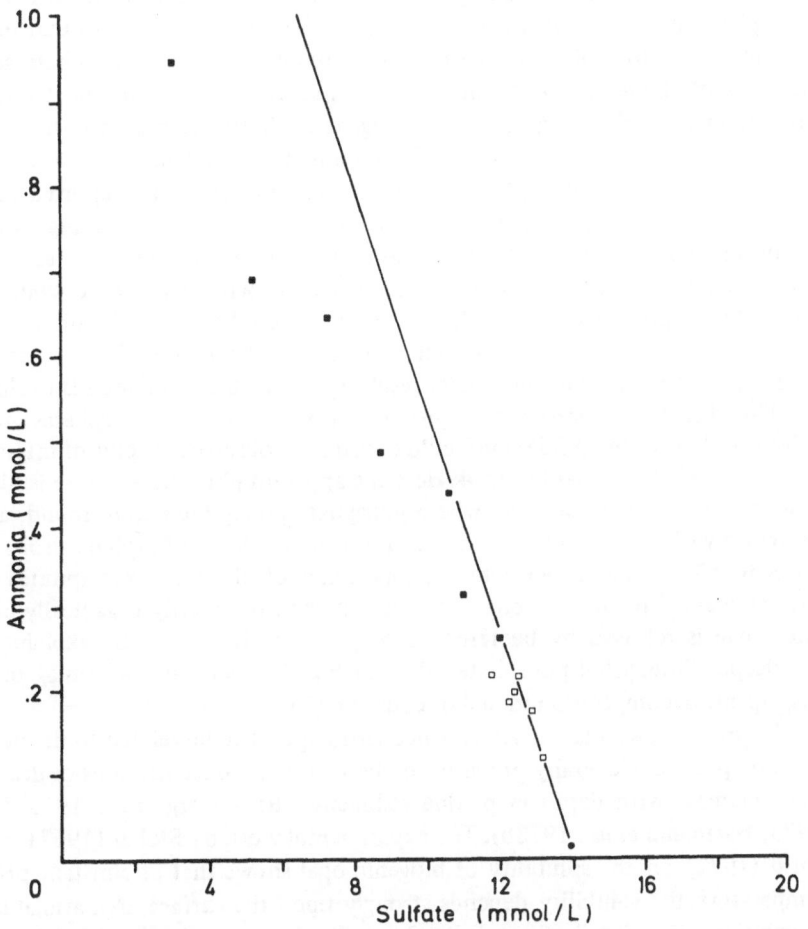

Fig. 2. Molar concentrations of dissolved ammonia and sulfate in the interstitial waters of Bornholm Basin sediments. The depth of each core is from 2–30 cm below the surface; the line indicates molar ratio of $NH_4^+ : -SO_4^{2-} = 53 : 8$. Stations 10191–1 = ▪ ; Station 10179–1 = □ ; ● = bottom water.

plain the low ammonia concentrations only in the deeper parts of the core (Fig. 2). But the deviations there could as well be due to further changes in composition of the model substance. In addition, it appears that the oxidation state of the model substance in equation (1) provides a better fit to the interstitial water data than that used in equation (2), which might be due to the difference between free water and interstitial water environment.

The data plotted in Fig. 3 suggest that authigenic iron–manganese–phosphate precipitates near the sediment surface are the source for the increase in interstitial phosphate. The acid-soluble iron and manganese are shown together with phosphate. The manganese and phosphate have a molar ratio from which a distinct phase may be formed ($Mn:PO_4 = 1:2$), but it is also thought that there is an acid-soluble iron phase which probably contains adsorbed phosphate and increases with increasing phosphate and manganese contents of the sediments. Formation of such authigenic iron–manganese–phosphate phases in the oxidizing-reducing transition zone of lake sediments is well known (Stumm and Morgan, 1970) and Fe(III)- and Fe(II)-phosphates have been reported as well (Tessenow, 1973). The reduction of Fe(III)-phosphates with decreasing sediment depth, i.e., with stronger reducing conditions, could, for example, be responsible for the excess of dissolved interstitial phosphate. With depth they would thus gradually disappear from the sediment, as is indeed the case with sediment samples from below 16 cm depth which contain only "background" concentrations of iron, manganese, and phosphate, resulting from other compounds (Table 1).

This distribution strongly suggests that iron–manganese–phosphates are dissolved by hydrogen sulfide and could explain the observed surplus of interstitial phosphate, as illustrated in Fig. 4. Here the apparent phosphate release is highest close to the sediment surface, where phosphate precipitates were found, and it decreases with depth where the molar ratio of sulfate to phosphate drops from $53:3$ to $53:1$ together with the disappearance of phosphate precipitates from the sediment. This ratio is close to that expected when only organically-bound phosphate is released by bacterial decomposition. This then also explains why the deepest interstitial phosphate values in Fig. 1 appear to conform again with the organic decomposition model of equation (1).

Figure 5 shows that dissolved silica also appears to be related to the decomposition process. Generally, it has been observed that concentrations of dissolved silica increase with depth in marine sediments (Rittenberg et al., 1955; Hurd, 1973; Hartmann et al., 1973b). The experimental work by Stöber (1967) and by Hurd (1972) on the solubility of biogenic opal shows that at constant pH and temperature the solubility depends strongly upon the surface area available for dissolution. The distribution of dissolved silica as shown in Fig. 5 might therefore indicate that an increased decomposition of organic matter in the sediment makes available more surface area of biogenic opal for dissolution. Whether this is due to the removal of protective coatings, such as shown to exist for Baltic Sea

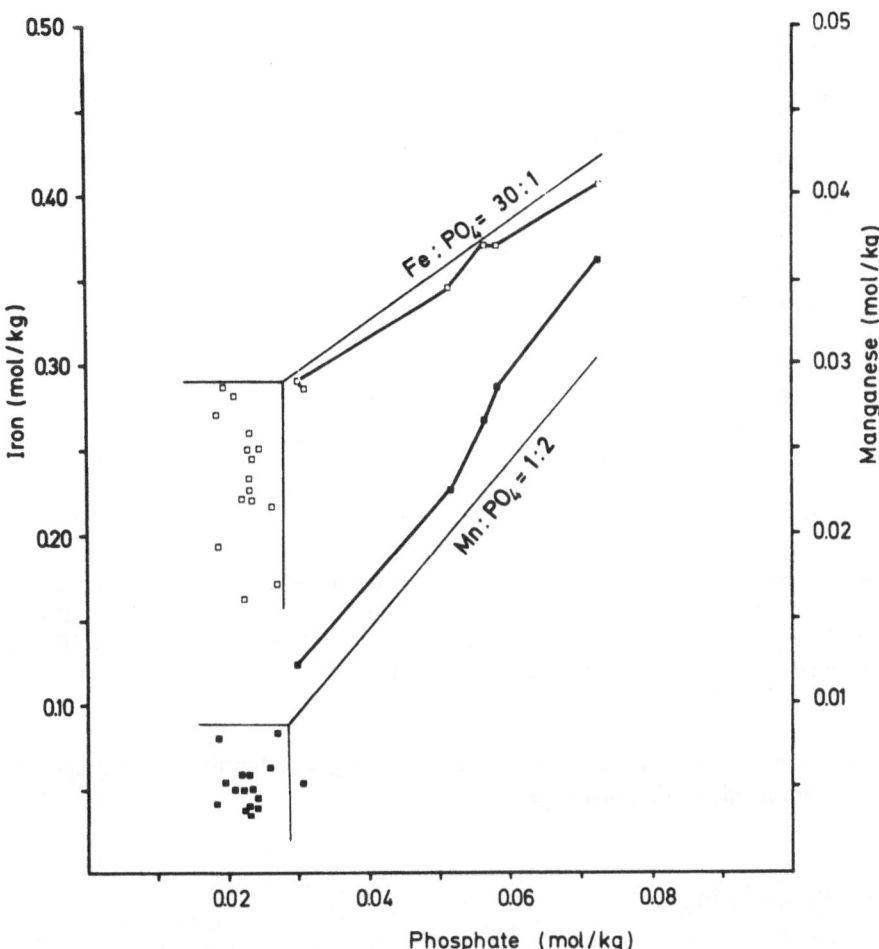

Fig. 3. HCl-soluble phosphate, manganese, and iron in near-surface sediments from the Bornholm Basin. The lines indicate the molar ratios of Fe : PO_4 = 30 : 1 and Mn : PO_4 = 1 : 2. The random distribution of phosphate, manganese, and iron below about 0.03 moles PO_4/kg, 0.03 moles Mn/kg, and 0.30 moles Fe/kg, respectively, is probably due to a mixture of soluble iron sulfides, hydroxides, and desorption or exchange processes between phosphate, iron, and clays, whereas above these values manganese–iron–phosphate and oxide phases appear to determine the element distribution in the HCl-extract. Stations 10193–2, 10191–1, and 10179–1.

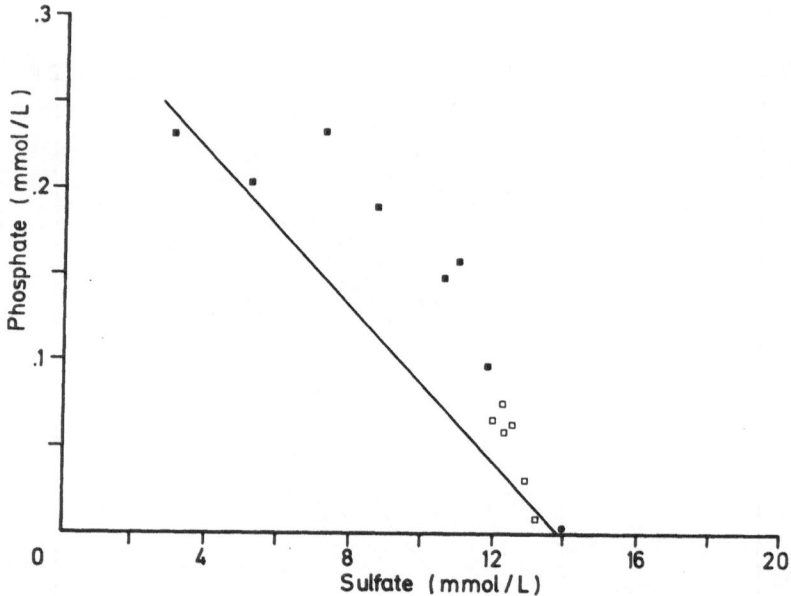

Fig. 4. Molar concentrations of dissolved phosphate and sulfate in the interstitial waters of Bornholm Basin sediments. The line indicates a molar ratio of $PO_4^{3-}:-SO_4^{2-} = 1:53$; samples deviating from this ratio contain additional phosphate from dissolution of manganese–iron–phosphates. Station 10191-1 = ■; Station 10179-1 = □; ● = bottom water.

diatom frustules (Schrader, 1971), or to pH changes within the sedimentary environment needs further investigation.

Continental Slopes

In a different case, on the West African continental slope, the hydrography does not provide such restricted and stagnant conditions, and consequently neither "lost nutrients" in the form of iron–manganese–phosphates nor "byproducts" in the form of metal sulfides are significant constituents of the sediment. But the interstitial water does contain accumulations of various nutrients. The distribution of these nutrients in the sediments shows that they may be supplied to the bottom near the foot of the continental slope because of diffusive transport along steep concentration gradients.

Figures 6A, 6B, 6C, and 6D illustrate respectively changes in interstitial sulfates, ammonia, carbon dioxide, and phosphate concentrations along an East-West profile at about 16°N (north of Dakar, Senegal). The length of the profile

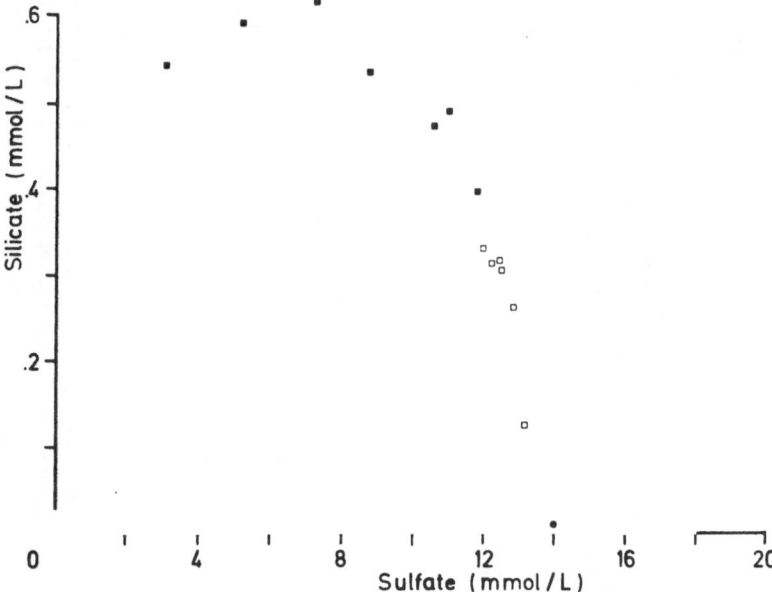

Fig. 5. Molar concentrations of dissolved silica and sulfate in the interstitial waters of Bornholm Basin sediments. The apparent correlation might indicate increasing dissolution of biogenic opal due either to the Ph changes or to the increasing surface area which accompanies the decomposition of organic matter from siliceous tests. Stations 10191–1 = □; Station 10179–1 = ▪; ● = bottom water.

extends from approximately 17°W to 22°W. The spacing of the lines of equal concentrations indicate steep gradients for these constituents near the foot of the slope, and hence a possible increase in molecular diffusion or other types of transport. The shelf, slope, and adjacent submarine topography are shown above the concentration profiles. The flux of nutrient constituents from the sediment to the bottom water needs to be calculated and compared with the regeneration of nitrate and phosphate in the water column to evaluate the relative contributions. Müller (1975) has calculated the flux of ammonia by diffusive transport to be 10^{-7} moles cm^{-2} y^{-1} although there are inherent difficulties in such an estimation (see Lerman, 1975; Berner, 1974). Nevertheless, the rate of ammonia flux appears reasonable when compared with the rate of nitrogen input to the sediment, as estimated independently (Müller, 1975). Over a depth profile of 5 m about 40% of the nitrogen input appears to be lost via diffusive transport as NH_4^+. It is then evident that even under the most favorable conditions—long residence time of the bottom water and little mixing—no significant nutrients supply can be expected. The steep concentration gradients at this

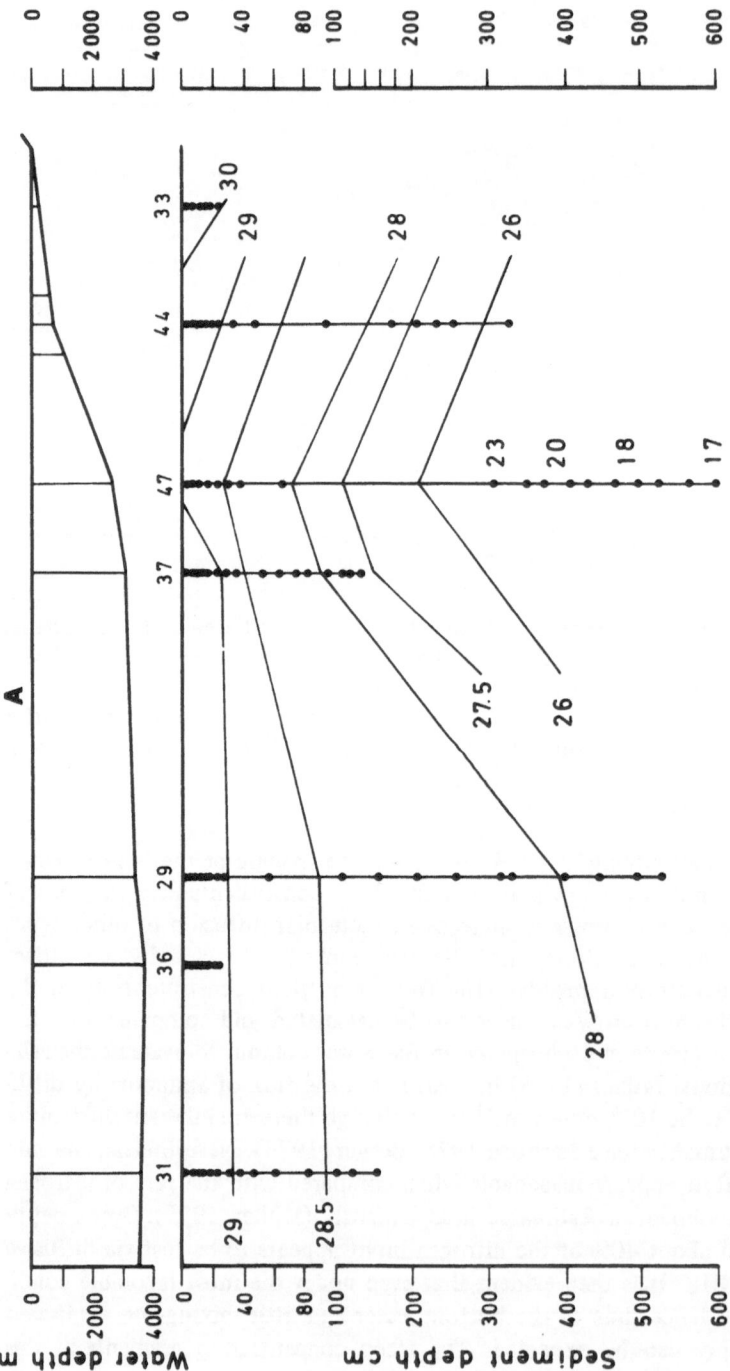

Fig. 6A. Changes in interstitial sulfate reduction, in mmol SO₄/L, with depth in the sediments along an East-West profile extending from 17°W to 22°W off the coast of Senegal. The shelf, slope, and adjacent submarine topography are shown above each profile. Numbers refer to stations. The steep concentration gradients of the interstitial water constituents of the continental slope are of particular significance. *Meteor* Cruise: M 25/71. Stations: 12338–1; 12344–1, 2; 12347–1, 2; 12337–4, 5; 12328–4, 5; 12329–4, 5; and 12336–1. (From Hartmann *et al.*, 1976.)

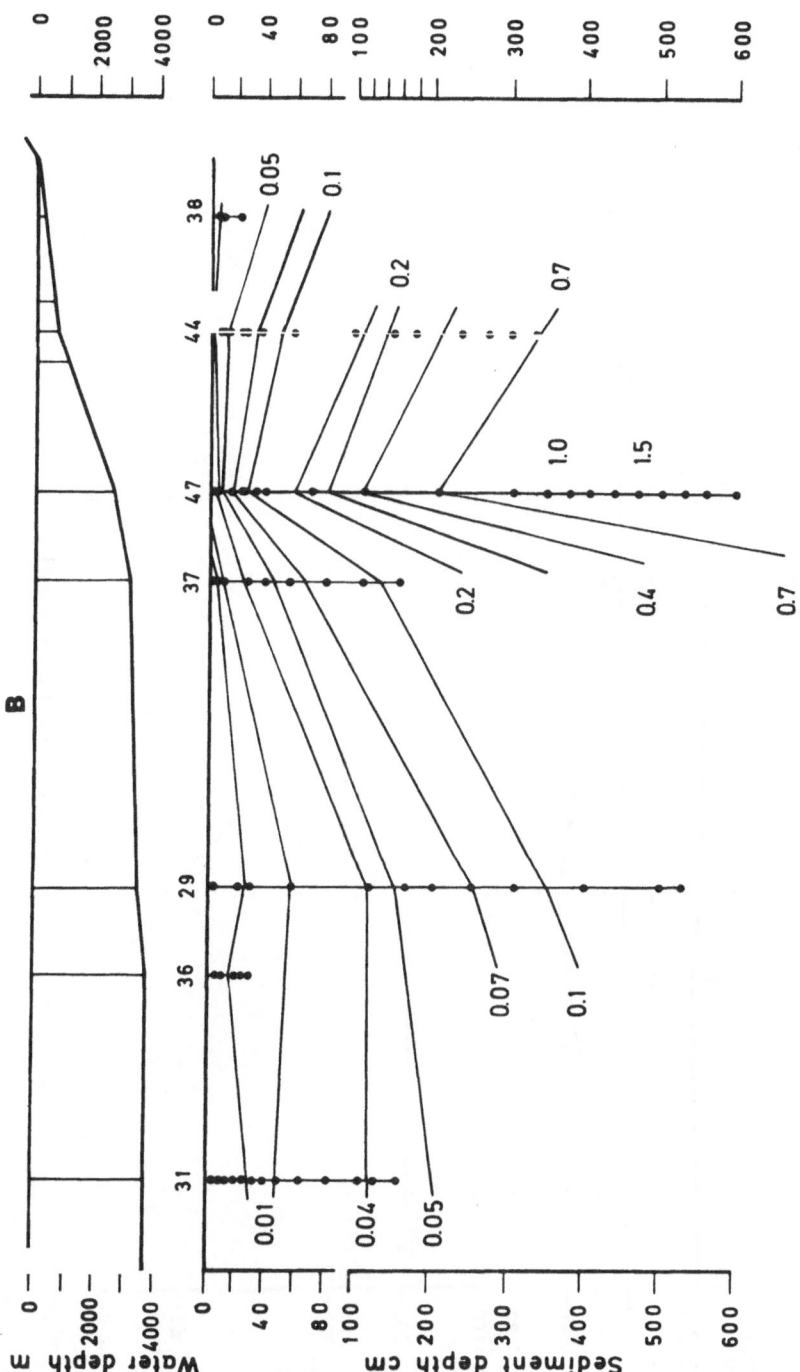

Fig. 6B. Ammonia, in mg-at. N/L.

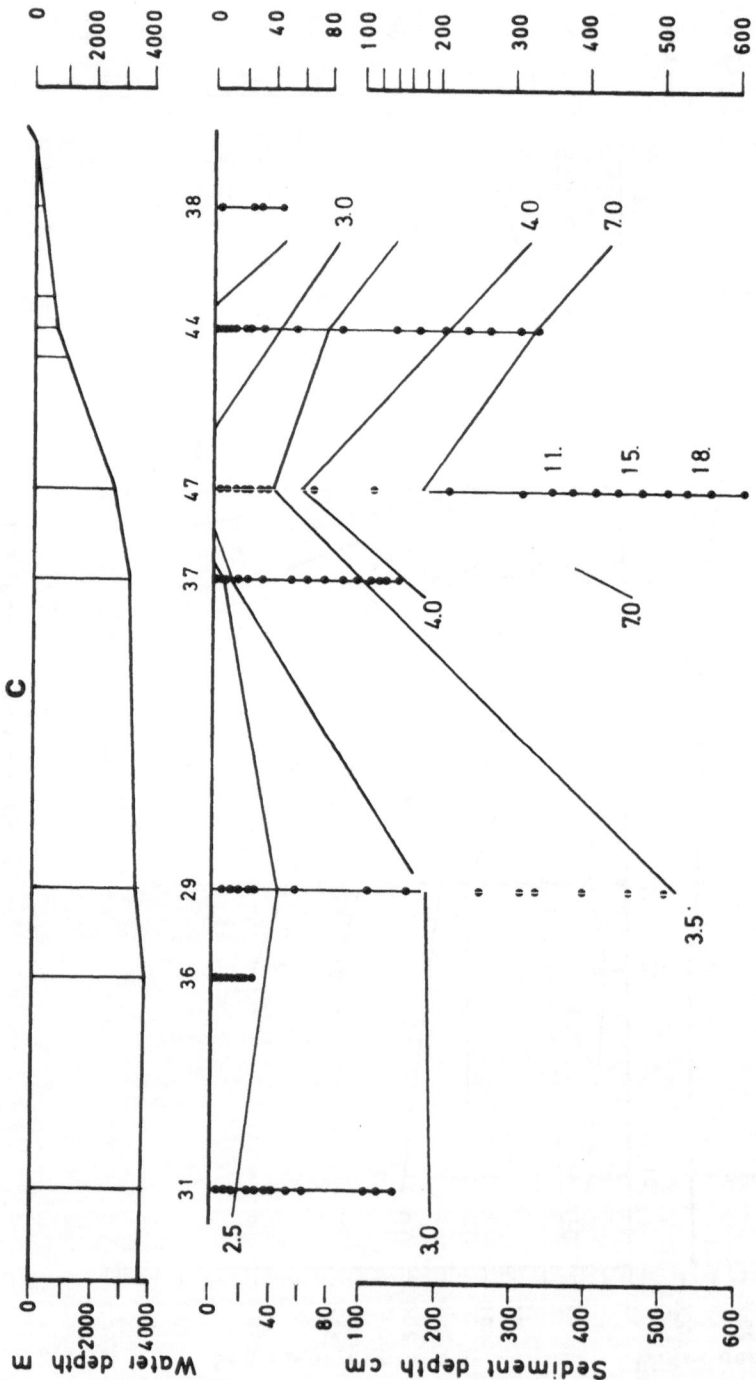

Fig. 6C. Alkalinity, in meq/L.

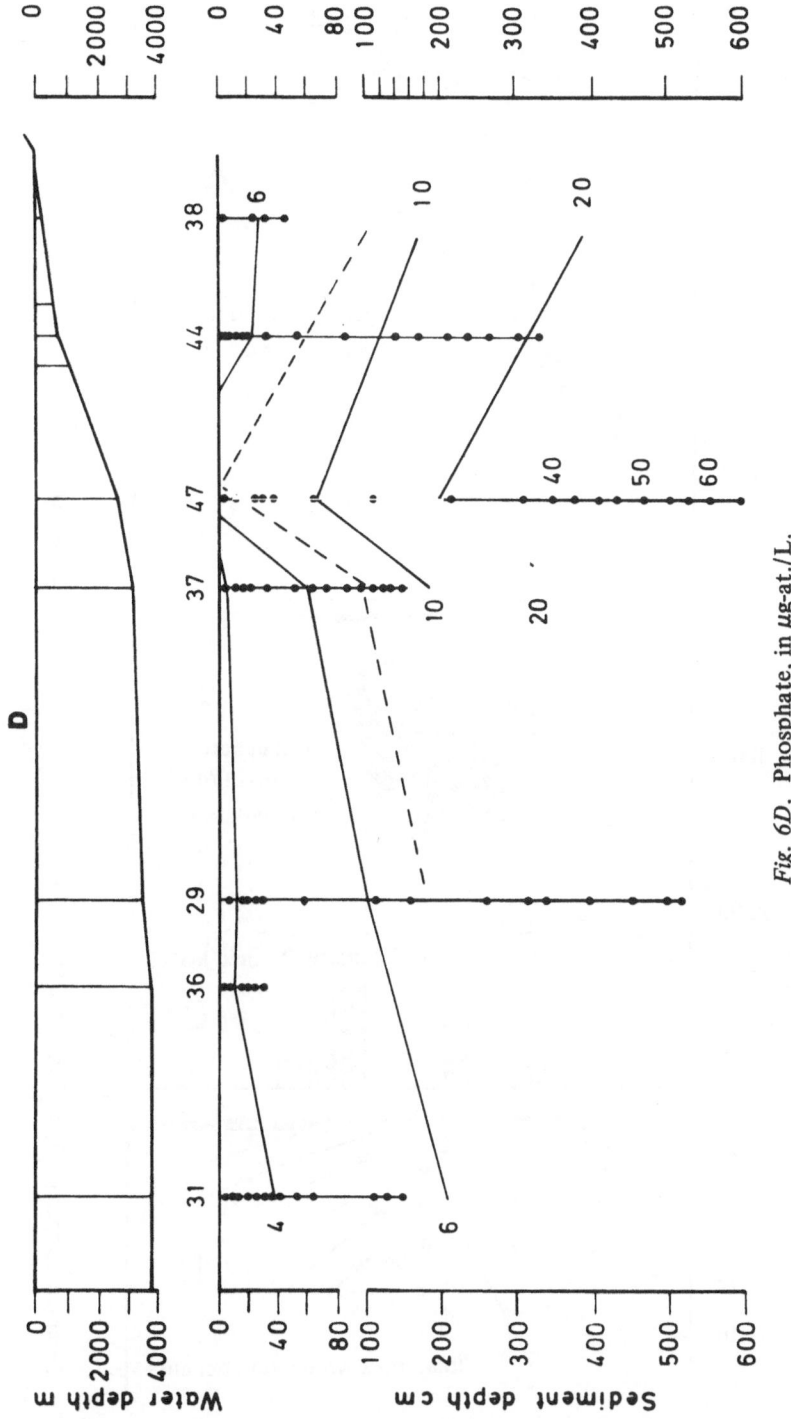

Fig. 6D. Phosphate, in μg-at./L.

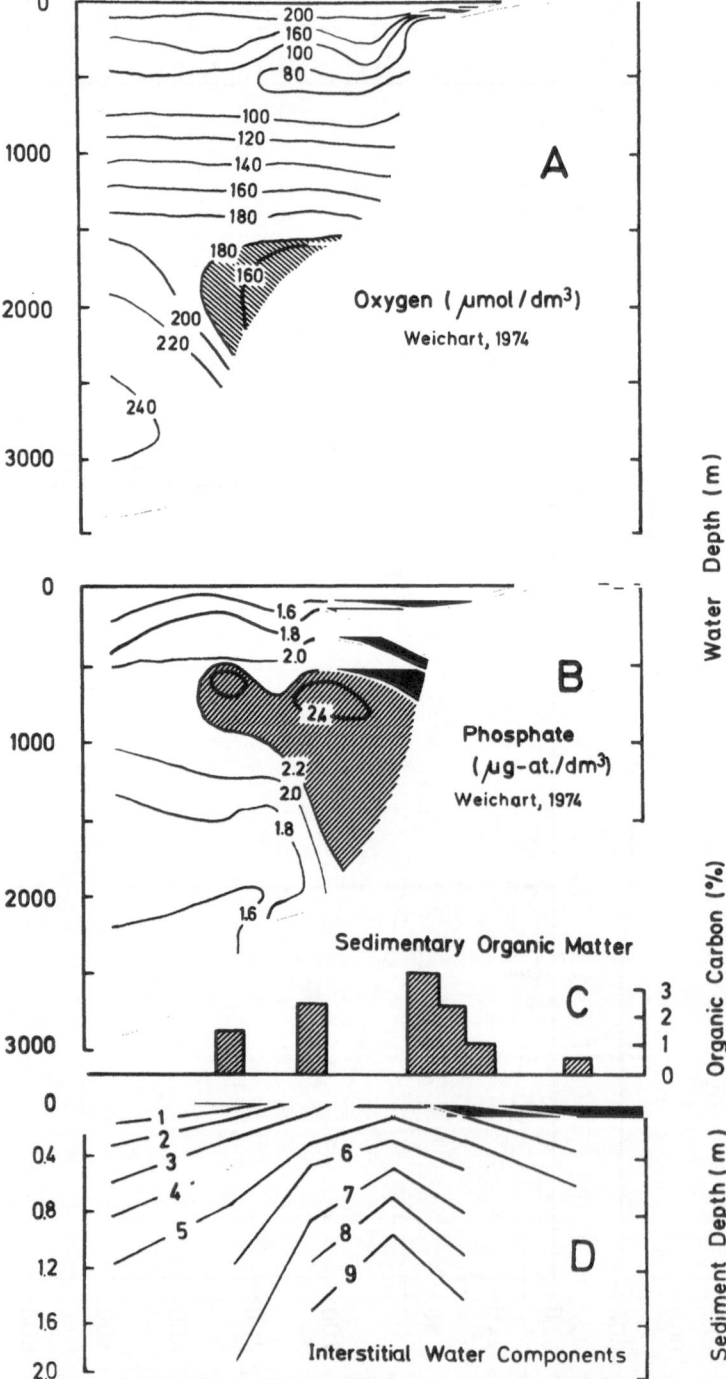

particular location suggest, however, that increased regeneration of nutrients may also take place at the sediment surface because the water depth at which they are found (between 700 and 2000 m) coincides with a phosphate maximum in the free oceanic water off West Africa and at other similar locations (Wattenberg, 1938; Riley, 1951; Redfield *et al.*, 1963; Wyrtki, 1971; and others). Perhaps the fact that the oceanic nutrient maximum deviates from the oxygen minimum layer close to the continents might be attributed off West Africa to the influence of increased regeneration of nutrients from the sediment surface and from the interstitial water. This possibility has recently been expressed by Weichart (1974), who suggests that accumulating organic matter on the ledges of the continental slope off West Africa leads to an increased regeneration of phosphate from the sediment surface. He has also reported an increased oxygen consumption in these bottom waters, but the related phosphate increase he reports is much larger than the stoichiometric model of oxygen consumption would predict. From this, it would appear likely that some additional sulfate reduction might be involved; this of course could only take place at or just below the sediment surface. There are several uncertainties inherent in the estimation of nutrient fluxes across the depositional interface, i.e., whether molecular diffusion properly describes this transport, or whether other more effective mechanisms such as bioturbation or tidal pumping are operating; also, the rate of input of mineralizable organic matter to the sediment surface, in contrast to input to the sediments (burial), is unknown. In view of these uncertainties, the possibility should be considered that significant nutrient regeneration might take place at the benthic boundary layer even under the hydrographic conditions encountered at the continental slope.

Unfortunately, there are presently not sufficient data, on either the nutrient concentrations from bottom waters at the West African continental slope or on the rates of input of mineralizable organic matter to the sediment surface to further substantiate this possibility. A diagram is presented in Fig. 7 to suggest this hypothetical relationship. This diagram shows the organic carbon content of surface sediments off West Africa. High organic carbon coincides with the steep nutrient concentration gradients of the interstitial water and with the oceanic nutrient maxima. In sediments from the northern Pacific Ocean, the northern Gulf of Mexico, and from many other areas, high organic carbon is found at

---

Fig. 7. Schematic illustration of the relationships among interstitial nutrient concentrations (D), amount of organic matter at the sediment surface (C), phosphate (B), and oxygen (A) in continental slope waters. Sediments with high organic carbon are not found at the same water depth at which the oxygen minimum layer impinges on the bottom, but at the depth of nutrient maxima. It is suggested that the phosphate maximum is influenced by flux from the sediment. Lines 1–9 show increasing nutrients and decreasing sulfate with depth; (A) and (B) are redrawn using data from Weichart (1974).

those depths where the oceanic oxygen minimum-layer impinges on the bottom (Thompson *et al.*, 1934; Richards and Redfield, 1954; and others); but off West Africa, high organic carbon is found deeper than the oxygen minimum-layer. Therefore, high organic matter in the sediment might not only be a simple question of preservation due to lack of oxygen, but might be a feedback process where nutrients are regenerated at an increasing rate on the continental slope and then transported upward. This would lead to increased productivity and to increased input to the sediment surface, and therefore to increased accumulation of organic matter.

In Figs. 7A and 7B the shaded areas show the depths of the additional oxygen minimum and the phosphate maximum layers in the ocean water close to the continental slope. These are thought to be influenced by mineralization processes according to the organic carbon distribution at the sediment surface (Fig. 7C) and by nutrients from within the sediment. This is illustrated in Fig. 7D by the concentration gradients of lines 1-9 in the upper 2 m of the sediment and by the organic carbon contents of the surface sediments. The lines of equal concentration 1-9 correspond to ammonia, phosphate, or alkalinity increase or to sulfate reduction, as shown respectively in Figs. 6A, 6B, 6C, and 6D. More data on interstitial concentration gradients from the continental slope are needed, as are detailed measurements of nutrients, oxygen, and particulates in near-bottom water samples, in order to estimate the significance of any nutrient regeneration from the sediment and from the sediment surface.

Deep Pacific Ocean

The hydrographic conditions characteristic of deep oceanic sedimentary environments, differing from those of the continental slope and stagnant basins, result in a different nutrient relationship at the benthic boundary layer. This is discussed here with data from Central Pacific Ocean sediments and their interstitial water constituents obtained from the vicinity of the Clarion Fracture Zone. Other important characteristics of deep ocean environments which affect this relationship are the slow rate of sedimentation, the low input of mineralizable organic matter, and perhaps the rate of mineralization. The distribution of organic matter with depth is illustrated for these sediments in Fig. 8A. The organic carbon and organic nitrogen contents decrease exponentially within the first 60 cm and remain essentially constant at greater depths. From these sediments, and from others in the immediate vicinity, soil mechanical properties and interstitial water chemical constituents have been determined (Hartmann *et al.*, 1973b, 1974; Hartmann and Müller, 1974). The ammonia and nitrate concentration changes with depth are shown in Fig. 8B and the total carbon dioxide

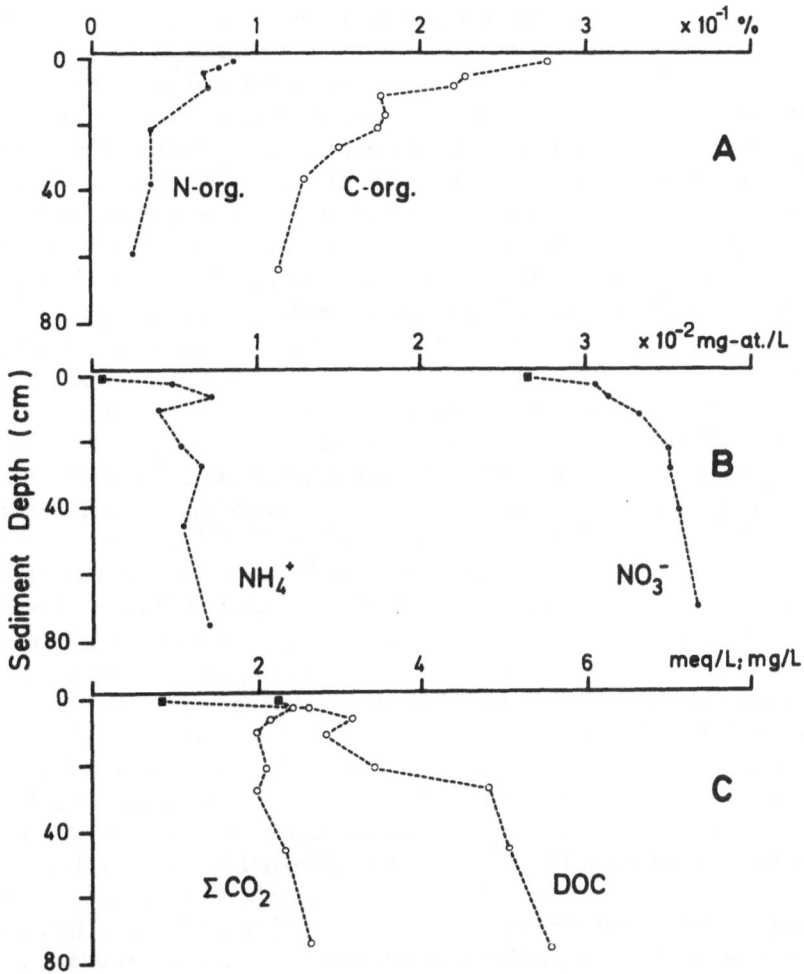

Fig. 8. Organic carbon and organic nitrogen from sediments of the Central Pacific Ocean (A), dissolved nitrogen (B), and dissolved carbon (C) constituents in the interstitial waters and sediment depth. It is suggested that mineralization of organic matter with time determines organic carbon and nitrogen depth distribution; concentration gradients of dissolved ammonia, nitrate, carbon dioxide, and organic carbon (DOC) suggest flux by diffusive transport; (*Valdivia* Cruise VA 05/4-1973; Stations 10127–1 and 10131–1; ■ = bottom water).

and dissolved organic carbon contents in Fig. 8C. All the dissolved constituents exhibited a more or less steep concentration gradient toward the sediment surface.

This organic carbon and nitrogen distribution is characteristic for all sediments that have been analyzed from that general area. Assuming that this distribution is determined by the loss of completely mineralized or partially solubilized intermediate products of organic matter, the flux of nutrient constituents and other products were estimated. A sedimentation rate of 3 mm/1000 y was used, radiometrically determined from the excess activity of protactinium ^{231}Pa for the upper 50 cm of the sediment column (Mangini, 1975); a bulk density of 1.15 g cm^{-3} was used and an average water content of 250% of dry weight (Kögler, 1974). Accordingly, the nitrogen flux of 5 X 10^{-9} mole cm^{-2} y^{-1} and the carbon flux of 20 X 10^{-9} mole cm^{-2} y^{-1} are to be expected from the recorded decrease of sedimentary organic nitrogen and carbon.

For comparison, the rates of flux could be calculated from the concentration gradients shown in Figs. 8B and 8C and from the appropriate sediment parameters, but because of the absence of any diffusion coefficients for dissolved organic carbon compounds and other uncertainties in such an estimation, it is sufficient to point out that diffusive ammonia transport, as listed by Lerman (1975), ranges from 1.3 X 10^{-3} mole cm^{-2} y^{-1} to as low as 4.8 X 10^{-9} mole cm^{-2} y^{-1}. The low values of 4.8 X 10^{-9} mole cm^{-2} y^{-1} and 18 X 10^{-9} mole cm^{-2} y^{-1} are reported from sediments of the Indian Ocean and represent deep oceanic conditions comparable to those of the Central Pacific Ocean. The agreement between the nitrogen flux in the Indian Ocean sediments as ammonia and in the Pacific Ocean sediments presumably as nitrate, is quite good, and accordingly the four-times-higher carbon flux also appears to be a reasonable estimate. Kroopnick (1974) considered the dissolved carbon dioxide released by *in situ* oxidation of organic matter within the deep Pacific water. He concluded that about 70% of the increase in total CO_2 was due to these processes and about 30% was due to dissolution of calcium carbonate. The average rate of oxygen consumption determined by Kroopnick ranged from 0.03–0.14 (μmol kg^{-1} y^{-1}). Accordingly, the rates of nitrate and carbon dioxide regeneration in the water should be between 0.004–0.017 (μmol kg^{-1} y^{-1}) and 0.02–0.11 (μmol kg^{-1} y^{-1}), respectively, using the Redfield-Ketchum-Richards model substance.

It is again not possible to compare the relative significance of the flux from within the sediments and from the sediment surface with regeneration of nutrients in the water column because of a lack of knowledge about residence time and about water-mixing close to the bottom. But, postulating a 1-m-thick bottom-water layer with no significant exchange with the overlying water masses, the nutrient flux from within the sediment would build up to measurable concentrations over a period of months. These speculations about the nutrient relationship near the deep oceanic sediment water interface might sound unreasonable, but they could help to clarify the space- and time-scales of meters and months over which parameters must be monitored in order, eventually, to evaluate fluxes from the benthic boundary layer.

Summary

The foregoing examples emphasize the need for further study of the relative contribution of nutrient regeneration from the depositional interface. The following questions are proposed as a focus for this study:

1. Why do the phosphate and ammonia concentrations actually released by sulfate reduction from the sediment deviate from the predicted values?

2. How does decomposition of organic matter in the sediment affect the solubility of biogenic opal?

3. Which nutrient-flux model is applicable to interstitial water conditions off West Africa and in the Central Pacific Ocean sediments?

4. What is the flux of nutrients from the sediment surface and what is the particulate input?

5. Is the nutrient maximum in the continental slope waters really influenced by increased nutrient flux from the sediment or is oceanic circulation more important?

6. Why is the relationship between organic carbon in the sediment and the oceanic oxygen minimum layer different off West Africa from that of other areas?

7. Is the loss of organic matter within the upper 60 cm of deep Pacific Ocean sediments due to mineralization?

8. Can the flux of nitrate and dissolved organic matter from the sediment to the bottom water there be responsible for that much loss?

9. What is the flux of dissolved organic carbon and why does removal of carbon from the sediment in this form appear to be quantitatively more important than removal in the form of carbon dioxide, as is the case in other sedimentary environments?

10. What is the ratio of oxygen consumption to nutrient release in Pacific Ocean bottom waters, and will flux of nutrients from the sediment affect this ratio?

11. What determines the rate of decomposition of organic matter, i.e., what are the rates of production of nutrients?

Acknowledgments

The data on which these discussions are based were collected during various cruises by the research vessels *F. S. Meteor*, *F. S. Valdivia*, and *F. S. Planet* to the East Atlantic, the Central Pacific, and the Baltic Sea, respectively. I would like to gratefully acknowledge the help and expertise of the crews during our sampling operations; financial support for the various investigations was supplied by the Deutsche Forschungsgemeinschaft, Bundesministerium für Forschung und Technologie, and the Fraunhofer Gesellschaft. Many discussions with colleagues from the Geologisch-Paläontologisches Institut, Kiel, and with participants in the NATO Science Committee Conference on the Benthic Boundary Layer have contributed to the above presentation of problems; among these I would particularly like to thank S. Calvert, J. Gieskes, M. Hartmann, and P. Müller. This paper was contribution No. 75 of the Sonderforschungsbereich 95, "Interaction Sea-Seabottom," at the University of Kiel.

References

Almgren, T., Danielsson, L. G., Dryssen, D., Johannson, T., and Nyquist, G., 1974, Release of inorganic matter from sediments in a stagnant basin, *Thalassia Jugoslavia* (in press).

Berner, R. A., 1974, Kinetic models for early diagenesis of nitrogen, sulfur, phosphorus, and silicon in anoxic marine sediments, *in: The Sea*, Vol. 5 (E. Goldberg, ed.), Wiley and Sons, New York, pp. 427–450.

Billen, G., 1974, Nitrification in the Scheldt estuary, Belgium and the Netherlands (submitted for publication).

v. Bodungen, B., v. Brockel, K., Smetacek, V., and Zeittschel, B., 1974, Ecological studies on the plankton in Kiel Bight. I. Phytoplankton, *Merentutkimuslaitoksen Julkaisu Havsforskningsinstitutets Skrift No.* 239, Helsinki.

Hallberg, R. O., Bagander, L. E., Engvall, A. G., and Schippel, F. A., 1972, Method for studying geochemistry of sediment-water interface, *Ambio, 1:* 71–72.

Hartmann, M., and Müller, P., 1974, Geochemische Untersuchungen an Sedimenten und Porenwassen, *Marine Technology, 5:* 201–202.

Hartmann, M., Müller, P., Suess, E., and van der Weijden, C. H., 1973a, Oxidation of organic matter in recent marine sediments, *"Meteor" Forschungs-Ergebnisse, C, 12:* 74–86.

Hartmann, M., Kögler, F. C., Müller, P., and Suess, E., 1973b, Preliminary results of geochemical and soil-mechanical investigations on Pacific Ocean sediments, *in: International Symposium on the origin and distribution of manganese nodules in the Pacific and prospects for exploitation, Honolulu,* July 23/25, 1973.

Hartmann, M., Kögler, F. C., Müller, P., and Suess, E., 1974, Untersuchungen zur Genese von Manganknollen (unpublished Annual Report to the Bundesministerium für Forschung und Technologie, Bonn).

Hartmann, M., Müller, P., Suess, E., and van der Weijden, C. H., 1976, Chemistry of late quaternary sediments and their interstitial waters from the NW African continental margin, *"Meteor" Forschungs-Ergebnisse, C., 24* (in press).

Hurd, D. C., 1972, Factors affecting solution rate of biogenic opal in sea water, *Earth and Planetary Science Letters, 15:* 411–417.

Hurd, D. C., 1973, Interaction of biogenic opal, sediment and sea water in the central equatorial Pacific, *Geochimica et Cosmochimica Acta, 37:* 2257–2282.

Kögler, F. C., 1974, Sediment-physikalische Eigenschaften von drei Tiefseekernen des zentralen Pazifischen Ozeans, *Marine Technology, 5:* 199–201.

Kroopnick, P., 1974, The dissolved O_2-CO_2-^{13}C system in the eastern equatorial Pacific, *Deep Sea Research, 21:* 211–227.

Lerman, A., 1976, Migrational processes and chemical reactions in interstitial waters, *in: The Sea*, (E. D. Goldberg *et al.*, eds.), Vol. 6, Wiley and Sons, New York (in press).

Mangini, A., 1975, Thorium- und Uran-Isotopenanalysen an Tiefseesedimenten, Institut für Umweltphysik der Universität, Heidelberg (unpublished).

Müller, P., 1975, Diagenese von stickstoffhaltigen organischen und anorganischen Verbindgungen in Sedimenten des westafrikanischen Kontinentalrandes und des zentralen Pazifik, Ph.D. Dissertation, University of Kiel.

Nehring, D., 1974, Untersuchungen zum Problem der Denitrifikation und Stickstoffentbindung in Tiefenwasser der Ostsee, *Beiträge zur Meereskunde, 33:* 135–139.

Redfield, A. C., Ketchum, B. H., and Richards, F. A., 1963, The influence of organisms on the composition of seawater, *in: The Sea*, (M. N. Hill, ed.), pp. 26–77.

Richards, F. A., 1965, Dissolved gases other than carbon dioxide, *in: Chemical Oceanography, 1:* (J. P. Riley and G. Skirrow, eds.), Academic Press, pp. 197–225.

Richards, F. A., and Redfield, A. C., 1954, A correlation between the oxygen content of seawater and the organic content of marine sediments, *Deep Sea Research, 1:* 279–281.

Riley, G. A., 1951, Oxygen, phosphate and nitrate in the Atlantic Ocean, *Bulletin of Bingham Oceanographic Collection, 13:* 1–126.

Riley, G. A., 1956, Oceanography of Long Island Sound 1952–1954. IX. Production and utilization of organic matter, *Bulletin of Bingham Oceanographic Collection, 15:* 325–344.

Rittenberg, S. C., Emery, K. O., and Orr, W. L., 1955, Regeneration of nutrients in sediments of marine basins, *Deep Sea Research, 3:* 23–45.

Schrader, H. J., 1971, Fecal pellets: Role in sedimentation of pelagic diatoms, *Science 1974:* 55–57.

Sen Gupta, R., 1973, Nitrogen and phosphorus budgets in the Baltic Sea, *Marine Chemistry, 1:* 267–280.

Stöber, W., 1967, Formation of silicic acid in aqueous suspensions of different silica modification, *in: Equilibrium Concepts in Natural Water Systems* (R. F. Gould, ed.), American Chemical Society (Washington) Publication 67, pp. 161–182.

Stumm, W., and Morgan, J. J., 1970, *Aquatic Chemistry*, Wiley Interscience, New York, pp. 514–563.

Tessenow, U., 1973, Lösungs-, Diffusions- und Sorptionsprozesse in der Oberschicht von Seesedimenten. II. Rezente Akkumulation von Eisen (II) phosphat (Vivianit) im Sediment eines meromiktischen Moorsees (Ursee, Hochschwarswald) durch postsedimentäre Verlagerung, *Archiv für Hydrobiologie/Beihefte, 42* (2): 143–189.

Thompson, T. G., Thomas, B. D., and Barnes, C. A., 1934, Distribution of dissolved oxygen in the North Pacific Ocean, *in: James Jonston Memorial Volume*, University Press, Liverpool, pp. 203–234.

Wattenberg, H., 1938, Die Verteilung des Sauerstoffs und des Phosphats im Atlantischen Ozean, *Wissenschaftliche Ergebnisse der Deutschen Atlantischen Expedition "Meteor" 1925–27, 9* (1) und (2): 1–179.

Weichart, G., 1974, Meereschemische Untersuchungen in nordwestafrikanischen Auftriebsgebiet 1968, *"Meteor" Forschungs-Ergebnisse, A., 14:* 33–70.

Wyrtki, K., 1971, *Oceanographic Atlas of the International Indian Ocean Expedition*, National Science Foundation Publication, Washington, D.C., pp. 1–531.

5
The Benthic Boundary Layer from the Point of View of a Biologist

P. A. W. J. de WILDE

The principal components of zoo benthic communities may be classified as macrofauna, meiofauna, and microorganisms. The ratio of meio- to macrofauna is extremely variable, ranging from 3% to >50% by biomass. This is found in "well-known" areas which comprise <5% of the world ocean. There is thus a profound need for more extensive investigation of the benthic fauna of the world ocean. Studies of energy flows are extremely revealing. The various sources and sinks of energy in benthic communities need to be examined and quantified for a number of situations. This will involve studies of the energy consumption in food ingested, energy converted into flesh, tissues, gametes, etc., metabolic energy loss, and energy loss in excreta. While a start has been made on measurement of these features in shallow water, few good estimates have yet been made for deep-water benthic communities. The use of these measurements in the construction of energy budgets is one of the objectives of this type of work, and this field of study overlaps with the work of chemists and sedimentologists who are examining other aspects of benthic processes. Quantification of the influence which organisms have on the benthic environment is another important but largely unrealized goal of the biologist.

Introduction

In discussing the biology of the benthic layer we are still mainly restricted to knowledge of the shallow parts of the ocean bed. Owing to technical and organizational problems, research in the deep sea has not kept pace with research in shallow and coastal waters on the continental shelf.

P. A. W. J. de WILDE ● Netherlands Institute for Sea Research, Texel, Netherlands.

As the author is mainly concerned with energy flow in benthic communities in a tidal flat area, the present view on the benthic boundary layer should be considered only as a special view on the subject. Intertidal areas probably are among the best-known marine habitats in the world. They are uncovered by the sea for a certain period each day. This unique feature makes them more accessible than any other part of the sea bed, greatly facilitating the possibilities for research.

Although tidal flats represent only a very small part of the total sea bed, they are among the most dynamic and productive parts of the ocean floor. In addition, it is postulated here that a large number of the basically important biological processes occurring near, on, and in the tidal flats are essentially closely related to those occurring in most of the other parts of the sea bed. Even as far as this "best-known" part is concerned, our knowledge is far from complete about the details of its benthic communities and their life processes.

Structure and Components of Benthic Communities

The principal components of benthic communities may be grouped under phytobenthos, macrofauna, meiofauna, and microorganisms. To describe a living benthic community one must begin by listing and counting its members and noting their features. With the probable exception of parts of the deep sea, and of meiofauna, the discovery of new organisms is now thought to be terminated. There are numerous fine examples of single-species investigations of ecology and distribution as well as qualitative and quantitative inventories of macrofaunal assemblages. The majority refer to near-shore bottoms in European and North American waters, but other parts of the ocean—tropical, polar, and parts of the deep sea—are also included. This is not the place to give an enumeration of the literature in this field of research. An exception must be made here for the book by Menzies et al. (1973). However, in my opinion, there is a need for more conveniently arranged data on geographical distribution and for a rough indication of abundance in important macrofauna species, such as given already in the serial atlases of the marine environment issued by the American Geographical Society (e.g., Korringa, 1969). Several marine biologists have tried to interrelate species composition and biotic and abiotic elements of a certain environment (e.g., Peterson, 1913; Thorson, 1957), thus emphasizing the existence of identifiable biocoenoses or communities. Marine environments offering more or less comparable or similar ecological conditions, are inhabited by the same or related groups of animal species, after Thorson (1957) 'life-forms', together forming the animal community. Similar inventories on meiofauna have received considerably less attention, but here, too, there are examples provided by Holme and McIntyre (1971), Gerlach (1971, 1972), Lasserre (1971), McIntyre (1969, 1971), and Wieser (1960). The relation between meio- and macrofauna may change from one locality to another. In general we are poorly informed about this. For the Heligoland Bight, Gerlach (1972) found that meiofauna biomass is only 3% of macrofauna biomass. In other places, however, it may equal macrofauna. There are now methods for estimating meiofauna biomass by determination of ATP content, without going into species composition (see Ernst, 1970).

Qualitative and quantitative studies in microorganisms are carried out, but probably are without significance. In broad terms, the traditional statement of Beyerinck that "everything is present everywhere and the environment will select" applies here. Most important are the intensities and kinetics of the biological processes, not the number of microorganisms.

Unfortunately, most data on the structure of meio- and macrofauna populations are only of small value for a proper understanding of the coastal and shallow-water ecosystems, because population dynamics in place and time are nearly always ignored.

Even in a rather large and homogeneous habitat such as the Dutch Wadden Sea, in the northern part of the Netherlands, covering some 2000 km^2, considerable variations in species composition and species abundance occur from place to place. In addition, there are large seasonal and annual variations in growth and consequently in biomass. An illustration is given by the work of Beukema (1974, and unpublished manuscript) on a single macrofauna species *Macoma balthica* in a tidal flat area. The abundance of the species in the Western Wadden Sea points to some mosaic distribution. Growth here is limited to the period March–July and consequently biomasses are high in the early summer and low during the late winter. Finally, during the winter, a remarkable redistribution of the O-group takes place. Data on other macrofauna species show similar changes in distribution and biomass. Research on meiofauna in this area is as yet insufficient, but will probably show the same trends.

Extending our boundaries to include the North Sea, we find several places in which the benthic fauna is locally well investigated. Most research, however, concerns near-shore waters, tidal flats, estuaries, lochs, and fjords. In the central and northern part of the North Sea knowledge about faunal composition, growth, production, and population dynamics is small, or absent. And our knowledge about the benthic subsystem of the North Sea, generally considered to be a well-known part of the ocean, is highly unsatisfactory.

Considering the ocean floor, in spite of recent investigations in offshore waters, probably less than 5% has received more than superficial attention.

The principal factors that control the performance of benthic organisms and populations are known. They include such diverse factors as light, temperature, salinity, availability of food, oxygen saturation, pressure, properties of the substrate, grazing, predation, and some other factors of minor importance. There is a lack of data from long-term investigations of benthic communities and environmental factors with adequate discrimination and coverage in space and time. Thus even the ranges of natural fluctuations are imperfectly known. Moreover, as many of the macroinvertebrates are long-lived species, at least several decades of careful work will be needed for an insight into systematic trends.

Proper understanding of the dynamics of benthic populations represents an essential basis for the understanding of the mechanisms of the complete systems. In addition, an understanding of the benthic world is the prerequisite for proper management of the natural resources of the sea (e.g., in the control of fish stocks).

Planktonic Communities

In contrast to our understanding of the benthic fauna, work on the variability of planktonic organisms has made considerable progress.

Extensive surveillance of the plankton in the North Sea and eastern North Atlantic, with the continuous Plankton Recorder, covering an uninterrupted period of over 25 years, provided remarkable insight in long-term variations in average abundance, biomass, and duration of the seasonal cycles throughout the year (Colebrook, 1972, 1973; Glover et al., 1972). Knowledge about annual and seasonal variations in plankton abundance and biomass is incomparably more advanced than knowledge about benthos. Notwithstanding their results from long-term investigations—pointing to systematic trends occurring in abundance, biomass, and seasonal cycle—the authors are extremely reserved in giving explanations and in trying to relate their findings to environmental factors. The paucity of environmental data and the inadequacy of research into this kind of time series, and the difficulties of interpretation even after apparent correlations have been found, are responsible for this reservation.

Energy Flows

Next to my plea for long-term investigations of benthic life, two other important aspects of the benthic system should be considered, namely, *energy flow* and *productivity*.

As in other biosystems, the benthic system is maintained by a constant flow of energy to all its living members. Each member, as it ingests food, is linked to the others according to its status as consumer or object of consumption. In this way definite food chains or food webs occur. Planktonic and benthic algae, because of their ability to photosynthesize, are always at the beginning of the food chain.

Studies of ecological energetics are an essential tool with which the importance of different species populations in their contribution to the structure, productivity, and functioning of communities can be evaluated. Theoretically, ecological energetics are extensively studied, but there is still a need for more field investigations. Energy budgets, either for a single specimen, for a population of a certain species, for a group of related species, or even for a community, conform to the following equation:

$$C = P + R + F + U$$

in which C is the energy content of the food ingested; P is the energy converted into flesh, tissues, gametes, etc.; R is the energy lost due to own metabolism; and F and U are energy lost by food remaining in the feces, urine, and other exudates. A great amount of basic knowledge exists in this area, in particular about details or about single terms of the equation. More recent work has concerned total budgets for species and populations and attempts have even been made to include complete communities. Two almost classical examples are studies by Hughes (1970) and Pamatmat (1968).

Energy uptake in bottom animals immediately introduces a sequence of other aspects. First, the exact status of a species as herbivore, carnivore, parasite, detritus feeder, feeder on microorganisms, or even feeder on dissolved organic compounds. Secondly, the exact relation between availability and quality (food value) of the food items and its actual utilization by the benthic fauna.

Food and Feeding

The importance of food chain studies is generally understood. Even in well-known and commercially important species such as cod and plaice there are still interesting aspects which need further research, as may be illustrated by some recent observations.

A species such as cod, for example may drastically affect macrobenthos populations. Young cod feed predominantly on benthic fauna. Some years ago, a single, extremely large year class destroyed the commercial shrimp populations in extensive parts of the coastal area of the North Sea (Boddeke, 1971). Any effect on other fauna elements by such a sudden disturbance of the benthic community was not recognized or studied. Now, after a subsequent drastic reduction of the cod stock by fisheries, the *crangon* stock has recovered and in the meantime has enlarged its area in an unexpected way (Boddeke, personal communication)

A second example concerns the recently discovered habit in some flat-fish species of feeding partially on certain appendages of bottom organisms: arms of ophuroids, siphons of bivalves, palps or tail ends of polychaete worms, etc. This type of harvesting of the benthos, without actually breaking into the populations, defies analysis by a normal approach of production studies. Its importance is still a matter of doubt, but in any case will give rise to underestimation of secondary production.

A third example concerns the effects of modern fishing gear on the benthic communities. Bottom trawls, loaded with the so-called "ticklers," may plough extensive parts of the sea bed to depths of 75 cm, and in certain areas, as many as several times a year. The effects on the bottom fauna are largely unknown.

Marine benthos is characterized by large numbers of sessile or relatively inactive animals. Some live attached to or move on the surface and are called epifauna, and some dig into the substrate or construct tubes or burrows and are called infauna. A second rough partition can be made based on the mode of feeding: Suspension- or filter-feeders that feed on suspended matter; and deposit-feeders, that take their food directly from the sediment. In both groups, ingenious morphological adaptions are found that enable them to collect sufficient amounts of food even in an environment relatively poor in food. Deposit-feeders are commonly said to be related to silty or muddy substrates, whereas filter-feeders would tend to predominate on or in firm and sandy substrates. However, in my opinion, it would be better to relate them to the availablity of food, as being more important than the properties of the sediment. Filter-feeders then are dependent on a constant supply of food, carried by water currents. Quiet water favors deposition of suspended particulate matter and thus is favorable to deposit-feeders. Sandy or muddy bottoms ought to be considered to be of a secondary nature. Examples of typical feeding behavior are given in the scientific film "Deposit-Feeding" by NIOZ (1973).

It must be emphasized here that filter- and deposit-feeding do not represent sharply separated mechanisms, because in almost all types of deposit-feeders there also occurs one or the other type of filtering. In both suspended and deposited organic matter three principal food components may be distinguished:

1. Living organic material: algae cells; sexual products, eggs and sperm; tiny organisms; etc.
2. Detritus: dead organic matter of both plant and animal origin, fecal pellets, etc.
3. Microorganisms: mainly heterotrophic bacteria

Perhaps a fourth source, dissolved organic matter, could be mentioned here. Recent investigations (Schlichter, 1975; and others) on uptake of dissolved matter gave evidence that certain macroinvertebrates (Actinians) are able to absorb this with quantitative significance. As dissolved organic matter seems to be an unlimited reservoir in the oceans, it would be wise to concentrate attention on this food source.

In general, living algae cells are to be expected only in the photic zone. Potentially, however, various algae remain viable during extremely long periods in complete darkness. Cadée (unpublished manuscript) found an almost instantaneous primary production by benthic diatoms, when exposed to sunlight after they had been buried for several months in the sediment.

Living cells, detritus, and microorganisms are thus considered to be the main food sources of the benthos in shallow water. In deep water only detritus and associated microorganisms, and perhaps dissolved matter, play a part.

To what extent they are useful as food for different organisms is a matter of debate. This will depend on the properties of the food as well as on the feeding mechanisms, retention, selection, adequate digestion, etc., in the consumers. Living algae cells, for example, are often to be found almost intact in the feces of different invertebrates. In molluscs this is ascribed to intracellular phagocytosis and the absence of certain enzymes. Conversely, certain deposit-feeding Tellins are able to crack diatom scales and digest them. Growth in some species is related to the occurrence of sufficient amounts of living algae.

The value of detritus, then depends mainly on its origin and age. Analysis of detritus in Danish waters, particularly the remnants of eel grasses, pointed to large percentages of compounds indigestible by macrobenthos. It is generally accepted that much of the particulate and dissolved organic material is relatively inert. In some cases the detritus itself shows a high food value, but the significance of microorganisms forming aggregates or being attached to the surface of detrital or even to inorganic material has also been emphasized. Particles would then be "coated" by a layer of microorganisms. In this view, soft bottoms with small particle size are characterized by large amounts of microorganisms.

Experiments by Newell (1965) in which gastropods and bivalves were repeatedly fed with incubated fecal pellets showed only the digestion of adhering microorganisms and the gradual reduction of the organic carbon source of the feces. The question has not been answered as to what extent the microorganisms supplied the energy requirements of the animals, finding expression in growth, reproduction, etc.

A similar process of breakdown of organic matter from the upper water layers apparently takes place during its way downward in the ocean. This has the effect of decreasing food contents of the continually recycled food particles (fecal pellets, etc.) entering deeper water layers. In accordance with this is the suggestion that depth of the ocean floor ranks first among the factors controlling standing stocks of the benthos. However, this is doubted by Menzies et al. (1973), who believe that the origins of water masses and bottom currents are of more significance (see Hollister, this volume). Thus the fertility and productivity in the euphotic zone is important in this respect. In a given body of water a distinct linkage between primary production and the metabolic rate of its benthic component is generally to be expected. In areas with a high primary production (or with a large supply of food from elsewhere) high standing stocks of the benthos are often found.

Much of our knowledge on detritus and its role in aquatic ecosystems is summarized in the proceedings of the IBP–UNESCO Symposium held in Pallanza, 1972, and I think we will all agree with the opinion in Mann's (1972) introductory remarks that "we shall not have a good understanding of functioning of aquatic ecosystems until we know a great deal more about the organic matter, with its associated micro-organisms, which we call detritus."

In tidal flats we encounter a part of the sea bed in which all three food components are amply available. In the Dutch Wadden Sea we are well informed on annual energy input into the system. Organic food is partially derived from *in situ* primary production; in channels and gullies (Postma and Rommets, 1970) by phytoplankton; and on the flats (Cadée and Hegeman, 1974) mainly by microphytobenthos. An annual production of 120 g C m^{-2} is available. Moreover, there is a large additional supply of allochthonous living and dead organic matter from the North Sea, transported by tidal currents and ultimately deposited on the flats. This amount was estimated by Postma (1954) to be 80 g C m^{-2}. Now, due to eutrophication of the coastal waters by nutrient-loaded river discharges, the supply of allochthonous organic matter in the Western Wadden area has probably doubled or even tripled during the last two decades (de Jonge and Postma, 1974). A direct influence on the standing stocks of macrobenthos, however, has not been established. As the meiofauna is probably of minor importance in this area it is postulated here that the surplus of food favors only the microorganisms. Apparently both meiofauna and microorganisms are a dead end in the food chains.

This example emphasizes the sometimes extremely indistinct interrelations between different trophic levels. (However, relations in the aphotic zone, in which algae are absent, are probably more simple.) Further research into energy transfer from one to the other trophic level is needed. We will return to this point at the end of this paper.

Suspended food particles are never present in a homogeneous mass, but consist of mixtures containing all kinds of living algae, small zooplankton, detritus, and microorganisms, as well as particles of inorganic origin. In deposited material the proportion of organic matter is sometimes small in comparision with that of silt and sand. Thus an important field of investigation includes the characterization and analysis of the separate components; we mention here: pigment analysis, determination of chlorophyll content, determination of ATP content, determination of size spectra by Coulter-counter analysis, microscopic examination, staining techniques, filtration or sedimentation techniques, determination of nitrogen–phosphorus ratio, and determination of organic carbon. Of great interest are techniques which evaluate the caloric content of the ingested food.

Often, because of admixtures of large amounts of inorganic material, the caloric content of the food is low, and thus normal determination in a bomb calorimeter is impossible. Other methods, such as wet combustion, are then used. Determination of food value after some biological test would be a most useful contribution.

Energy Stored as Production

Let us now return to the energy equation and consider the term for *energy stored as production*. Here we need information on the energy content of the tissues added to the specimen, to the population, or to the community due to growth, recruitment, and elimination. In natural populations, as mentioned before, detailed insight in population dynamics is needed. That is, insight into mortalities, recruitment, and growth, as well as into the variabilities of the population.

The *energy content stored in the development of sexual products* is energy that from time to time is liberated as spawned gametes. In invertebrates we are usually poorly informed on this sort of production. As it often equals flesh production, this represents a considerable deficiency in our knowledge.

Energy Consumed in Metabolism

The next term of the equation deals with the energy lost by the *energy requirements of the organisms themselves*. This will absorb the largest part of the energy ingested. In general, determination by the indirect method of measuring oxygen consumption is applied. New methods by direct calorimetry (Pamatmat, personal communication; Spaargaren, 1975) are in progress. However, the bulk of our present knowledge is derived from oxygen consumption (Winkler method, Clark electrodes, polarographic methods, Warburg methods, Cartesian divers, etc.), which are to be converted in calories by application of oxycalorific coefficients.

Much basic information is available on O_2 uptake in relation to certain species and to body size, temperature, salinity, food conditions, behavior, stress, diurnal and seasonal rhythms, etc. The most important determinants of the respiration rates of an organism, however, are size and temperature.

Unnatural measuring conditions are probably of minor importance in microorganisms and meiofauna. However, in more organized and larger macroinvertebrates, large errors are involved in extrapolating respiration rates, as measured in the laboratory, to the situation in the natural habitat. Even invertebrates will display rather complex patterns of behavior which are related to different levels of oxygen uptake. Measurements on infauna ought to be made in the sediment. An adequate apparatus for measuring O_2 uptake in burrowing animals is now used at our Institute (de Wilde, 1973). Even so, one must be extremely cautious in using experimental data to predict what happens in nature. For example, laboratory studies on the behavior of *Macoma balthica*, kept in aquaria and in the measuring vessels of a respirometer, showed indeed the same elements of behavior, but the patterns were rather different, as was oxygen consumption.

For community respiration, *in situ* methods, in which a small part of the sea bed is enclosed by a transparent bell jar, box, or clock, are used. The O_2 consumption is measured by electrodes. By treatment of the enclosed environment inside the box with formalin and bacteriocides, the separate consumption of microorganisms, higher invertebrates, and chemical oxygen demand is analyzed. Finally, the benthic fauna inside the box is sampled and analyzed. Here we can agree with the opinion of Brinkhurst *et al.* (1972) that the attempt to recreate energy or material budgets for communities from laboratory studies of isolated components is less promising than the attempt to study communities *in situ*. For the latter, however, complicated technical problems need to be solved. In addition, *in situ* measurements require long-term periods of observation to include seasonal changes.

Besides the biological oxygen demand, the sediment itself uses oxygen for chemical oxidation processes; thus blanks must be subtracted. Here again, numerous other aspects of oxygen uptake and respiration ought to be discussed.

Depending on sediment properties, organic matter content, permeability, and animal activity, the oxygen saturation in deeper sediment layers may be reduced, and thus higher forms of animal life will find difficulties in inhabiting this region. Muddy substrates are frequently anoxic below 1–5 cm, and infrequently even the water over the sediment is anoxic. Higher animal life can cope with these types of environment by construction of well-ventilated burrows, which make contact with the outside world; others possess blood pigments that highly facilitate oxygen binding; still others reduce metabolic activity or switch temporarily to anaerobic forms of metabolism.

Numerous types of anaerobic bacteria, some protozoa and perhaps a few nematodes, can live in the completely reduced zone. In anaerobic respiration some inorganic compound is used as electron acceptor. In tidal flat areas, sulfur bacteria of the genus *Desulfovibrio* are well known, reducing sulfate to sulfide and ultimately forming the black sulfide layer.

Excretion

The last terms of the energy equation concern the *energy lost with feces, urine, and other exudates*. Animals are able to extract only part of the energy accumulated in organic food. Thus feces may be a valuable food material for the same or other organisms. There is some doubt concerning the large energy losses from exudates described by several authors, which points to considerable leakage from the outer membranes of animals and algae. We have strong indication that such leakage is caused by improper handling of the organisms. Some publications mention losses of over a third of the energy ingested (Johannes and Satomi, 1967; Hargrave, 1971).

Energy Budgets

Although the construction of energy budgets is a laborious affair, the usefulness of measuring metabolic activities and calculating energy flow is obvious. Not only will they give direct information on numerous biological questions, but they are also one of the principal and most proper ways of characterizing whole biological systems. These data make possible a direct comparison of totally different communities, independent of structure, geographical localization, and depth. Expressing the activity in oxygen consumption per square meter per hour, we will observe highest values in coral reef communities—up to 750 ml O_2—and probably the lowest values in places on the abyssal plains (on the order of 1 ml O_2). A fair estimate for tidal flat areas, including the Dutch Wadden Sea is on the order of 100 ml O_2 m^{-2} h^{-1}.

Closely tied to the one-way path of energy in the ecosystem, ultimately being dispersed into heat, is the cyclic pathway of materials. In spite of the great importance of this subject we will go into it only briefly here. It has become increasingly obvious that the benthic layer plays a dominant part in recycling processes. Tidal flat areas that receive large amounts of organic matter are analogous to chemical factories, producing large amounts of valuable nutrients as phosphate, nitrate, and silicate, which are liberated in such a well-proportioned ratio and in such a place, that they again become available immediately for primary production. Other important cycles deal with sulfur and carbon. From a geochemical point of view, specialized microorganisms, yeasts, fungi, algae, and protozoa are important in the transformation of numerous elements, especially those which are of biological significance, such as C, H, O, N, S, P, Fe, I, B, Mn, Co, etc.

In the ocean, with its relatively long food chains and low productivity, higher organisms are mainly responsible for mineralization; bacteria play a minor part. In productive areas, upwelling zones, coastal areas, and tidal flats, food chains are shorter. Dead and living organic matter is abundant, sedimentation rates become large, oxygen supply for aerobic decomposition may be limited, and thus the sediment becomes anaerobic. There is also both fermentation and respiration of compounds such as nitrate, sulfate, and carbon dioxide, on which nitrogen, sulfide, and methane are produced.

Not only do microorganisms serve recycling processes directly, but they also may change the environmental conditions in such a way that solution of other compounds is accelerated, as, for example, in silica and calcium.

Often the interstitial water serves as a primary place of reception. From here it disappears, by diffusion processes, wave or tidal pumping, and bioturbation, into the overlying water. It has recently become increasingly obvious that there may be large differences between the concentrations of dissolved compounds in the interstitial water and in the overlying water.

Particularly in polluted areas this may have serious consequences. Pollutants are transported with the suspended particles and are deposited in the sediments of the tidal flats or elsewhere on the sea bed. Here they are liberated and ultimately arrive in the interstitial water (Duinker *et al.*, 1974). In the interstitial waters of the Western Wadden Sea, Duinker found concentrations of chlorinated hydrocarbons on the order of 10 times as high, and of heavy metals in the order of 100 times as high, as those in the overlying sea water. Its influence on the infauna is totally unknown.

Space does not permit discussion of other biological aspects of the sea bed, among which are (1) aspects dealing with the often vicious circle of the organism–sediment relationship, in which organisms change their environment, after which the changed environment becomes suited for other organisms, and (2) the effects of organisms on the sediment, causing either consolidation or erosion of the sea bed. In general, the significance of bioturbation is highly underestimated. Often the upper sediment layers to a depth of 5–10 cm or even more are turned over completely by a varied assemblage of bottom organisms such as crustaceans, worms, molluscs, fish, etc. They will simultaneously circulate tremendous volumes of water through burrows or simply through the sediment, and thus act upon the exchange of dissolved and particulate matter between bottom sediment and overlying water. Investigation of quantitative methods for examination of bioturbation would be of great interest.

I will close this paper with a single remark on the large-scale, fully controlled, indoor basin as a useful tool for the study of the biology of the sea bottom. Such a basin, measuring 10 X 5 m and containing a tidal flat system, was recently constructed at the Netherlands Institute of Sea Research on Texel, the Netherlands. The principal aim will be investigation of the interrelations between different trophic levels under controlled environmental conditions. The prospects are very encouraging.

References

Beukema, J. J., 1974, Seasonal changes in the biomass of the macrobenthos of a tidal flat area in the Dutch Wadden Sea, *Netherlands Journal of Research 8:* 94–107.

Boddeke, R., 1971, The influence of the strong yearclasses of cod 1969 and 1970 on the stock of brown shrimp along the Netherlands coast in 1970 and 1971, *International Council for the Exploration of the Sea, C. M. 1971/K:* 32 (mimeo).

Brinkhurst, R. O., Chua, K. E., and Kayshik, N. K., 1972, Interspecific interactions and selective feeding by tubificid olegochaetes, *Limnology and Oceanography, 17:* 122–133.

Cadée, G. C., and Hegeman, J., 1974, Primary production of the benthic microflora living on tidal flats in the Dutch Wadden Sea, *Netherlands Journal of Research 8:* 260–291.

Colebrook, J. M., 1972, Changes in the distribution and abundance of zooplankton in the North Sea, 1948–1969, *Symposium of the Zoological Society of London, 29:* 203–212.

Colebrook, J. M., 1973, Geographical distribution and the abundance of plankton, *International Council for the Exploration of the Sea, CM 1973/L:* 18, p. 11 (mimeo).

Duinker, J. C., van Eck, G. T. M., and Nolting, R. F., 1974, On the behaviour of copper, zinc, iron, and manganese, and evidence for mobilization processes in the Dutch Wadden Sea, *Netherlands Journal of Sea Research 8:* 214–239.

Ernst, W., 1970, ATP als Indikator für die Biomassa marine Sedimente, *Oecologia (Berlin) 5:* 56–60.

Gerlach, S. A., 1971, On the importance of marine meiofauna for benthos communities, *Oecologia (Berlin) 6:* 176–190.

Gerlach, S. A., 1972, Die Produktionsleistung des Benthos in der Helgoländer Bucht, *Verhandlungsbericht der Deutschen Zoologischen Gesellschaft 65:* 1–13.

Glover, R. S., Robinson, G. A. and Colebrook, J. M., 1972, Plankton in the North Atlantic—an example of the problems of analyzing variability in the environment, *in: Marine Pollution and Sea Life,* FAO and Fishing News Books Ltd., pp. 439–445.

Hargrave, B. T., 1971, An energy budget for a deposit-feeding Amphipod, *Limnology and Oceanography 16:* 99–103.

Holme, N. A., and McIntyre, A. D., 1971, *Methods for the study of marine benthos,* IBP Handbook No. 16, Blackwell, Oxford, 334 pp.

Hughes, R. N., 1970, An energy budget for a tidal-flat population of the bivalve *Scrobicularia plana, Journal of Animal Ecology 39:* 357–379.

Johannes, R. E., and Satomi, M., 1967, Measuring organic matter retained by aquatic invertebrates, *Journal of the Fisheries Research Board of Canada 24:* 2467–2471.

Jonge, V. N. de and Postma, H., 1974, Phosphorus compounds in the Dutch Wadden Sea, *Netherlands Journal of Sea Research 8:* 139–153.

Korringa, P., 1969, Shellfish of the North sea, in: *Serial atlas of the marine environment, American Geographical Society, Folio 17:* 1–7, 8 plates.

Lasserre, P., 1971, Oligochaeta from the marine meiobenthos: taxonomy and ecology, *Smithsonian Contributions to Zoology 76:* 71–86.

Mann, K. H., 1972, Detritus and its role in aquatic ecosystems. *Memoire Istituto Italiano Idrobiologico 29* (Suppl.): 13–16.

McIntyre, A. D., 1969, Ecology of marine meiobenthos, *Biological Reviews 44:* 245–290.

McIntyre, A. D., 1971, Observations on the status of subtidal meiofauna research, *Smithsonian Contribution to Zoology 76:* 149–154.

Menzies, R. J., George, R. Y., and Rowe, G. T., 1973, *Abyssal Environment and Ecology of the World Oceans,* John Wiley and Sons, N.Y. London, 488 pp.

Newell, R., 1965, The role of detritus in the nutrition of two marine deposit feeders, the prosobranch *Hydrobia ulvae* and the bivalve *Macoma balthica, Proceedings of the Zoological Society of London 144:* 25–45.

NIOZ, 1973, *Deposit Feeding,* scientific film, Netherlands Institute for Sea Research, Texel, Netherlands.

Pamatmat, M. M., 1968, Ecology and metabolism of a benthic community on an intertidal sand flat, *Internationale Revue der gesamten Hydrobiologie 53:* 211–298.

Peterson, C. G. J., 1913, Valuation of the sea. II. The animal communities of the sea bottom and their importance for marine zoogeography, *Reports of the Danish Biological Station 21:* 1–44.

Postma, H., 1954, Hydrography of the Dutch Wadden Sea, *Archives Néerlandaises de Zoologie 10:* 405–511.

Postma, H., and Rommets, J. W., 1970, Primary production in the Wadden Sea, *Netherlands Journal of Sea Research 4:* 470–493.

Schlichter, D., 1975, The importance of dissolved organic compounds in sea water for the nutrition of *Anemonia sulcata* Pennant (Coelenterata), *Proceedings of the 9th European Marine Biology Symposium:* 395–405.

Spaargaren, D. H., 1975, Heat production of the shore-crab *Carcinus maenas* (L.) and its relation to osmotic stress, *Proceedings of the 9th European Marine Biology Symposium:* 475–482.

Thorson, G., 1957, Bottom communities (sublittoral and shallow shelf), in: (J. N. Hedgepeth ed.), *Treatise on Marine Ecology and Paleoecology, Geological Society of America Memoir 67* (1): 461–534.

Wieser, W., 1960, Benthic studies in Buzzards Bay. II. The meiofauna, *Limnology and Oceanography 5:* 121–137.

Wilde, P. A. W. J. de, 1973, A continuous flow apparatus for long-term recording of oxygen uptake in burrowing invertebrates with some remarks on the uptake in *Macoma, Balthica, Netherlands Journal of Sea Research 6:* 157–162.

6
Metabolic Activities of Benthic Microfauna and Meiofauna
Recent Advances and Review of Suitable Methods of Analysis

PIERRE LASSERRE

Some of the more significant studies of the metabolic activities of microbenthos and meiobenthos are reviewed, and technical and theoretical difficulties in such investigations are noted.

Introduction

The "interstitial fauna" of sand, or "mesopsammon" (Remane, 1940), and the mud "meiofauna" (Mare, 1942), have been studied very extensively in their morphology, distribution, and ecology (see Delamare Deboutteville, 1960; Swedmark, 1964; and McIntyre, 1969, for reviews along these lines).

Mare (1942) divided the benthic fauna into *macrofauna*, *meiofauna* (including metazoans smaller than 1 mm, foraminifers, and juvenile stages of macrofauna: the "temporary" meiofauna), and *microfauna* (bacteria, protophytes, and ciliated protozoans, excluding the foraminifers). The meiofauna is composed of

---

PIERRE LASSERRE ● Institut de Biologie Marine, Université de Bordeaux, Arcachon, France.

metazoans passing a 1.0 or 0.5 mm sieve. According to Fenchel (1969) "sieving is a poor way to sort the animals according the size since, for example, worm shaped animals may pass a sieve in which an equally large animal of another shape is retained." However, reliable methods for sampling and sorting of these small benthic organisms have been described and discussed (see Hulings and Gray, 1971; Uhlig *et al.*, 1973, for reviews).

Sensu stricto, the size distinction of 1 mm has no biological and ecological meaning and many groups retained by a 1.0-mm sieve display meiobenthic characters. This is true for small nematodes, oligochaetes, polychaetes, archiannelids, and ascidians. Furthermore, it is more and more established that meiofauna comprises small metazoans which have acquired a high degree of specialization; this is related to a convergent mechanism of evolution.

Efforts concentrating on the questions of population density are relatively recent, and the need for more quantitative information was stressed in an important review published by McIntyre (1969). The purpose of his review was to assess the role of meiofauna in the ecology of the marine ecosystem. This idea was also developed by Fenchel (1969) in his comprehensive work on the ecology of microbenthos, with special reference to ciliated protozoans.

The recognition of benthic communities of ciliates and meiofauna with extremely high individual numbers is recent and is of steadily increasing interest. The most numerous taxonomic groups are ciliates, nematodes, harpacticoid copepods, oligochaetes, turbellarians, gastrotrichs, tardigrads, archiannelids, and rotifers.

Intertidal and subtidal benthic communities may contain from 55,000 to 1,000,000 meiofauna individuals square meter (McIntyre, 1969). If the majority of meiofauna concentrates in the upper centimeters, abundant populations are present far down in sandy and muddy sediments. In medium-coarse calcareous sand, 40 m deep, collected in Mururoa atoll, up to 1,300,000 meiofauna specimens were found in the superficial stratum at the sediment–water interface; however, in a stratum 15 cm below the interface, 360,000 specimens were also present in an equal quantity of sediment (Salvat and Renaud-Mornant, 1969). Fenchel (1967) has found more than 10,000,000 ciliates per square meter in Scandinavian sediments, representing up to 2.3 g wet weight per square meter.

Standing stocks of the meiofauna between 2 and 20 g wet weight per square meter are frequent in littoral and sublittoral sand and mud. For example, in the Baltic, the standing crop of the meiofauna above the halocline has been estimated to be 5–20 g wet weight square meter (Elmgren, quoted in Jansson, 1972). Thiel (1972) states that the meiofauna biomass in the Iberian deep sediments (3500 meters) is between 0.24 and 2.8 g wet weight per square meter.

In shallow-water areas, meiofaunal and microfaunal groups have short generation times, and their production: biomass ratios tend to be considerably higher than for macrofauna (e.g., McIntyre, 1969; Fenchel, 1969; Gerlach, 1972). In the deep-sea benthos, the available data suggest a year-round reproduction pattern (Rokop, 1974) and the abyssal meiofaunal community has been estimated to be 1.5 to 3.9 times as abundant as the macrofauna (Hessler and Jumars, 1974).

The striking contrast between the community structure of these comparatively stable and permanent sets of species of great diversity—covering probably several trophic levels—and that of macrobenthos is now a widely accepted feature.

It is more difficult to assess the exact role of ciliates and meiofauna within the sand and mud ecosystem. The need for more information on the interaction of microbenthos and meiobenthos with lower and higher elements in the food chain becomes urgent. Some aspects regarding feeding and predation have been considered recently (e.g., Fenchel, 1969; Thane-Fenchel, 1970; McIntyre *et al.*, 1970; McIntyre and Murison, 1973). Furthermore, recent improvements of our knowledge about the metabolic role played by the microfauna and meiofauna within the benthic boundary layer have resulted from measurements of total bottom metabolism and from more physiological and biochemical studies on the metabolic solutions which have been developed by different species to succeed in oxidized or anoxic sediments. These and other related problems have been discussed recently (Lasserre and Renaud-Mornant, 1975). The following arguments on the metabolic activities of benthic microfauna and meiofauna are based on data published mainly during the last decade.

Metabolic Studies on Single Taxa

The most evident characteristic of both microfauna and meiofauna lies in the very small size of the animals. This feature necessitates acute specifications with regard to the dimensions of the techniques that will be used in the metabolic approach. For example, the direct calorimetric measurement of the loss of energy, in the form of heat resulting from oxidations and other metabolic activities, is impracticable for these minute organisms. Conversely, it is possible to measure with a sufficient degree of accuracy and reproducibility the oxygen uptake and to convert this into the energy equivalent of the food that is presumed to have been oxidized.

It is also interesting to evaluate part of the energy releases through anaerobic processes. Little is known of the possible magnitude of such effects. The significant works which have been done recently by different authors tend to encourage future studies.

Maintenance of Standard Metabolism

Oxygen consumption has two components: one active rate related to activity (e.g., ciliary activity, locomotion) and which increases logarithmically with temperature; and one low "standard" rate which is relatively independent of activity. A high level of "standard" metabolism has been found in meiofauna living in oxidized conditions. When one compares different species or different groups of species, it is apparent, for example, that harpacticoid copepods are active and oligochaetes are more sluggish. This external activity may influence the level of O_2 consumption, a fact which is important to consider in the ecological utilization of respiratory measurements. If locomotion is inhibited by anesthetics, it appears that the rates of oxygen consumption are lowered around 10% only in the case of the mystacocarid *Derocheilocaris remanei*, and 5% only in oligochaetes *Marionina achaeta* and *Marionina spicula* and in the polychaetes *Stygocapitella subterranea* or *Hesionides arenaria*. This phenomenon is not so clearly visible in other groups, such as the archiannelid *Protodriloides symbioticus* and other fast-moving species, which have a more important active rate of O_2 consumption but no more than 50% of the total respiratory rate.

Therefore, for constant and nonstressful conditions, the oxygen consumption is maintained at the same rate. Locomotion is not the primary cause of a high metabolic rate, which is generally apparent in meiofauna living normally in oxidized sediments (Lasserre, 1971a). A comparable pattern of respiration was noted by Zeuthen (1947b) in larval forms of echinoderms and polychaetes.

Influence of Body Size and Metabolic Levels

The possibility of subdividing microbenthic and meiobenthic species into groups of species presenting different rates of oxygen consumption provides stimulating perspectives. However, for the sake of comparison between animals different in size and weight, it is necessary to account for the equation $Q = aW^b$ which relates body weight (W) in wet-weight or dry-weight, to oxygen consumption (Q) per unit time. The coefficient a is the level of metabolic expenditure per time, and b the regression coefficient which defines the rate of change of metabolism with body weight. The regression analysis, assuming a linear relationship, is carried out on the data after a double logarithmic transformation ($\log Q = \log a + b \log W$). The respiration body weight regressions are compared by covariance analysis.

In ciliates and meiobenthic metazoans, oxygen consumption is clearly weight-dependent (Lasserre, 1970, 1971a; Atkinson, 1973b; Ott and Schiemer, 1973; Vernberg and Coull, 1974a). The value of the regression coefficient (b) falls within the range 0.6 to 0.9, supporting the values published for other poikilotherms (Zeuthen, 1953; Hemmingsen, 1960). A common regression line with a slope $b = 0.75$ can relate oxygen uptakes of ciliates, meiofauna, and macrofauna. This overall value represents only an estimate given for comparison of metabolic levels between different benthic groups. It is likely that the slope b relating oxygen consumption and body weight can be modified by both endogenous and exogenous factors (Lasserre, 1971a; Atkinson, 1973a).

From Fig. 8 it appears that the ciliate *Tracheloraphis sp.*, studied by Vernberg and Coull (1974a), has an extremely high respiration rate, ranging from 2350 to 7000 μl O_2 h^{-1} g^{-1} wet weight, and averaging 4500 μl O_2 h^{-1} g^{-1}. Meiofauna respiration rates range from 400 to 3000 μl O_2 h^{-1} g^{-1} wet weight in *group I "high."* This group is composed of species displaying a clear tendency to live—in nature—aerobically, and comprises species living intertidally and subtidally in well or reasonably oxygenated sediments. Their metabolic rate is approximately 5 to 100 times more than that of macrobenthic species.

Meiofauna species classified in *group II "low"* have respiration rates ranging from 90 to 400 μl O_2 h^{-1} g^{-1} wet weight.

This is found for several nematodes (Ott and Schiemer, 1973; Schiemer and Duncan, 1974; Wieser *et al.*, 1974), two gnathostomulids (Schiemer, 1973) and the polychaete *Stygocapitella subterranea* (Lasserre and Renaud-Mornant, 1973). The low levels of respiration observed by these authors, using two different techniques of microrespirometry (stoppered and standard divers), may reflect an ecophysiological adaptation. These species live in oxygen-deficient or anoxic marine and limnic sediments. The metabolism of these species could be partially anaerobic even when oxygen is available. However, this does not seem always the case: the ciliate *Tracheloraphis sp.* lives in nature in low-oxygen or anaerobic conditions (Vernberg and Coull, 1974b) and displays very high respiratory rates. Moreover, several "euroxybiontic" species live in nature, or in the laboratory, in low oxygen tensions (Lasserre and Renaud-Mornant, 1973, and later in this review), and also possess a high respiration rate.

In nematodes, the oxygen consumption was found to be directly proportional to the size of the buccal cavity. Species with a large buccal cavity respire at the highest rate (Wieser and Kanwisher, 1961; Teal and Wieser, 1966). Similar observations were made in a more recent work (Wieser *et al.*, 1974); in addition, oxygen consumption was found to be inversely proportional to sediment depth. Ott and Schiemer (1973), Wieser *et al.* (1974), concluded that, in nematodes, the differences in respiratory rates could be correlated with different trophic habits, in oxygenated or in reduced habitats.

Considering the data obtained so far, respiration rates obtained in the Cartesian diver and oxygen electrode respirometers do not seem different from those which presumably take place in the natural environment. High and low rates of oxygen uptakes are very probably related to specifically different modes of ecophysiological adaptation. The choice of the Cartesian diver method is

highly advisable if we admit that the meiofauna animal is able to behave nat
rally in the respiration chamber.

To apply such results to estimate the energy loss by respiration in differe
benthic ecosystems, more data on different species at different weight (and ag
classes are required.

Strategies of Ecophysiological Adaptation of Micro- and Meiobenthic Organisms

A high degree of tolerance to many environmental factors has been shown
many microbenthic and meiobenthic species. Salinity, temperature, and oxyg
availability are the main factors which have been considered.

Many marine ciliates are euryhaline and can survive at temperatures fro
0°C to a little below 30°C (Fenchel, 1969).

Behavioral reactions, tolerance, and preference limits have been studied e
perimentally on several meiobenthic species (e.g., Remane, 1933; Noodt, 195
Wieser and Kanwisher, 1961; Jansson, 1962, 1967; Boaden, 1962, 1963; Gra
1965, 1967; Lasserre, 1969, 1971b; Coull and Vernberg, 1970; Lasserre ar
Renaud-Mornant, 1971a,b, 1973; Vernberg and Coull, 1975; Wieser *et a*
1974). These studies focused on physiological correlations which possibly exi
between environmental features and animal distribution.

This possibility of physiological adaptation was underlined by Swedma
(1964) in his critical review on the interstitial fauna of marine sand. Jansson h
shown that interstitial animals are "guided in the field both by their toleran
reactions and preference reactions." Tolerances were tested by the lethal effe
limit of 50% and the preference experiments were carried out employing sma
experimental chambers where the animals are placed in a gradient of the fact
which is studied. These experiments give very useful information. However, the
ecological interpretations should be considered with prudence.

Within the complex picture of all the adaptive features acquired by mei
fauna and microfauna in order to succeed in different habitats, we should al
try to define physiological and biochemical adaptations, in other words, th
cellular or molecular mechanisms that will explain at least part of their evol
tionary adaptation.

Metabolic Compensations

The metabolic compensations represent the possibility for organisms 1
exhibit similar metabolic rates under widely different environmental factor
such as temperature, salinity, oxygen pressure, etc. The molecular mechanisn
underlying these compensatory responses probably include a number of impo

tant enzymatic changes. Metabolic compensations have been described in differ-, ent invertebrates and fish (Bullock, 1955; Prosser, 1955, 1958, 1967; Fry, 1958; Newell and Northcroft, 1967; Rao, 1967) and the physiological and biochemical basis of such mechanisms of compensation are now widely recognized as a "strategy of biochemical adaptation" (Hochachka and Somero, 1973).

Metabolic compensations have been found in meiofaunal organisms: oligo-chaetes (Lasserre, 1969, 1970, 1971a), polychaetes (Lasserre and Renaud-Mornant, 1973), mystacocarids (Lasserre and Renaud-Mornant, 1971a,b), harpacticoid copepods, and archiannelids (Lasserre and Renaud-Mornant, 1973).

Measurements of O_2 consumption were performed on single individuals, using the Cartesian diver technique. This analytical approach was used in order to provide information on the homeostatic capabilities of two meiobenthic oligochaetes, *Marionina achaeta* and *Marionina spicula*. Differences in species distribution have been explained in terms of expenditure of energy—the most adapted species consuming less energy—inside a specific compensation range. The overall O_2 consumption was found to be statistically constant over "respiration adaptive plateaus," the limits of the plateau being different for each species. Different "respiration adaptive plateaus" are shown in Figs. 1–5.

Lasserre (1971a) found that the enchytraeid oligochaete *Marionina achaeta* could tolerate salinities somewhat above 25%$_{oo}$, but only with the expenditure of considerably more metabolic energy; it would therefore be at a disadvantage on the lower shore where the salinity is normally above this level. Therefore *Marionina achaeta* is an euryhaline species with a clear tendency to be more adapted to live in a low salinity environment. *Marionina spicula*, on the other hand, could tolerate salinities above 20%$_{oo}$ without the expenditure of metabolic energy, but was unable to survive at the high temperatures (above 30°C) which have been recorded during summer in the upper part of the range of *Marionina achaeta*. Lasserre concluded that "while both species were well adapted for life in the fluctuating intertidal environment, each had its own specific ability to succeed in different temperature or salinity ranges" (quoted in McIntyre, 1969).

With the mystacocarid *Derocheilocaris remanei*, the temperature compensation was less effective than the salinity compensation and it was concluded that the respiratory metabolism of *Derocheilocaris remanei* was not strongly influenced by rapid changes in salinity. It is therefore adapted to living in fluctuating gradients of salinity (between 10%$_{oo}$ and 36%$_{oo}$), but was limited to living in the lower part of the beach were temperatures are more stable than in the higher part of the beach. Some regulatory respiration curves were obtained after acclimation to different oxygen tensions (Fig. 4). The molecular mechanisms underlying these compensatory responses include a number of important enzymatic changes.

In a detailed study on the osmotic capabilities of *Marionina achaeta*, Lasserre (1969, 1975) found an energetic relationship between ionic regulation and

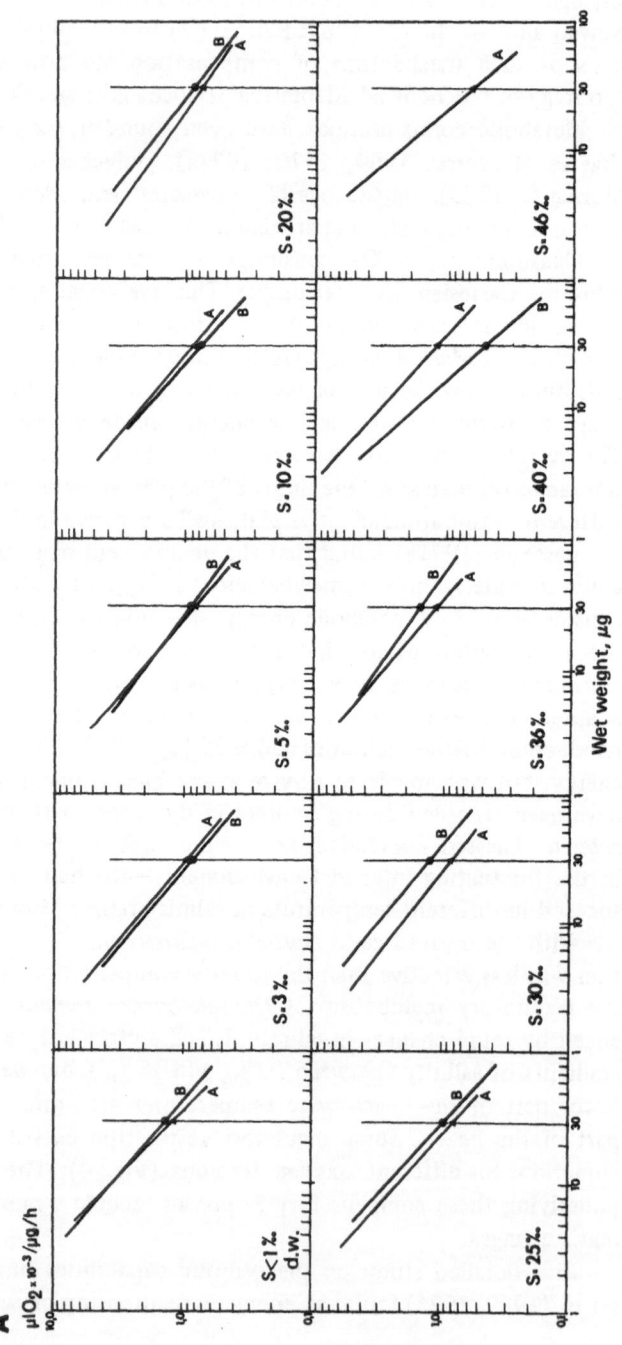

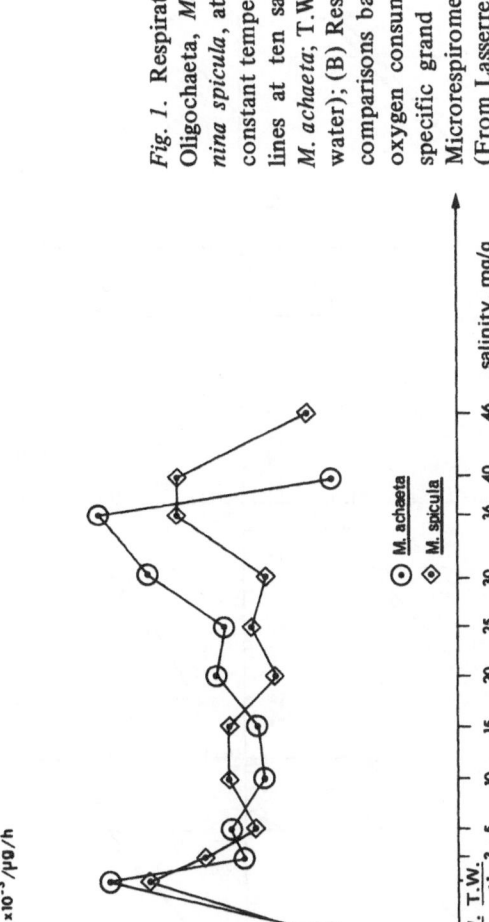

Fig. 1. Respiration of two meiobenthic Oligochaeta, *Marionina achaeta* and *Marionina spicula*, at different salinities and for a constant temperature (19°C). (A) Regression lines at ten salinities (A = *M. spicula*; B = *M. achaeta*; T.W. = tap water; D.W. = distilled water); (B) Respiration rates at 12 salinities: comparisons based on estimates from mean oxygen consumption adjusted to the inter-specific grand mean wet weight of 30 μg. Microrespirometer: standard Cartesian diver. (From Lasserre, 1971a.)

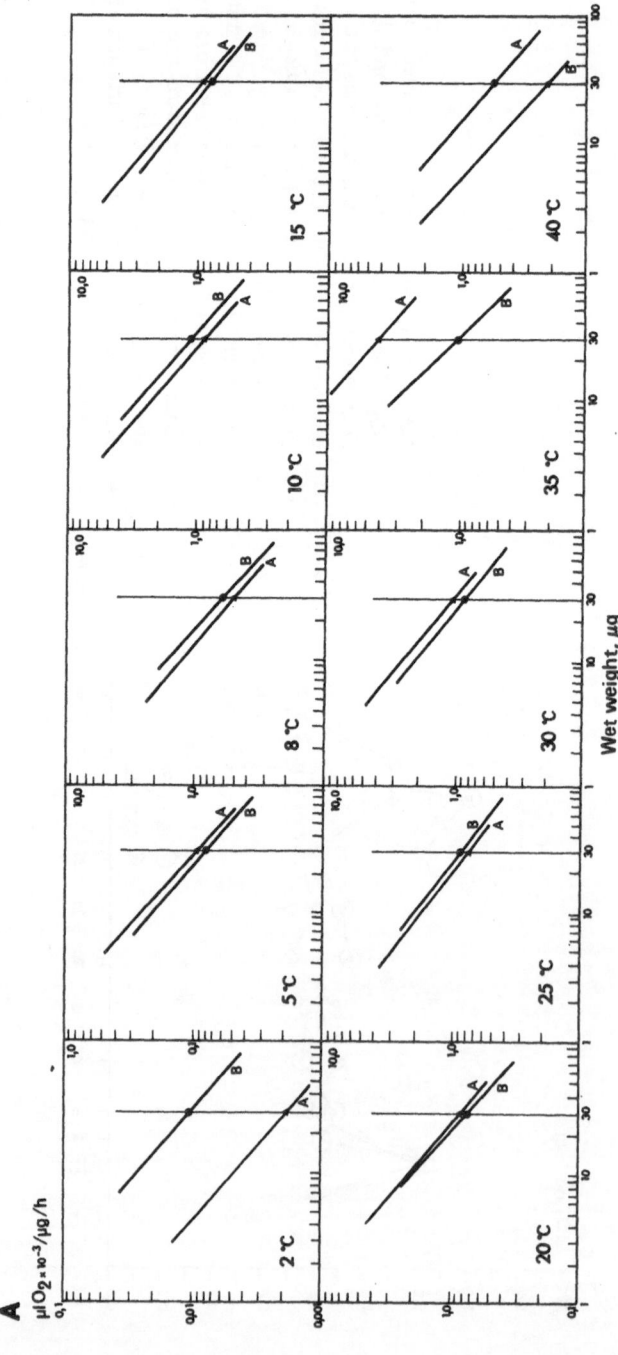

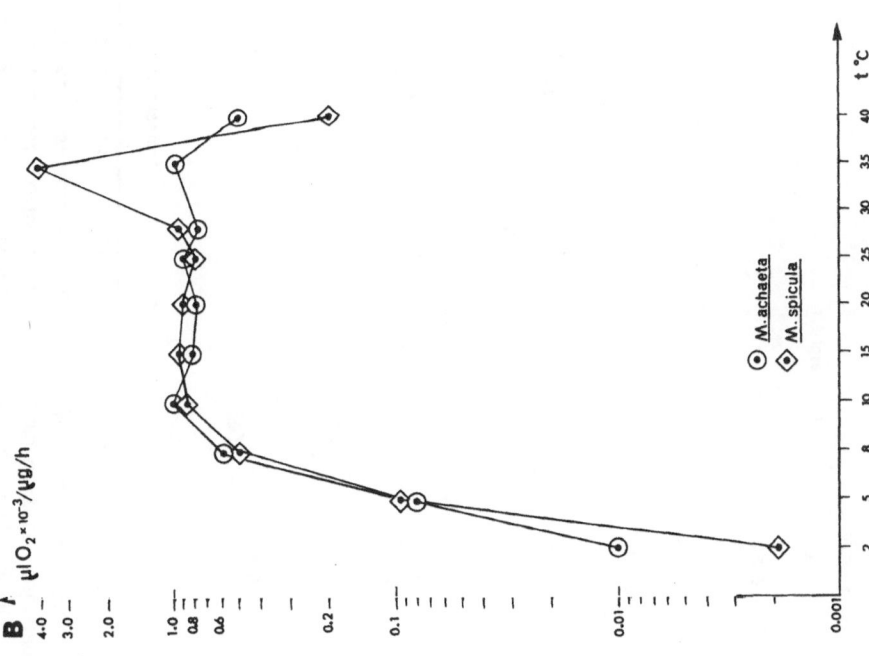

Fig. 2. Respiration of two meiobenthic Oligochaeta, *Mari.-nina achaeta* and *Marionina spicula*, at different temperatures and for a constant salinity of 15⁰/oo. (A) Regression lines at ten temperatures (A = *M. spicula*: B = *M. achaeta*). (B) Respiration at ten temperatures; comparisons based on estimates from mean oxygen consumption adjusted to the interspecific grand mean wet weight of 30 μg. Microrespirometer: standard Cartesian diver. (From Lasserre, 1971a.)

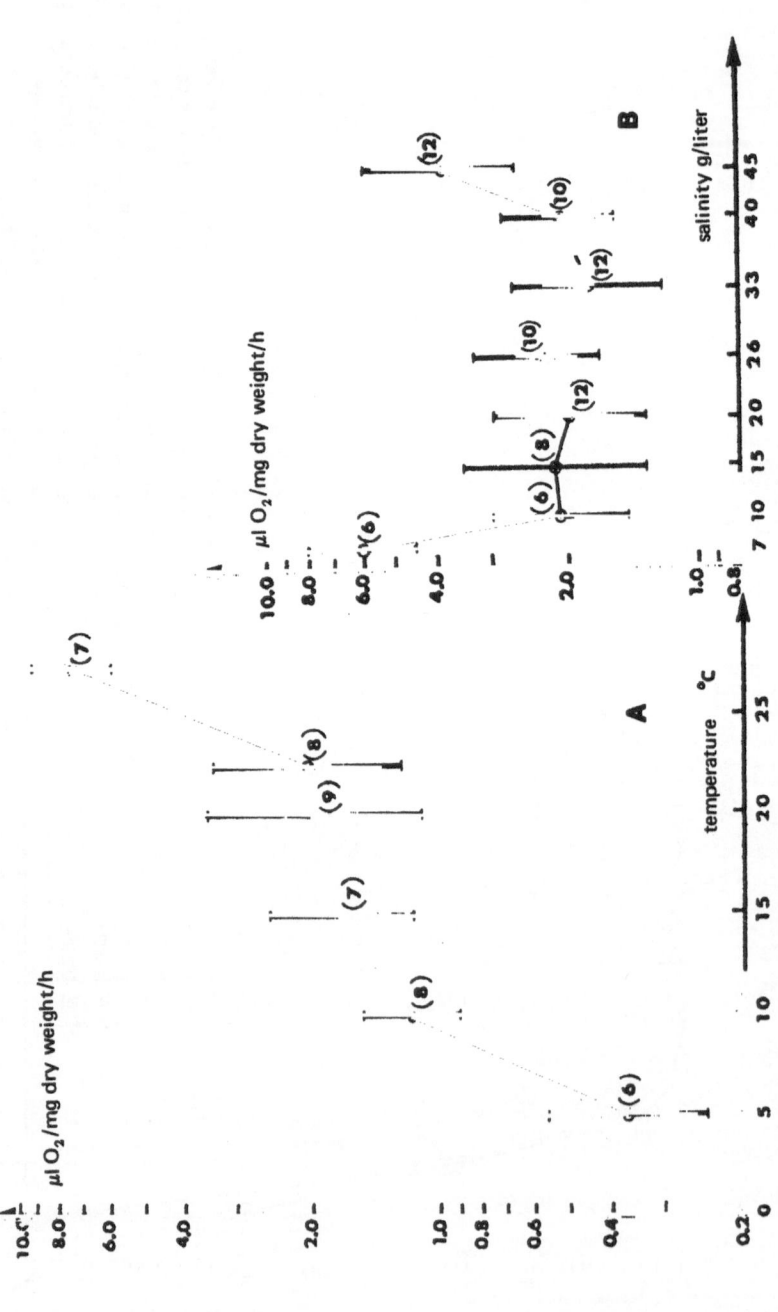

Fig. 3. Respiration of *Derocheilocaris remanei* (*Crustacea, Mystacocarida*) at six temperatures and eight salinities. Comparisons based on estimates from mean oxygen consumption adjusted to the grand mean dry weight of 0.43 µg. Limits of the standard error of adjusted means. Microrespirometer: standard Cartesian diver. (From Lasserre and Renaud-Mornant, 1971b.)

respiratory rates. Some of the energy liberated by the respiratory metabolism in the form of ATP is probably utilized by a sodium–potassium pump which maintains ionic concentration gradients in the animal body. In the presence of ouabain, which is known as a specific inhibitor of $(Na^+ - K^+)$ ATPase activity, oxygen uptake is depressed. This phenomenon suggests that a significant part of oxygen consumption is linked to some hydrolyzing $(Na^+ - K^+)$ ATPase system, the latter being implicated in some active process of ionic regulation.

Ecophysiological Adaptation to Oxygen-Deficient and Anoxic Conditions

Fluctuating Oxygen Regimes. Differences in metabolic responses to fluctuating oxygen regimes have been detected on meiofauna living in "semistable" or protected and oceanic sandy beaches of the Atlantic (Lasserre and Renaud-Mornant, 1973). A typical pattern of species composition was chosen and served as background in this study. On the protected beach of Eyrac the percolation conditions of the sediment are complex. The pore system provides good conditions for oxygenation; however, high density of decaying sea weeds and zostera can produce redox discontinuity layers which limit the oxidized biome. On the other hand, at Mimizan, the entire interstitial climate of this oceanic-type beach is under the influence of very strong current dynamics and the sand is fully oxidized. In this "high-energy beach" (terminology of Riedl, 1971) two interstitial crustaceans, one mystacocarid and one harpacticoïd copepod, are strictly bound to the retention and resurgence zones, deep in the sand and 10–20 cm above the subsoil water table at high levels.

Considering the tolerance experiments to show O_2 deficiency effects alone, the species fall into three groups: one in which the capacity to live in very poorly oxygenated habitats (< 0.5 ml O_2 per liter) is small (*Paraleptastacus spinicauda; Derocheilocaris remanei*), 50% of the individuals being dead after approximately 2–5 h; a second in which the species remain alive much longer, the 50% mortality occurring at 48–78 h (*Hesionides arenaria; Protodriloides symbioticus*); and a third group which can endure longer exposures (5 days for *Stygocapitella subterranea* and approximately 3 days for *Marionina achaeta*). Oxygen deficiency and the simultaneous presence of H_2S were fatal for all species. However, the resistance of *Stygocapitella subterranea* declined less sharply, a phenomenon that was probably concomitant with a low rate of respiration.

The effects of changes in the imposed oxygen regime have been studied in terms of respiratory physiology. Oxygen uptakes were determined using the Cartesian diver microrespirometer, a special technique for filling divers with different gas mixtures being utilized. The respiration profiles determined give a very accurate and reliable estimation of the different degrees of physiological adaptation to fluctuating oxygen regimes, and the results obtained so far show

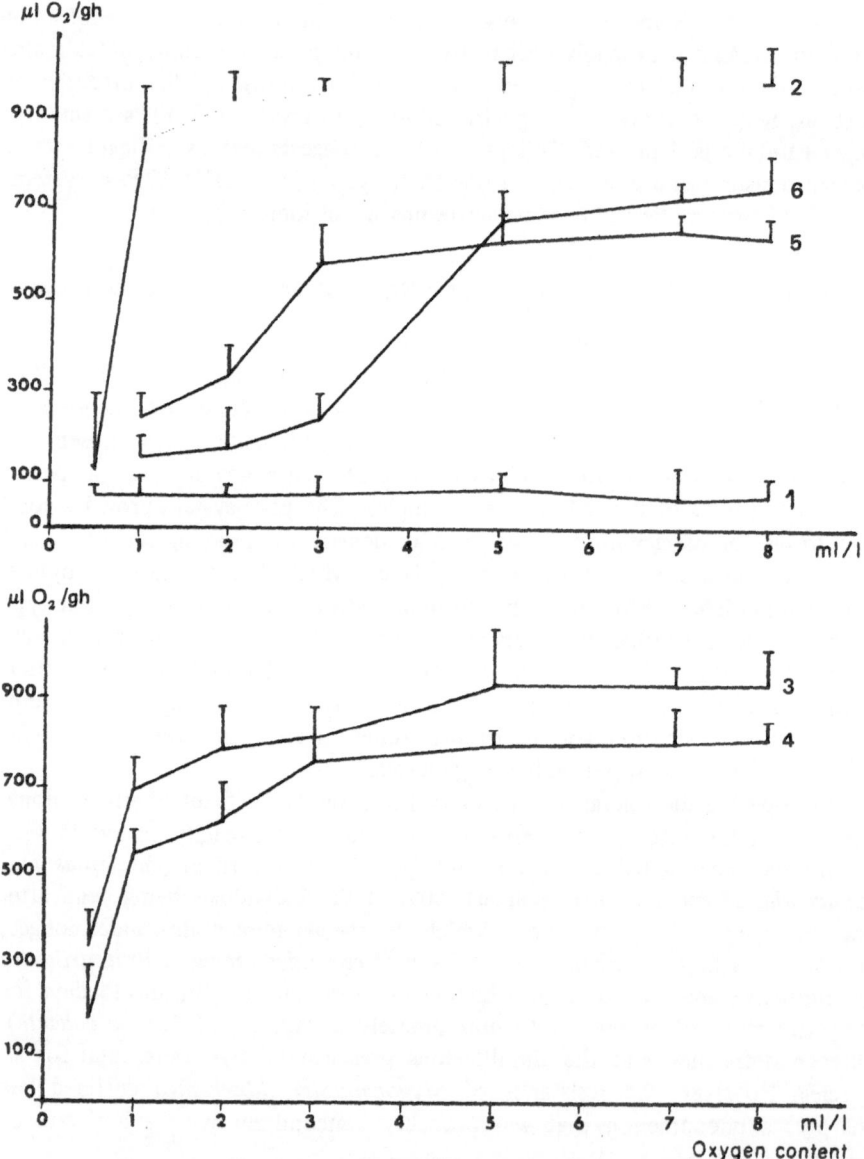

Fig. 4. Respiration of six meiofauna species at different oxygen concentrations. Polychaeta: 1, *Stygocapitella subterranea*; 4, *Hesionides arenaria*. Oligochaeta: 2, *Marionina achaeta*. Archiannelida: 3, *Protodriloides symbioticus*. Copepoda Harpacticoida: 6, *Paraleptastacus spinicauda*. Mystacocarida: 5, *Derocheilocaris remanei*. Species 1 and 2 are euroxybionts, 5 and 6 are stenoxybionts, 3 and 4 have complex metabolic profiles. Microrespirometer: standard Cartesian diver. (From Lasserre and Renaud-Mornant, 1973.)

very good correlations with the different distribution patterns of the selected species of meiofauna, these patterns following largely (but not exclusively) the chemophysical gradient of oxygenation.

The oxygen consumption was found to be independent of the ambient oxygen down to a critical O_2 concentration (P_c). Below this critical level the respiration rates were more or less directly proportional to variations in ambient oxygen content (Fig. 4). The following classification has been proposed (Table 1):

1. *Euroxybiontic regulators* (*Stygocapitella; Marionina*). These species show the highest tolerance to low O_2 tension ($P_c < 0.5$ ml O_2 per liter). This adaptive feature enables the species to live in oxygen-deficient habitats.

2. *Stenoxybionts* (*Derocheilocaris remanei; Paraleptastacus spinicauda*). Oxygen uptake is near the line of a strictly dependent relationship between O_2 tension and O_2 consumption, below 5 ml O_2 per liter and 3 ml O_2 per liter of oxygen content, respectively.

3. *Complex metabolic types* (*Protodriloides; Hesionides*). The relationships between O_2 consumption and ambient oxygen content are sometimes more complex. A regulator or conformer type of metabolism is not clearly distinguishable.

The euroxybiontic meiofauna species (*Stygocapitella* and *Marionina*), showing highest tolerance to low oxygen tension, live in oxygen-deficient sandy sediments.

The stenoxybiontic and conformer species (*Derocheilocaris; Paraleptastacus*)

Table 1. Metabolic Rates, Critical O_2 Concentrations (P_c), and Respiration Profiles of Six Meiobenthic Species

Species		Resistance to O_2 deficiency	Metabolic rate	P_c, ml O_2/liter	Respiration profile[a]
S. subterranea	1	Euroxybiont	Low	0.5	R
M. achaeta	2	Euroxybiont	High	1	R
P. symbioticus	3	Euroxybiont	High	5	R above $P_c = 5$
				1	R + C for $1 < P_c < 5$
H. arenaria	4	Euroxybiont	High	3	R above $P_c = 3$
				1	R + C for $1 < P_c < 3$
D. remanei	5	Stenoxybiont	High	3	R above P_c C under P_c
P. spinicauda	6	Stenoxybiont	High	5	R above P_c C under P_c

[a]R: metabolic regulator; C: metabolic conformer; R + C: complex type. (From Lasserre and Renaud-Mornant, 1973.)

require a fully oxidized biotope, the "high-energy beach" being the appropriate habitat.

The complex metabolic types (*Protodriloides; Hesionides*) are well adapted to resist in poorly oxygenated habitats; they tend nevertheless to be more adapted to live in a fully oxidized or in fluctuating oxygen regimes. This is the case in the lower part of the semistable beach where the two species are distributed.

The effects of fluctuating oxygen regimes have also been studied on the marine enoplids *Enoplus brevis* and *Enoplus communis* by Atkinson (1973a,b).

One of these free-living nematodes, *Enoplus brevis*, occurs in estuarine muds where there are changes in the environmental O_2 concentration while *Enoplus communis* is found in habitats where oxygen is continuously available. The oxygen uptake, measured with an oxygen electrode microrespirometer before acclimation, is reduced by each lowering in the imposed oxygen tension (135, 75, 35, and 12 Torr). However, after prolonged exposure to low oxygen tension (35 Torr), the level of O_2 consumption was appreciably higher. Compensatory mechanisms in the metabolism would appear in the region of 30–40 Torr. This range of oxygen tensions are nearly the lowest at which these acclimated nematodes are able to maintain a full range of metabolic activity (Fig. 5). Considering the effect of exposure to oxygen-free sea water on the subsequent oxygen uptake at air saturation, it would seem that the levels of oxygen consumption are not appreciably altered after exposure to anaerobic conditions (Fig. 6).

The Sulfide Biome. Ecological evidences of the presence of a characteristic fauna in marine and limnic anoxic sediments have been well documented, particularly when free-living nematodes and ciliates are concerned (Wieser and Kanwisher, 1959, 1960; Fenchel, 1969; Ott and Schiemer, 1973; Vernberg and Coull, 1975). According to Fenchel and Riedl (1970), the sulfide system, characterized by anoxic conditions and containing toxic compounds such as hydrogen sulfide, nevertheless has a highly diverse metazoan fauna. This sulfide biome contains "more than 12 phyla, ranging from blue-green algae to metazoans with many new groups and new functional and organizational types" (Fenchel and Riedl, 1970). Boaden and Platt (1971) have coined the term "thiobios" to describe the biome of sulfide-rich zones, and various examples of "thiobiotic" species have been described (Boaden, 1974). However, physiological proof for anerobic metabolism has not been clearly established for such metazoans. Prolonged anoxia apparently induces cryptobiosis in several species and neither glycogen nor lipid stores are utilized on such occasions (Atkinson, 1973a). The data for two gnathostomulid species and for benthic nematodes living in strongly reduced sediments (Ott and Schiemer, 1973; Schiemer, 1973) indicate a general low level of oxygen consumption (Fig. 7) in comparison with animals from better-oxygenated habitats, and the authors suggest that some of

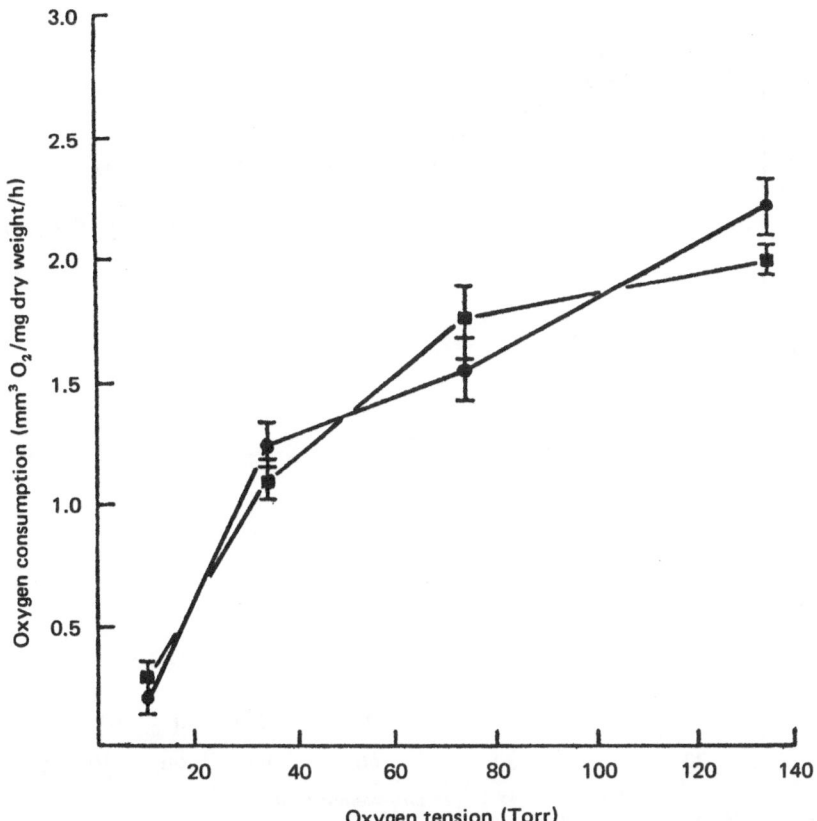

Fig. 5. Respiration of *Enoplus brevis* (solid circles) and *E. communis* (solid squares), benthic Nematoda, at four oxygen tensions. Comparisons based on estimates for mean oxygen consumption adjusted to the interspecific grand mean dry weight 45.5 μg. Microrespirometer: Clark-oxygen electrode (Radiometer). (From Atkinson, 1973a.)

the metabolic energy is released mostly by glycolytic processes. A very low oxygen uptake (Fig. 4) has been found also with the interstitial polychaet *Stygocapitella subterranea* (Lasserre and Renaud-Mornant, 1973). This species shows a very good tolerance to low oxygen concentrations (less than 0.5 ml O_2 per liter), and also a limited tolerance to simultaneous presence of H_2S.

Recently, Wieser *et al.* (1974) considered that the nematode *Paramonhystera n. sp.* "is the first marine metazoan in which it can be shown that a specific biological process is favorably affected by anoxic conditions if compared with the situation at normal pO_2."

The metabolism of these strict or facultative anaerobes has not yet been

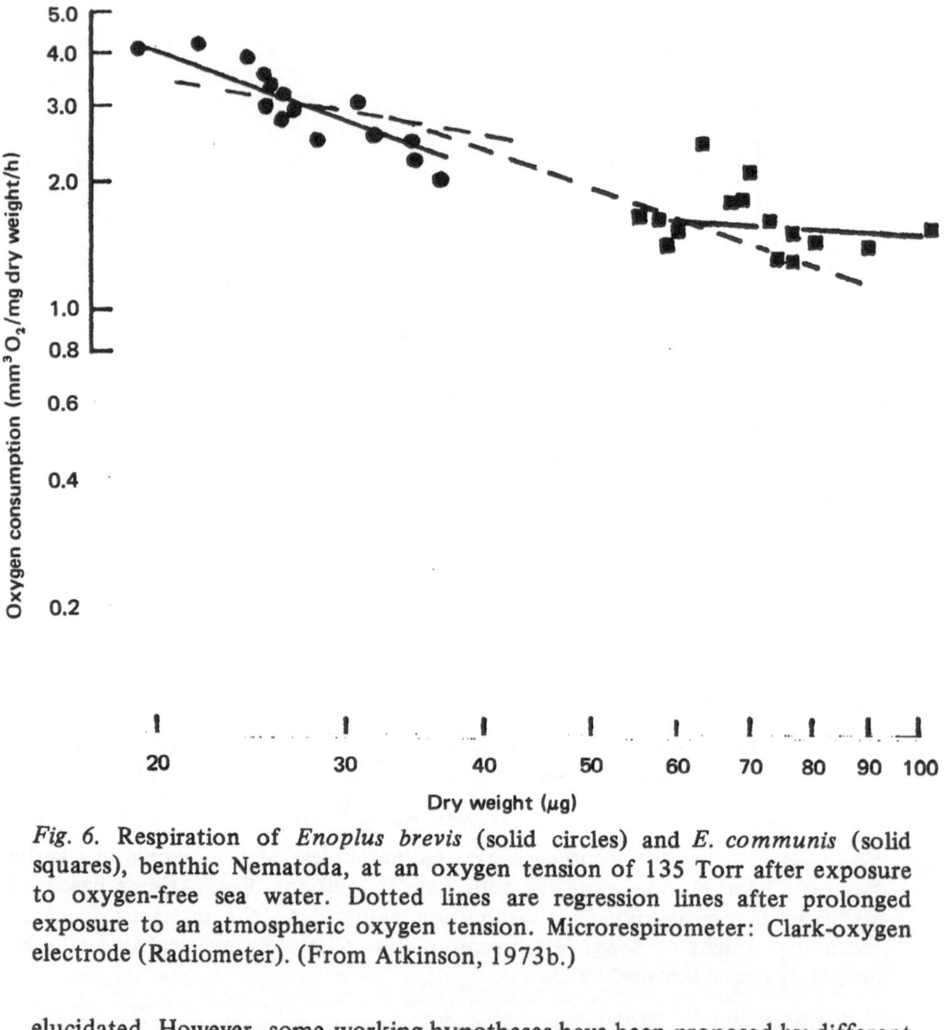

Fig. 6. Respiration of *Enoplus brevis* (solid circles) and *E. communis* (solid squares), benthic Nematoda, at an oxygen tension of 135 Torr after exposure to oxygen-free sea water. Dotted lines are regression lines after prolonged exposure to an atmospheric oxygen tension. Microrespirometer: Clark-oxygen electrode (Radiometer). (From Atkinson, 1973b.)

elucidated. However, some working hypotheses have been proposed by different authors (Atkinson, 1973a; Wieser *et al.*, 1974; Maguire and Boaden, 1975).

In the nematode *Enoplus brevis*, a high-affinity respiratory pigment (hemoglobin) seems to function at low pO_2 (Ellenby and Smith, 1966; Atkinson, 1973a). In *Paramonhystera n.* sp., "changes of coloration which may be due to a new pigment system" have been observed (Wieser *et al.*, 1974, p. 246).

The "thiobiotic" gastrotrich, *Thiodasys sterreri*, survives completely anaerobic conditions and seems able to fix carbon dioxide (Maguire and Boaden, 1975). According to these authors, a reverse Krebs cycle sequence might be involved. This assumption is based on incubation experiments in ^{14}C-labeled sodium-bicarbonate; autoradiography and chromatography techniques have been

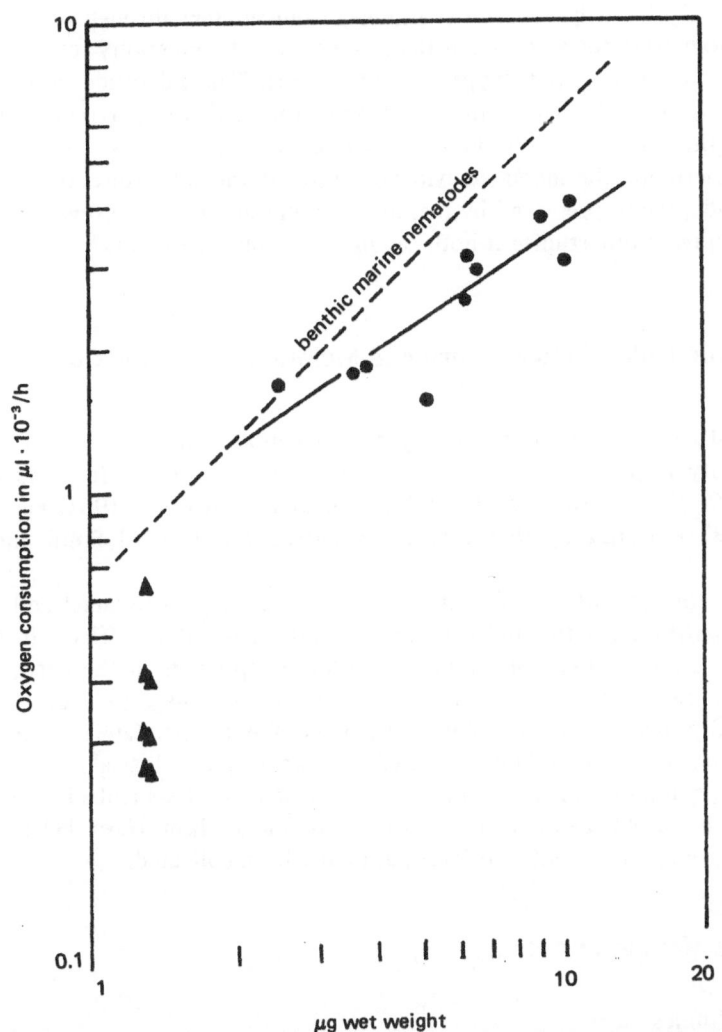

Fig. 7. Respiration of two meiobenthic Gnathostomulida, *Haplognathia* sp. (circles) and *Gnathostomula* sp. (triangles). Dotted line: regression line for benthic marine nematodes. Microtechnique: stoppered Cartesian diver. (From Schiemer, 1973.)

used in conjunction electron microscope observations of tissues (mitochondria are similar to those of other anaerobic species). However, to confirm the authors' conclusions a more precise identification of radioactive components is necessary.

Concluding Remarks. In a shallow stressed area off Sapelo Island, Georgia, seasonal respiration of the meiofaunal, microfaunal, and microfloral component remained significantly constant (and high) from summer through winter and no correlations were found between this component of community respiration and temperature or dissolved oxygen (Smith, 1973). The independence of *in situ* respiration over the environmental temperature and oxygen ranges strongly suggests the existence of physiological metabolic compensations.

Furthermore, the increased evidence of many original features of ecophysiological adaptation presented by meiobenthic and microbenthic organisms tends to encourage future studies at both cellular and molecular levels.

Metabolic Role of Microfaunal and Meiofaunal Benthic Components

Results of *in situ* and laboratory methods used to measure overall benthic community respiration have been discussed by several authors (e.g., Kanwisher 1959, 1962; Kanwisher *et al.*, 1974; Teal and Kanwisher, 1961; Pamatmat, 1968, 1971a; Hargrave, 1973; Smith *et al.*, 1972; Smith, 1973; Smith and Teal, 1973).

With the present insufficient knowledge regarding meiofaunal and microfaunal contributions to total community metabolism, it is difficult to draw a precise picture of the role of these benthic components in the energy flow. Furthermore, an accurate estimation of the relative magnitude of the total community respiration contributed by macrofauna, meiofauna, microfauna, microflora, and bacteria should be based on more ecophysiological investigations, including studies on metabolic profiles at a single taxon level and a better understanding of aerobic and anaerobic parts of the metabolism. Nevertheless, in the last few years, some significant information has been collected.

Aerobic Metabolic Activity

In Situ Studies

In experimental sand ecosystems composed of columns filled with sand and provided with sea water flowing down the column at a constant rate, McIntyre *et al.* (1970) report oxygen demands in excess of the organic matter input they can account for. If we admit that respiratory rates of meiobenthic communities are high, we would expect the organic carbon to be lowered in the sediments.

Marshall (1970) measured organic input from primary production sources and found approximately 300 g C m^{-2} year^{-1}. He concluded that, at maximum rates, respiration by microfauna and meiofauna would account for all of this decrease. In a salt marsh where nematodes were exceptionally abundant and had

Table 2. Summary of Total Monthly Community Respiration and Component Parts[a]

Month	Total community respiration	Macrofaunal respiration	%	Bacterial respiration	%	Meiofauna and micro-fauna-flora respiration	%	Sediment chemical demand
July	92.7	5.1	5.5	55.7[b]	60.1	31.9	34.4	8.5
September	89.5	7.2	8.1	49.9 ± 3.0	55.7	32.4	36.2	7.1
October	81.4	6.2	7.6	49.1[b]	60.3	26.1	32.1	8.1 ± 0.1
November	74.1	3.8	5.2	37.7 ± 1.7	50.8	32.6	44.0	6.4 ± 0.0
December	69.9	4.3	6.1	29.1[b]	41.7	36.4	52.2	4.1
January	53.9	6.7	12.4	16.1 ± 2.4	29.9	31.1	57.7	2.8 ± 0.0
February	59.1	8.1	13.7	17.8[b]	30.2	33.2	56.1	3.1
March	65.3	16.8	25.8	26.4 ± 0.8	40.4	22.1	33.8	4.5 ± 0.0
April	73.0	15.7	21.5	37.9[b]	52.0	19.4	26.5	5.6
May	85.1	19.4	22.8	44.7 ± 2.5	52.5	21.0	24.7	7.6 ± 0.0
June	92.3	12.5	13.5	50.1[b]	54.3	29.7	32.2	7.3 ± 0.1

[a] Percentage of each component is given (values in ml O_2 m^{-2} h^{-1}). Standard errors included where replicate samples were obtained.
[b] Bacterial respiration values estimated from alternate month data. From Smith (1973).

an unusually high biomass (8.7–18.4 g wet weight per square meter), it was calculated that their respiration was 25–33% of the total sediment respiration, while in other marshes the corresponding value for respiration was only 3% of the total (Wieser and Kanwisher, 1961; Teal and Wieser, 1966).

Fenchel (1969) calculated that 20% of the total community respiration in Nivå Bay was due to meiofaunal respiration.

From recent studies made on shallow estuarine sediments (Smith, 1973), it is evident that a relatively high percentage of the total community respiration (25–58%) may be contributed by the meiofaunal, microfaunal, and microfloral components. Smith has indirectly evaluated this contribution by subtraction of the experimentally determined components of macrofauna and bacteria (Table 2).

Single Taxon Studies

Lasker *et al*. (1970) have partitioned carbon utilization of a sand dwelling copepod, *Asellopsis intermedia*, into growth, moulting, reproduction, and respiration: only 7% of the assimilated carbon occurred in growth, and over 90% was lost due to respiration.

More generally, respiration rates calculated from individual measurements on micro- and meioorganisms (Zeuthen, 1953; Wieser and Kanwisher, 1959, 1961; Lasserre, 1969, 1971a; Coull and Vernberg, 1970; Lasserre and Renaud-Mornant, 1971a,b, 1973; Vernberg and Coull, 1974) with aid of different microrespirometric techniques give oxygen uptakes 5–70 times higher than that of an equal biomass of macrofauna (Fig. 8). In other words, micro- and meiofauna consume 5–70 times more energy than the same biomass of macrofauna.

Critical Comments

Both indirect and direct data obtained *in situ* and in the laboratory show an energetic importance of aerobic microfauna and meiofauna.

In lagoons, as in other shallow, sheltered areas, deposit-feeding micro- and meiofauna utilize the primary production and are preyed on by carnivorous meiofauna such as turbellarians and by larger metazoans such as polychaetes and crustaceans (Renaud-Debyser and Salvat, 1963; McIntyre, 1969). However, this predatory activity seems to be comparatively small.

It is probable that the importance of meiofauna as food has been exaggerated (Tietjen, 1969). Moreover, very high densities of microfauna and/or meiofauna have been found in estuarine sediments: 10^5–10^7 organisms per square meter (Fenchel, 1969; McIntyre, 1969). In such areas a significant part of the primary production which is converted to micro- and/or meiofauna tissue is probably not involved in movement upward to higher trophic levels. As there is strong evidence that estuarine macrofauna can use organic matter directly (e.g., grey

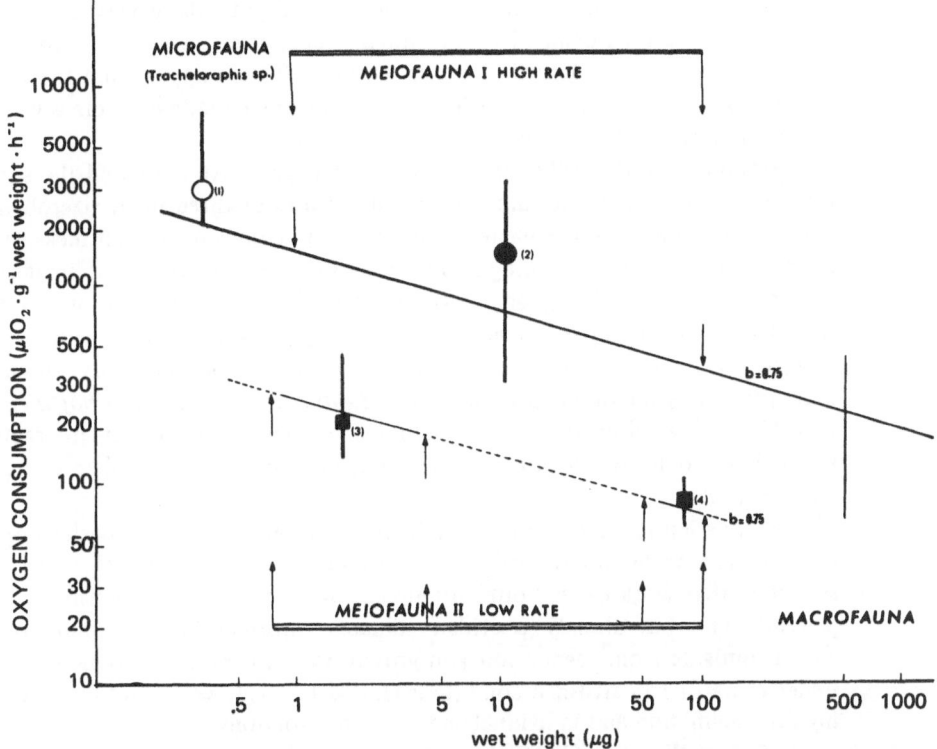

Fig. 8. Rates of respiration of the microbenthic ciliate *Tracheloraphis* sp., meiofauna (high and low rates), and macrofauna expressed on a unit weight basis.
Mean weight-specific rates of oxygen consumption have been transformed into
logarithmic values and corrected to a standard temperature of 20°C (correction
factor of Winberg, 1971). The two regression lines: $O_2/W = aW^{b-1}$ have a
similar mean slope $b = 0.75$, and different coefficient a (intercept of the Y-axis).
Vertical lines indicate minimum and maximum rates of oxygen consumption.
Data from (1) Vernberg and Coull (1974) for the ciliate *Tracheloraphis* sp.;
(2) Coull and Vernberg, (1970), Lasserre (1970, 1971a, and unpublished),
Lasserre and Renaud-Mornant (1971a,b, 1973), Ott and Schiemer (1973),
Wieser and Kanwisher (1960, 1961), Wieser *et al.* (1974); (3) Ott and Schiemer
(1973), Schiemer (1973), Wieser *et al.* (1974); (4) Lasserre and Renaud-Mornant
(1973) for the polychaet *Stygocapitella subterranea*.

mullets (*Mugilidae*) feeding from the first trophic level: Lasserre and Gallis,
1975), the amount left for these larger consumers would not be optimal. The
fishponds in the lagoon of Arcachon are portrayed as having a variable organic
input, generally high. Some of this organic supply is carried away from the area
by currents, some is utilized by herbivorous macrofauna, and there is a high loss

through respiration of the meiofauna since these potentially very active communities have a high demand for organic matter (Lasserre *et al.*, in preparation). While further work in this direction is required, it does appear that the role of meiofauna, ciliates, and bacteria in lagoons and other shallow waters is very important in terms of energy flow.

Brinkhurst *et al.* (1972), in a provocative paper, have evaluated the sum of respiration and growth in pure and mixed cultures of three small macrobenthic and limnic tubificid oligochaetes. The respiration rates have been measured in respiratory chambers containing 150 worms of each species or 50 specimens each of the three species (using an oxygen electrode). Compared with pure culture, the mixture of all three species produced a significant decrease in the rate of oxygen consumption and the presence of *Peloscolex multisetosus* stimulated more growth by the other species, *Tubifex tubifex* and *Limnodrilus hoffmeisteri*, than they achieved in its absence. These results suggest interspecific interactions which have to be taken into consideration in attempts to calculate energy or material budgets.

The possibility of metabolic interactions among associated species populations remains to be determined with microfauna and meiofauna. However, it is also clear that work carried out on pure cultures and isolated organisms is indispensable and has already provided valuable information. Determination of annual budgets from respiration and growth rates measured on organisms held under more or less artificial conditions should be made very carefully, accounting for acclimation and individual and seasonal variations.

Chia and Warwick (1969) have demonstrated, using autoradiographic techniques, the ability of meiobenthic nematodes to assimilate dissolved organic molecules: glucose and, to a lesser extent, amino acids. Glucose is taken up directly through the gut wall. Significant amounts of labeled glucose were present in the tissues after placing batches of animals in filtered sterilized seawater containing 10^{-5} M D-glucose-6-^{3}H.

Reproducing populations of meiofaunal animals can be kept in the laboratory (McIntyre, 1969). Gerlach (1971, 1972) and Tietjen and Lee (1972) are studying growth and reproductive rates of nematodes under laboratory conditions. The influence of such variables as temperature, salinity, and selected food are taken into consideration.

Anaerobic Metabolic Activity

Pamatmat (1971a,b) and Pamatmat and Bhagwat (1973) have pointed out that the rate of total oxygen uptake by the sediment surface underestimates benthic community metabolism. The recent data obtained on heterotrophic bacteria and the existence of anaerobic meiofauna and microfauna indicate that

Table 3. Mean Rates of Oxygen Uptake and of Sulfur Cycle in
Upper 10 cm of a Marine Sediment Model System
(from Jorgensen and Fenchel, 1974)

	Period, days			
	7–19	19–41	41–76	76–215
O_2 uptake	24	61	111	112
SO_4^{2-} reduction	66	72	76	41
SO_4^{2-} uptake (gross)	21	38	53	–
SO_4^{2-} uptake (net)	17	21	14	–
$H_2 S$ reoxidation	4	17	39	65

[a]Units: mg eq $\cdot$ m^2 $\cdot$ day^{-1}.

anaerobic metabolism can represent a higher fraction of the total metabolic budgets of benthic communities, aerobic respiration being predominant in sandy bottoms exposed to surf and currents.

Keeping in mind the energetic disadvantages of anaerobic metabolism, the dimensions and the conservative functions of the sulfide system must be underlined. The complexity and the importance of this anaerobic environment underlying the oxidized layers of porous sea bottoms has been established by Fenchel and Riedl (1970). More recently, Jørgensen and Fenchel (1974) have studied the sulfur cycle of a marine sediment experimentally. This model system was followed for seven months. Rates of oxygen uptake and of sulfate reduction in surface sediment have been measured and it was found (Table 3) that the sediment "rapidly changed from oxidized to strongly reducing conditions, and then slowly began to reoxidize again. . . . SO_4^{2-} were equally as important as O_2 for transporting oxidation equivalents from the water to the sediment" (Jørgensen and Fenchel, 1974).

Concluding Remarks

There is more and more evidence that the energetic differences and resemblances between communities living in stressed littoral and shallow sediments are reflected notably by differences and resemblances between the strategies of physiological and biochemical adaptation developed by the organisms to succeed in these areas. It is also very probable that different evolutionary strategies have been adopted by micro- and meiobenthic organisms in physically stable environments, e.g., in cave habitats (Poulson and White, 1969) and in the deep sea (Grassle and Sanders, 1973). The first *in situ* measurements of oxygen consumption were made recently by Smith and Teal (1973) on undisturbed deep-sea benthic communities. They found that oxygen uptake at 1850 m on the lower slope of New England was about two orders of magnitude less than at shallow shelf depths. It is not possible to say if the specific respiration rates per weight of meiofauna are significantly lower than in the shallow bottom sediments.

A recent attempt has been made to correlate dehydrogenase activity (DHA) with adenosine triphosphate (ATP) contained in marine sediments. The DHA:ATP ratio increased with depth of sediment layer (Pamatmat and Skjoldal, 1974). Ernst and Goerke (1974) have found decreasing values of ATP with increasing depths (from 0.96 μg ATP ml^{-1} wet sediment, at 252 m, to 0.07 μg ATP ml^{-1} wet sediment, at 5510 m). The ATP content of several nematodes was determined and found to be correlated with tissue weight. The deep-sea nematodes have less ATP per weight (1°/$_{oo}$ of the wet weight) than littoral nematodes (1.8^o/$_{oo}$). However, deep-sea nematodes can be damaged, and it is impossible to be certain that ATP content per animal weight in them is significantly lower than in nematodes living in shallow areas.

At the present time, no definite interpretations can be made from these interesting approaches. More estimates relating ATP and total organic carbon in single organisms are necessary.

It is now evident that a good understanding of metabolic activities and production of microfauna and meiofauna only can be achieved by means of parallel *in situ* and laboratory studies. A good knowledge of physiological and biochemical profiles of the organisms is also indispensable. It is necessary to increase our knowledge of the dynamic aspects of adaptative adjustments—with ecological bias—of closely related species which exhibit differential distributions. Physiological periodicity, temporal shifts due to localized abiotic and biotic effects, and variation due to reproductive conditions should all be taken into consideration in the future.

Review of Suitable Micromethods Applicable to Metabolic Studies on Microbenthic and Meiobenthic Organisms

Introduction

One of the most evident characteristics of meiofaunal and microfaunal groups is the very small size of the animals. According to Muus (1967) and Fenchel (1969), individual wet weight of microbenthic ciliates ranges between 10^{-9} and 10^{-5} g. The meiofauna is composed of all the benthic metazoans weighing less than 10^{-4} g wet weight and comprising, notably, nematodes, turbellarians, oligochaetes, polychaetes, harpacticoid copepods, ostracods, gastrotrichs, tardigrads, rotifers, archiannelids, and juvenile stages of larger animals—the "temporary" meiofauna. Benthic metazoans weighing more than 10^{-4} g wet weight are considered as macrofauna. A number of classical techniques used in physiological biochemistry are not suited for work approaching meio- and microfaunal dimensions. In the small-volume works, special problems of obtainable sensitivity, reproducibility, surface tension, evaporation, and diffusion, etc., are quite important. Considering these inherent difficulties, it is necessary to define the theoretical aspects of such an approach and to investigate only with appropriate methods. Since the last decade, the cellular physiologists have been most successful in their efforts to improve the microscale techniques. The recent manual by Neuhoff (1973) can be consulted with profit. It is quite important to preserve unaltered the function of organisms during laboratory experiments and to know as well as possible the microenvironment both in nature and in the experimental vessel.

Laboratory Microtechniques

Microrespirometers

Micromanometric Techniques: The Cartesian Diver Respirometer and Its Various Modifications. Among the techniques available for respiratory physiology, the Cartesian diver respirometer, because of its versatility, is becoming of more universal use. A growing number of researchers are using this technique. Gas volume changes smaller than 0.1 μl per hour with an accuracy of at least 5% can be recorded. The measurements performed in a Warburg apparatus are on a scale several thousand times larger (error larger than 1 μl).

The Cartesian diver technique has found application in the determination of respiratory rates of planktonic and benthic marine organisms (Zeuthen, 1947b; Lasker, 1966), free-living nematodes (Wieser and Kanwisher, 1961; Ott and Schiemer, 1973), gnathostomulids (Schiemer, 1973), interstitial oligochaetes (Lasserre, 1969, 1970, 1971a,b; Lasserre and Renaud-Mornant, 1973), the mystacocarids *Derocheilocaris remanei* (Lasserre and Renaud-Mornant, 1971a,b), interstitial polychaetes, one harpacticoid copepod, one archiannelid (Lasserre and Renaud-Mornant, 1973), and one ciliate (Vernberg and Coull, 1974).

The Cartesian diver has nothing to do with the philosopher Descartes; it was described by Magiotti in 1648. This pupil of Galileo published the first account of the so-called "Cartesian diver."

This principle was no doubt later called "Cartesian" as a synonym for anything "scientific" or mechanical and remained, without application, a philosophical toy until the present time. This lovely toy became, in the hands of the Danes Linderstrøm-Lang (1943), Holter (1943), and Zeuthen (1943), a unique and versatile technique to be applied in the fields of chemistry, cellular physiology, and the physiological ecology of very small organisms.

The principle of the diver technique lies in the fact that any change in the amount of gas in the diver, which is used as a reaction vessel, requires a corresponding change in the pressure necessary to hold the gas volume constant, so that the diver will remain submerged at a fixed level in the flotation medium surrounding it. Thus, the pressure changes become measures of the changes in amount of gas in the diver.

I will restrict this review to a discussion of diver techniques which are readily useful and not too elaborate, and which are easy to set up.

The principal difficulty lies in making and filling the divers. It is necessary to draw capillaries of convenient diameter (0.5–0.16 mm) from glass tubes of proper density (approximately 2.2 g ml^{-1}). A pipetting assembly (necessary to fill the divers) is composed of a series of pipettes placed in a holder with proper illumination. The diver is placed in a diver clamp stand. A "breaking" pipette is properly centered above the opening of the diver, and the latter is raised until the tip of the pipette almost touches the bottom. The solution containing the animal is gently blown out and the diver gradually lowered. The same procedure is observed to place successively the sodium hydroxide seal, the oil seal, and finally the flotation liquid seal (Fig. 9).

A technique for filling divers with different gas mixtures is described by Lasserre and Renaud-Mornant (1973).

Excellent reviews on the Cartesian diver respirometers including, both theoretical and practical aspects, have been published by Glick (1961) and Holter and Zeuthen (1966).

Standard Divers: (Linderstrøm-Lang, 1943; Holter, 1943). The standard Cartesian diver respirometer is shown in Figs. 9 and 10. The diver proper is made of a piece of capillary glass blown to form a bulb at one end and supplied with a tail of solid glass.

The diver is charged by the successive pipetting of three or four accurately measured volumes of fluid into the bulb and the neck. This of course requires accurate and special pipetting techniques. To facilitate such pipetting the inside of the diver is siliconed. In method A (Figs. 9 and 10), the diver contains a bottom drop with the animal and two neck seals, one made up by a CO_2-absorbing NaOH solution and the other by paraffin oil that prevents evaporation of water from inside the diver. The mouth seal contains a strong salt solution, the Holter's medium, present in the flotation vessel in which the diver is placed (Fig. 10: 1). The solubility of gases in this medium is very low so that the gas diffusion in the mouth seal reduces to an acceptably low level. It is an almost saturated solution of $NaNO_3$ and NaCl.

To prevent water from distilling from the reaction medium (e.g., a drop of

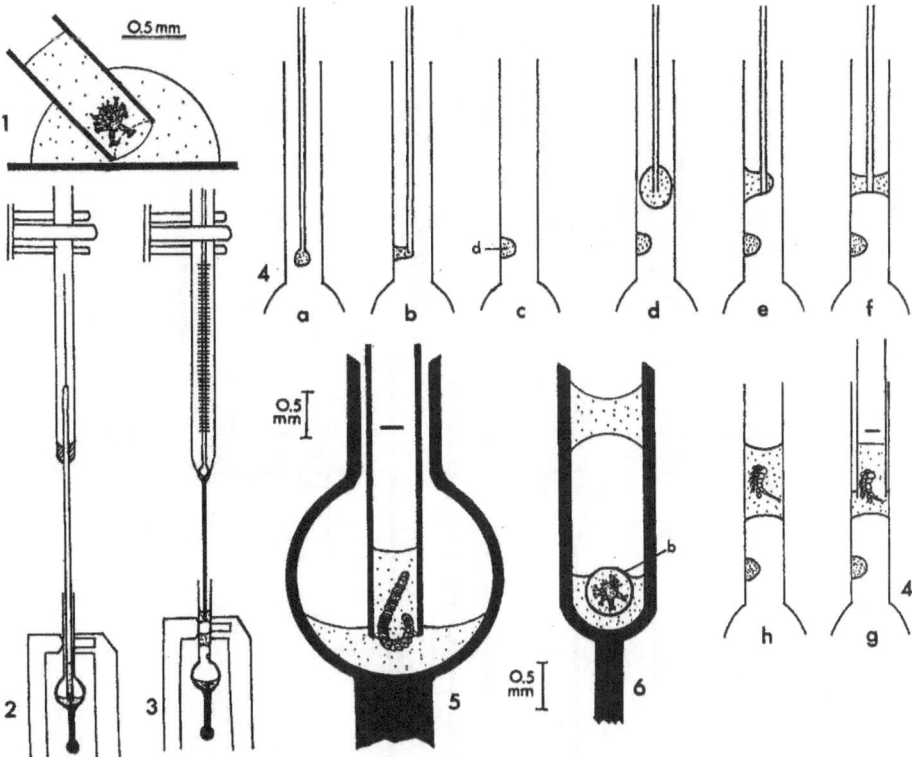

Fig. 9. Standard Cartesian diver. 1, 2, 3: different procedures for filling the diver; 4a, b, c: with formation of a neck drop (d); 4d–h: and neck seal containing one meiobenthic copepod; 5: standard diver; 6: cylindrical diver containing one glass bead (b).

sea water containing the animal) to the high salinity flotation medium in the mouth, a separate oil seal is introduced between the reaction medium and the mouth fluid.

The sodium hydroxide seal is made isotonic with the reaction medium; this is important in the case of sea water medium.

The filling method B (Fig. 10: 3, 5; Fig. 9: 4d–h) is used to modify, in the course of an experiment, the ionic concentration of the water where the animal is living and/or to add some drug which eventually modifies the oxygen uptake (e.g., ouabain: Lasserre, 1969). To mix the drug droplet with the seal containing the animal, a manometer pressure is applied to push down the seals until the seal with animal engulfs the different droplet (Lasserre, 1975).

By means of the connected manometer it is possible to vary the pressure

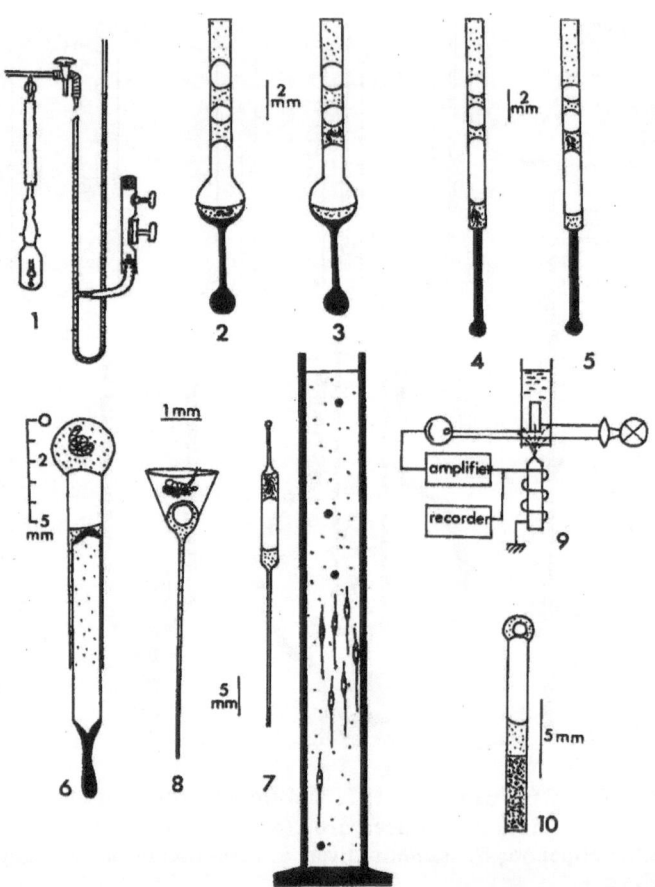

Fig. 10. 1: Flotation vessel with manometer; 2–5: standard Cartesian diver; 6: stoppered diver; 7: gradient diver, and a collection of divers in a density gradient; 8: diver balance; 9: principle of the automatic electromagnetic balance, with the corresponding diver respiration chamber (10). (9 and 10 from Løvtrup, 1973.)

above the surface of the flotation medium and consequently the location of the menisci of the seals in such a way that the compound density of the whole system becomes equal to that of the medium, enabling it to float.

If part of the enclosed oxygen is consumed, it is necessary to reduce the pressure to keep the diver floating. This pressure change, which is proportional to the oxygen disappearance, can be read on the manometer, calibrated in millimeters. The oxygen uptake per unit time is expressed as

$$\frac{dV}{dt} = (p - pl) \cdot V$$

where dV/dt is the O_2 consumption per hour in $\mu l \times 10^{-3}$ O_2 h^{-1}, $(p - pl)$ is the pressure change in atmosphere inside the diver read on the manometer, and V is the total volume inside the diver (in μl) at equilibrium (between 0.5 and 15 μl).

To prevent atmospheric pressure variation the open branch of the manometer is connected to a constant atmosphere maintained in a six-liter bottle immersed in the same thermostatic bath as the flotation vessels. The temperature is regulated to an accuracy of $\pm 0.01°C$.

Zeuthen has constructed a number of variants of this diver model, most of which are distinguished by an enhanced accuracy. However, the handling of such ultramicro divers is delicate and drift of the blank is considerable (Holter and Zeuthen, 1966).

In our laboratory we also use a modification of the standard diver. This cylindrical type (Fig. 9: 6; Fig. 10: 4, 5) consists of a narrow capillary tube 0.5 mm wide and about 15 mm long. The capillary is sealed at one end and a long glass extrusion "tail" is attached to the outer closed end, facilitating the balancing of the instrument. The sensitivity of the diver depends on the volume of the gas phase. For the standard diver in the volume range of 5–12 μl this results in a sensitivity of $1-10 \times 10^{-3}$ μl O_2 h^{-1}. The gas volume of the cylindrical standard diver has decreased to 0.5-5 μl and the accuracy has increased from 2 to 5 $\times 10^{-5}$ μl, permitting straight respiration measurements within the range 2×10^{-4} to 2×10^{-3} μl h^{-1}.

In both these standard divers, it is possible to place one small glass bead of the same density and frosted. This glass bead serves as a support to the meiofauna organism like a sand grain (Fig. 9: 6).

Stoppered Diver: (Zeuthen, 1950; Klekowski, 1971). With organisms suspended in a high-salinity medium, loss of water by distillation from the medium holding the animal can be omitted, without using an oil seal. The diver floats in an isotonic solution of Holter medium ($+0.1$ N NaOH isotonic to sea water) which penetrates deep into the diver's mouth. Into the mouth is inserted a hollow glass stopper which is less than 5% of the capillary's cross section open for diffusion of gases. Such stoppered divers are air tight and can be made tight even to CO_2 with which the diver is easily charged. Because of its simplicity of construction and calibration, and because it can be charged by hand, the stoppered diver has a high sensitivity and is therefore advantageous. However, for measurements of O_2 consumption lower than 1×10^{-3} μl per hour, the cylindrical standard diver and the ampulla diver are more suitable (on the order of 1×10^{-4} μl per hour). Two other limitations are that the liquid charge is inaccurately defined and that mixing of reactants inside the diver is not practicable (Fig. 10: 6 and Fig. 11: 5).

Concluding Remarks on the Standard Cartesian Divers and the Stoppered Diver. With both types of divers it is easy, without any damage, to introduce the organisms and remove them after the completion of the measurements. Because of this feature, one can repeatedly run the records of metabolism of the same individual within its developmental cycle, as well as of individuals subject to experimental treatment between measurements. At the end of one experiment the divers are rinsed and washed carefully with distilled water–acetone–toluene–acetone–distilled water and then dried in an oven at 100°C. Each dried diver is subsequently stored under cover, ready for use.

The Cartesian diver microrespirometer, with all its advantages, has also some unfavorable properties, which should be borne in mind when this technique is applied.

1. This is a respirometer of the "closed-vessel" type. In the course of measurement, the concentration of O_2 decreases and that of metabolites increases. This should be taken into account when planning the size of divers and the duration of measurements. However, with a seven hour run, 5–20×10^{-3} μl of oxygen is consumed, and no more than 10% decrease of inside oxygen tension is observed in a standard diver. This would not cause any significant change in respiration intensity of most meiofauna animals and CO_2 output can easily be measured when no seal of sodium hydroxide is present.

2. The oxygen consumption of the individual is measured while the animals move in a water film inside the diver (in the neck seal or in the bottom seal). How can this level of activity be compared with active life in the interstitial biotope? It is well established that physical work is accompanied by increased O_2 consumption, although the relationship need not necessarily be linear. The metabolism during rest is a convenient starting point. In very active crustaceans, such as mystacocarids, the oxygen consumption was found to be only about 10% higher than in immobilized or anesthetized animals (Lasserre and Renaud-Mornant, 1971a). In the case of annelids there is only a 5% augmentation. In an actively creeping archiannelid a 15–22% augmentation was observed (unpublished data).

We conclude that diver data obtained from meiofauna animals are quite applicable to natural conditions. Mystacocarids, oligochaetes, and polychaetes can live three or four days inside a standard diver without apparent effects on their activity.

The Gradient Diver Microrespirometer: (Løvlie and Zeuthen, 1962; Nexø, et al., 1972). The gradient diver is a non-Cartesian diver microgasometer in which the diver floats in a linear gradient made up with Na_2SO_4 in water. Manometric calibration is replaced by direct calibration based on the positions taken in the gradient by a series of density standards made of glass beads of known densities. Calculations according to previous original equations were cumbersome in practice. This difficulty was overcome after the equations were programmed in Algol, for solution by computer. The theory and practice were subsequently highly simplified (Nexø et al., 1972). With this procedure, gradient and diver characteristics are introduced into simple equations. The gradient diver (Fig. 10: 7) has only two fluid compartments (reaction medium and flotation medium +0.1 N NaOH). In making the diver, the narrow capillary of 500 μm is pulled with two fine tips forming a "tail" and a "shaft"; the "ampulla" constitutes the reaction chamber (Fig. 10: 7). This type of diver can be easily charged with any gas mixture. If no biological object is enclosed, the diver finds a steady equilibrium position in the gradient (control diver), but if an object is enclosed, it travels along down the gradient (experimental diver) due to contraction of the gas phase (Fig. 10:7). The diver displacement may be recorded visually or by photographing the gradient at constant intervals. Gradients need not even be thermostatically controlled. In such a case, control as well as experimental divers perform vertical migration due to changes in temperature and pressure; such migrations are corrected by measurement of change in distance from control to experimental divers. The control diver has the function of a thermobarometer (Fig. 11: 1-4).

The gradient diver lacks the versatility of the original standard Cartesian divers as designed and shaped by Holter (1943), and it cannot replace the stoppered diver, although it is highly sensitive. The reading accuracy in microliters is a function of the precision (in millimeters) with which the diver migrations are measured. When migrations of the divers are recorded photographically the precision is close to $\pm 4 \times 10^{-6}$ μl O_2; when a cathetometer is used the reading accuracy is 5×10^{-3} mm and the precision is close to $\pm 1 \times 10^{-6}$ μl O_2. Performances are elevated and oxygen consumption of 5×10^{-5} μl O_2 h^{-1} to 2×10^{-3} μl O_2 h^{-1} are measured with 1-μl gradient divers. The simplified procedure can be operated without a thermostat on the laboratory bench; however, we recommend a good temperature control (thermostatted bath) to limit the formation of temperature gradients.

The gradient diver has been used by Løvlie (1964) in studies of growth and photosynthesis in small pieces of normal and mutant *Ulva* algae. In this case there is an upward migration of the diver due to effects of photosynthesis.

Electromagnetic Diver: (Larsson and Løvtrup, 1966; Løvtrup, 1973). One of the most recent and sophisticated developments is the automatic diver balance described first by Larsson and Løvtrup (1966) and critically commented upon and reviewed by Løvtrup (1973). This automatically recording electromagnetic microrespirometer works upon the principle that a small glass body, the diver, having a density lower than some aqueous medium, is kept floating in a glass vessel filled with this medium by a magnetic force acting on a magnet enclosed in the diver. This force is generated by a coil located beneath the vessel. In the floating position, the diver cuts off a light beam until its buoyancy is compensated by the magnetic force (Fig. 10: 9). When loaded, the diver will sink until this force is reduced by an amount corresponding to the load. Since the change in the column of an enclosed air bubble in a submerged system entails a change

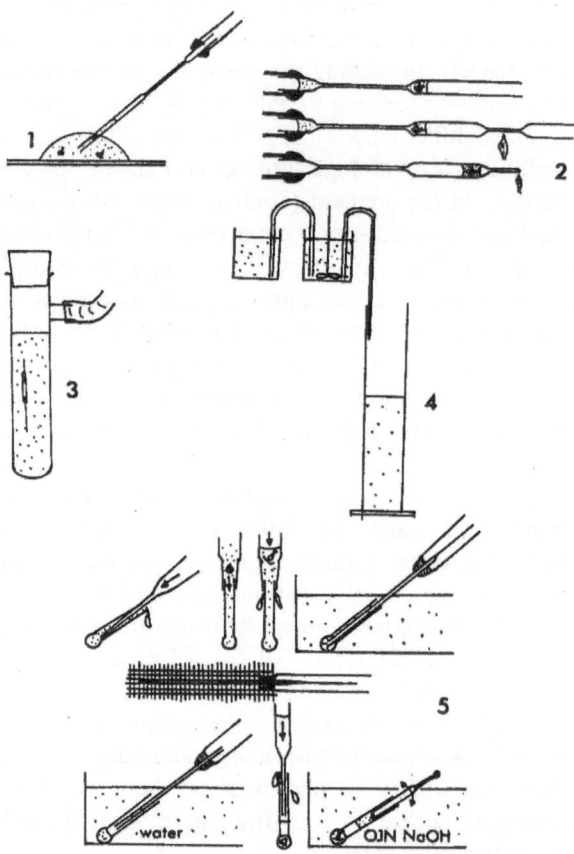

Fig. 11. Gradient diver. 1–4: procedures used in making and charging the gradient diver (1–2); 3: balancing the diver; 4: set-up for making the linear density gradient made up with Na$_2$SO$_4$ in water (Modified from Nexφ *et al.*, 1972); 5: stoppered diver, procedure for charging the diver (from Klekowski, 1971.)

in reduced weight (RW), a respiration chamber is placed on top of the balance in order to follow any manometric changes taking place. In fact, the diver is very similar to a standard Cartesian diver, but a true "tail" is missing. A fundamental difference is that Cartesian divers *sensu stricto* are constant volume respirometers, while the electromagnetic compensated diver is a constant pressure respirometer. The light reaching the photocell generates a DC current which is converted to a DC voltage in a preamplifier. It appears that when the RW of the diver increases (e.g., as result of oxygen consumption) it will move downward from a weaker to a stronger magnetic field. Consequently, a reduction in current will occur. This reduction would comprise two components, one due to the decrease in buoyancy of the floating system and the other due to the change in field strength. To prevent the change in field strength the diver does not move at all; this is given by an integrating amplifier. Therefore, the current measured is proportional to oxygen consumption. Any change in the volume of an enclosed air bubble gives rise to a change in the RW of the floating system equal to the weight of the same volume of water. If the density of the latter is 1.0, the oxygen uptake is expressed by

$$\Delta O_2 \, (\mu l \times 10^{-3}) = \Delta RW(\mu g)$$

The diver is calibrated by means of standard weights. For this purpose, weights are made with very thin (25-μm) platinum wire, they are weighed on a Mettler microbalance, and their RW is calculated from the relation

$$RW = W(1 - 1/\rho)$$

where W is the weight (μg) and ρ the density of the standard.

The diver itself is made from capillary glass and one end is shaped into a small bulb. The open end is painted with black epoxy varnish, containing about 25% carbonyl iron by weight (Fig. 10: 10). The filling is not different from the standard diver procedure. A difficulty is that this method gives considerable variation between individual divers as regards their stability. This parameter has to be checked before each experiment.

Clark-Type Oxygen Electrode Microrespirometers and pCO$_2$ Electrode. Oxygen gas in solution reacts with a negatively polarized metal surface (cathode) and becomes OH$^-$. Proteins and many other substances in the solution also are attracted to the surface and poison it or alter its reactivity with oxygen. Clark (1956) separated the unknown solution from the electrolyte–platinum cathode by means of a polyethylene membrane, through which oxygen diffused to the platinum surface. This original design consists of a platinum surface (diameter 2 mm) fused in glass (an electrolyte cavity with a pressure relief opening away from the active tip), and a silver reference electrode. A potential of about – 0.6 V with respect to the silver is applied to the platinum. The current through the cell is then a linear function of the oxygen tension in the solution bathing the electrodes.

The introduction of plastic membranes (teflon, polypropylene) selectively permeable to O$_2$ and to CO$_2$, but not to ions or water, has permitted the construction of specific electrodes for these gases, which have over the past decade been developed into sophisticated commerically available systems of adequate accuracy and stability. Such electrodes are well suited for respirometric work in general, and are convenient to use in conjunction with manometric methods.

A typical recent version mounted in a thermostatted cuvette is made by Radiometer Denmark (digital acid base analyzer with pO$_2$ and pCO$_2$ modules).

In the pO_2 electrode, marketed by Radiometer, the active metallic surface is the exposed end of a platinum wire 10–20 mm in diameter and sealed in glass. The electrodes are insensitive to turbulence in the gas or liquid in which they are immersed, and stirring is not necessary. The very low current ($\sim 10^{-11}$ A mm Hg^{-1}) is measured on a high-impedance electrometer. Radiometer electrodes have linear characteristics, and they have to be standardized only against an oxygen-free solution and a solution with known O_2 or CO_2 tensions. The amplified outputs of the electrodes are recorded on a strip-chart recorder. The effect of temperature is to change the current for a given O_2 tension. Temperature regulation of the respiratory chamber is therefore a necessity. As a control, it is convenient to measure the O_2 concentration in the respiratory vessel itself by means of the micro-Winkler titrimetric method of Fox and Wingfield (1938).

With the Radiometer system, there is a possibility of determining the partial pressure of CO_2 with a specially designed pCO_2 electrode using a CO_2-permeable membrane. However, the electrode does not work in sea water.

A constant-temperature microrespirometer, based on the Clark-type oxygen electrode, has been described recently by Atkinson and Smith (1973). The apparatus is capable of measuring oxygen consumptions of the order of 0.02–0.2 μl O_2 h^{-1}. This respirometer is a modification of the steel electrode housing provided by Radiometer (thermostatted cell type D616) for use with the pO_2 electrode E5046. The respirometer cell is maintained at constant temperature by a Radiometer water bath (VTS 13). The low current, flowing as the result of the reduction of oxygen at the cathode ($\sim 10^{-11}$ A $Torr^{-1}$ at 38°C), is measured with the high-gain stable amplifier of the Radiometer pH meter (PH27) connected to the oxygen monitor PHA 927b. The amplified output of the electrode is continuously recorded on a pen recorder. Permeability of the cell materials to oxygen and the absorption of the gas by the elastomer seals may be limiting factors. A gas-tight respiratory chamber has been obtained even at low oxygen tension using Vitron O rings.

A similar system is now being utilized in our laboratory. The steel electrode housing is a slight modification of that provided by Radiometer (thermostatted cell type D616); pO_2 electrodes are used and the electrode outputs are connected to the high-gain stable amplifier of the Radiometer pH meter PHM 72 (Digital Acid–Base Analyzer) connected to a pO_2 module (type PHA 934).

Conclusions on the Microrespirometers. There is no need to consider the gasometric and the electrometric methods as competitive; both have their own merits and their performances, and in many ways are complementary. Because of the reliability and the ease with which absolute measurements can be made, the manometric Cartesian diver methods fully justify the interest in them. The pO_2 electrodes of the Radiometer type are quite suitable for monitoring oxygen uptakes.

Quantitative Microdeterminations of High-Energy Compounds
by Bioluminescence Technique

A very sensitive quantitative assay for adenosine triphosphate (ATP) and other high-energy molecules (CPK, AMP, etc.) and their hydrolyzing enzymes is based on the firefly luminescent system.

The luciferin–luciferase enzyme is extracted from tails of the firefly *Photinus pyralis* (desiccated tails available from Sigma, U.S.A.). The preparation has the property of binding specifically to ATP over a wide range ($10^{-9}-10^{-14}$ mole) and of giving a bioluminescent compound:

luciferin–luciferase + ATP $\underset{\rightleftharpoons}{\overset{Mg^{2+}}{}}$ adenyl–luciferin +

pyrophosphate adenyl–luciferin $\xrightarrow{O_2}$ adenyl–oxyluciferin + *light*

The light emission is measured with a scintillation counter or any kind of high-performance photometer (e.g., Chem Glow, Aminco photometer, type 4-7441). Quantities of ATP as small as 10^{-14} mole in a 10 μl sample are detected with this method.

Because of the prominent role played by ATP in the isolated luminescent system of the firefly, numerous assay methods can be based upon this system. ADP, AMP, and, in fact, all other compounds which are directly or indirectly related to ATP should be capable of measurement in this way. The luciferin–luciferase system is applicable to determination of ATPase activity, coenzyme A, creatine phosphokinase, and DPN pyrophosphatase, all of which would bring about changes in ATP concentration and hence influence light emission in the firefly system.

Inherent problems of reproducibility in the preparation of the samples seem a major difficulty in utilizing the bioluminescence method. A detailed technique for the measurement of ATP in the ocean has been given by Holm-Hansen and Booth (1966). For the assays of ATP in sediments, the procedure of Ernst (1970) and its modifications (Ernst and Goerke, 1974; Pamatmat and Skjoldal, 1974) can be consulted with profit.

Measurement of Wet Weight, Dry Weight, and Reduced Weight

The problems associated with the measurement of very small amounts of biological samples are inevitably combined with serious manipulative difficulties.

Dry Weight with Microgram Balances. It is now possible to determine weight on commercially available fast-weighing microbalances. One of the best that we have used was the Mettler ME 22, with a sensitivity of ±0.1 μg and a reproducibility of ±0.4 μg.

To measure dry weight of meiobenthic individuals, one to twenty animals are allowed to dry on a microscope cover slip; activated silica gel is placed in the balance box as a drying agent and is replaced frequently. The animals had been previously dipped in distilled water with 10% formalin. They are deposited on a tared cover slip, the latter being transferred to the balance pan. Loss of weight is recorded on a pen recorder and constant weight is obtained in a short period; repeated observations over periods up to 24 h showed that no further loss of weight occurred, even after heating in an oven at 100°C. Dry weights determined by this direct method give very good estimates.

For the greater sensitivity necessary for individual measurements on smaller micro- and meiobenthic organisms, quartz fiber balances based on that described by Lowry (1944) give sensitivities of about ±0.03 μg and a reproducibility of 0.1 μg. The subject of the construction and the design of quartz fiber balances has been discussed in detail by Lowry and Passonneau (1972).

Wet Weight with Volume-Density Determinations. It is often necessary to obtain quantitative values for the wet weights of micro- and meiobenthic individuals. Wet weights can be derived from estimates of volume, based on body dimensions, and of density.

Volume. In the case of nematodes (Nielsen, 1949; Wieser and Kanwisher, 1960) and oligochaetes (Lasserre, 1970, 1971a) it is possible to determine the volume by calculating: $l(D/2)^2\pi$, where l is the length, and D the mean body diameter. These animals correspond reasonably well to imaginary cylinders. Similar estimations have been made on Mystacocarids and on other groups of meiofauna (Lasserre and Renaud Mornant, 1973).

Fenchel (1967, 1969) estimated the wet weights of ciliates on the same basis. Ciliates are pressed in the Roto-Compressor of Heunert and Uhlig (1966) until they have plane parallel sides. The thickness is then measured using a microscope and the volume is calculated from the outline traced by a camera lucida.

In many other cases the Holter method (1945) is recommended. The animal is drawn up into a capillary tube of known diameter, which is wide enough to avoid deformation of the animal, and dye solution (acid violet) of known density is poured into the capillary with the animal. The total length of the column containing the animal is measured and the dye is emptied into a known volume of water. The concentration of dye is determined colorimetrically in a microcuvette and is compared with a standard curve made from columns of solutions with dye alone. $V = A(L_t - L)$, where A is the cross-sectional area of the capillary, L_t is the length of the dye + object in the capillary, and L is the length of the dye alone. The value L corresponds to the colorimeter reading of the unknown dye solution containing the animal and is obtained from the standard curve.

Density (Specific Gravity). The density of a single meiofauna animal can be determined by flotation in a bromobenzene-xylene gradient column, following the method of Low and Richards (1952), modified by Richards and Berger (1961). For routine measurements, density can be assumed to be 1.0.

The procedure described recently by Neuhoff (1971) for fresh weight determination of very small tissue samples in the range 1–0.01 mg is also applicable.

Reduced Weight with the Cartesian Diver Balance. Zeuthen (1947a) described the Cartesian diver balance, which is a diver used not as a reaction chamber but as a balance. This technique permits gasometric determinations of "reduced weights" (RW). The so-called "reduced weight" is the weight of sample minus the weight of an equal volume of water. This parameter is independent of the water content of the sample. This technique is also applicable to determinations of absolute weights, densities, and volumes. (The diver balance is shown in Fig. 10: 8.) It consists of a very fine glass capillary, blown to form a bulb at one end, on top of which is placed a cup fused to the outside of the bulb with polyethylene.

The animal is placed in the cup and the diver is allowed to float in a medium which is optimal for the animal (sea water or dilute sea water). In the diver is an air bubble which is surrounded with fluid that is continuous, through the tail, of the external medium. The equilibrium pressures of the diver are measured before and after loading with an animal. The difference between the two pressures is a measure of the RW of the load. The manometer and the flotation vessel are the same as for the Cartesian diver respirometer. The balance diver is calibrated with very small polystyrene beads of known RW, previously determined in a density gradient. Standards made of very thin platinum wire (0.025 mm) can be weighed directly on a Mettler microbalance "ME 22" and their RW calculated from the formula: $RW = W(1 - 1/P)$, where W is the weight and P the density of the standard.

The diver balance can measure RW with an error of ±10–33ng. This method is rather difficult and needs good experience.

Total Nitrogen and Protein Quantitative Assays for Submicrogram Amounts

Oxygen consumption can be calculated per nitrogen content of tissue. This procedure was followed by Zeuthen (1947b). Total nitrogen was determined in this work with a micromethod of the order of micrograms developed in the Carlsberg laboratory by Bruel *et al.*, (1946). This very sophisticated method can be employed in analysis of samples containing 0.1-1.0 μg of nitrogen with an accuracy of 0.005 μg for the total analysis. However, it is a time-consuming procedure which required particularly stingent levels of cleanliness to avoid contamination. We have used at the Arcachon laboratory the rather simple micro-Kjeldahl method of Shaw and Beadle (1949). Very recently, a sensitive colorimetric protein assay, linear between 0.01 and 0.2 μg protein, has been described by Goldberg (1973).

Wet weights and dry weights are preferable in order to calculate respiration rates; however, total protein and/or nitrogen assays are fully indicated when enzyme activities (or concentrations in ATP) are calculated: it is preferable to refer specific activities to protein contents.

Ingestion and Assimilation of Food

Radioactively Labeled Food. Rates of food ingestion of a freshwater benthic nematode, *Plectus palustris*, have been determined using radioactively labeled bacteria cells ($[^{14}C]$ glucose). The number of bacteria ingested during t minutes is equal to the *mean cpm per nematode ÷ specific activity cpm per cell*. This assumes no loss of radioactive material and that no labeled bacteria adhered to the outside of the worms' bodies. Assimilation rates were not calculated since incorporated ^{14}C began to accumulate as body production and some was lost as $^{14}CO_2$ (Duncan *et al.*, 1974).

Assimilation of Dissolved Organic Molecules. Chia and Warwick (1969) have shown that the meiobenthic nematodes *Pontonema vulgare* and *Dorylaimopsis punctatus* can freely absorb simple sugars and probably also amino acids through the gut wall rather than the body wall. Adult living specimens are incubated at 8°C in 10^{-5} M D-glucose-6-^{3}H or in 6.10^{-6} M glycine-2-^{3}H dissolved in filtered and sterilized seawater. Batches of specimens are removed and fixed in hot Bouin after 30 min, 3 h and 24 h, 3 days, 1 week and 2 weeks. Autoradiographs are prepared from serial paraffin sections using liquid Ilford "K2" nuclear emulsion. Autoradiographic slides are developed after 4 weeks incubation. After 24 h, blackening of the emulsion was considerable and this was mostly confined to the intestinal epithelium. After 3 days, the intensity of labeling extended progressively to various tissues, and notably to somatic muscles, hypodermal cord, genital apparatus, and the pharynx. Oocytes, spermatocytes, and cuticle were not labeled (Chia and Warwick, 1969).

With aid of these techniques, in addition to data on respiration rates, weight, and growth and reproduction rates, valuable information will be obtained on daily energy budgets.

Direct Measurement of Total Aerobic and Anaerobic Metabolisms of Benthic Communities

The *in situ* measurements of aerobic and anaerobic metabolic rates require much pioneer research work. An interesting review was published by Crisp (in Holme and McIntyre, 1971).

Direct *in Situ* Respiration Rates

In situ measurements of sediment oxygen demand can be made with bell jar systems described notably by Pamatmat (1965, 1968), Pamatmat and Fenton

(1968), and Smith *et al.* (1972). The apparatus described by Smith *et al.* (1972) consists of an opaque Plexiglass cylinder (15 X 30.0 cm diameter) which is closed at the top. An oxygen electrode is inserted in the cylinder, and water is circulated within the bell jar by a small stirring motor and magnetic bar. The electrode is connected to a recorder (Rustrak 12VDC, 100 μA) which is enclosed in a water-tight camera housing.

In Situ Measurement of Anaerobic Metabolism

Jϕrgensen and Fenchel (1974) have made direct measurements of major components of the sulfur cycle. Rates of sulfate reduction have been calculated with an *in situ* radiotracer technique, in conjunction with other chemical and radioisotopic measurements, including sulfate, free sulfide, elemental sulfur, acid-soluble sulfide, total residual sulfur, total organic matter, rate of oxygen uptake in the dark and in the light, and redox potential.

Pamatmat and Bhagwat (1973) found a practical way to estimate total anaerobic metabolism with dehydrogenase assay of substrate sediment layers. In this quantitative assay, triphenyltetrazolium chloride (TTC) is used as an efficient hydrogen acceptor and in the presence of dehydrogenase forms a red compound of triphenylformazan (TPF) which can be measured colorimetrically.

Specific Gas- and Ion-Sensing Electrodes for *in Situ* Measurements

Oxygen Electrodes

Jansson (1967), using the platinum microelectrode method of Lemon and Erickson (1952), made interesting studies on "oxygen availability" in different sandy beaches. However, the cathode is composed of a bare noble metal (platinum, gold) and the current reading that is obtained in any particular situation may be influenced by a number of factors other than oxygen tension itself (Van Doren and Erickson, 1966). These factors include the effect of stirring, the presence of interfering substances which may form complexes with the electrode metal, and the mechanical alteration of the surface of the electrode (which is rather large: 200 μm to 1 mm in diameter) by the position of foreign substances. The presence of hydrogen sulfide may result in a rapid "poisoning" of the platinum surface and erratic behavior of the microelectrode may occur.

The introduction of plastic membranes (Teflon, polypropylene, etc.), selectively permeable to oxygen, has given rise to many commercial types of electrodes (Beckman Instruments, Yellow Spring Instruments, Ponselle, Orbisphere Laboratories). A type of oxygen electrode insensitive to hydrogen sulfide, even at high concentrations (350 ppm) was marketed recently by Orbisphere Laboratories (Geneva).

Ion-Sensing Electrodes

Ion-sensing electrodes have been developed during the last decade mainly by Orion (Cambridge, Mass., U.S.A.). These electrodes are selective to hydrogen, ammonia, calcium, chloride, fluoride, nitrate, nitrite, potassium, redox, sulfide, and sodium (see the booklet published by Orion Research: "Analytical Method Guide").

Conclusions

Important knowledge has been gained, notably from (1) measurements of *in situ* benthic metabolism, with better partition between the aerobic and the anaerobic parts of the metabolism, and (2) the ecophysiological approach which is a necessary and important complement of the *in situ* studies.

Further work in these two directions is still required and it appears very probable that meio- and microfauna, respectively, with their nonnegligible biomasses which sometimes equal those of macrofauna, may play an important metabolic role in the benthic boundary layer, from mangroves, lagoons, and shallow seas to the deepest part of the oceans.

References

Atkinson, H. J., and Smith, L. 1973, An oxygen electrode microrespirometer, *Journal of Experimental Biology 59:* 247-253.

Atkinson, H. J., 1973a, The respiratory physiology of the marine nematodes *Enoplus brevis* (Bastian) and *E. communis* (Bastian). I. The influence of oxygen tension and body size, *Journal of Experimental Biology 59:* 255-266.

Atkinson, H. J., 1973b The respiratory physiology of the marine nematodes *Enoplus brevis* (Bastian) and *E. communis* (Bastian). II. The effects of changes in the imposed oxygen regime, *Journal of Experimental Biology 59:* 267-274.

Boaden, P. J. S. 1962 Colonization of graded sand by an interstitial fauna, *Cahiers de Biologie Marine 3:* 245-248.

Boaden, P. J. S. 1963 Behaviour and Distribution of the archiannelid *Trilobodrilus heideri*, *Journal of the Marine Biological Association of the United Kingdom 43:* 239-250.

Boaden, P. J. S., and Platt, M. 1971, Daily migration patterns in an intertidal meiobenthic community, *Thalassia Jugoslavica 7:* 1-12.

Boaden, P. J. S., 1974a, Three new thiobiotic Gastrotricha, *Cahiers de Biologie Marine 15:* 367-378.

Brinkhurst, R. O., Chua, K. E., and Kaushik N. K., 1972, Interspecific interactions and selective feeding by tubificid oligochaetes, *Limnology and Oceanography 17:* 122-133.

Bruel, D., Holter, H., Linderstrøm-Lang, K., and Rozits, K., 1946, A micromethod for the determination of total nitrogen (accuracy 0.005 µg N), *Comptes Rendus des Travaux du Laboratoire Carlsberg, Série Chimie 25:* 289-324.

Bullock, T. H., 1955, Compensation for temperature in the metabolism and activity of poikilotherms, *Biological Reviews 30:* 311-342.

Chia, F. S., and Warwick, R. M., 1969, Assimilation of labelled glucose from seawater in marine nematodes, *Nature* (London) *224:* 720-721.

Coull, B. C., and Vernberg, W. B., 1970, Harpacticoid copepod respiration: *Enhydrosoma propinquum* and *Longipedia helgolandica, Marine Biology 5:* 341-344.

Clark, L., 1956, Monitor and control of blood and tissue oxygen tensions, *Transactions American Society for Artificial Internal Organs 2:* 41-48.

Crisp, D. J. 1971, Energy flow measurements, in: *Methods for the Study of Marine Benthos* (Chap. 12). Holme & McIntyre, eds. (IBP Handbook No. 16). Blackwell, Oxford, 334 pp.

Delamare Deboutteville C., 1960, Biologie des eaux souterraines littorales et continentales, *Herman*, Paris, 740 p.

Duncan, A., Schiemer F., and Klekowski, R. Z., 1974, A preliminary study of feeding rates on bacterial food by adult females of a benthic nematode, *Plectus palustris* de Man 1880, *Polskie Archiwum Hydrobiologii 21:* 249-258.

Ellenby, C., and Smith L., 1966, Haemoglobin in *Mermis subnigrescens* (Cobb), *Enoplus brevis* (Bastian), and *E. communis* (Bastian), *Comparative Biochemistry and Physiology 19:* 871-877.

Ernst, W., 1970, ATP als Indikator für die Biomasse mariner Sedimente, *Oecologia* (Berlin) *5:* 56-60.

Ernst W., and Goerke, H., 1974, Adenosin-5'-triphosphat (ATP) in Sedimenten und Nematoden der nordostatlantischen Tiefsee, *"Meteor" Forsch.-Ergebnisse C.18:* 35-42.

Fenchel, T., 1967, The ecology of marine microbenthos. I. The quantitative importance of ciliates as compared with metazoans in various types of sediments, *Ophelia 4:* 121-137.

Fenchel, T., 1969, The ecology of marine microbenthos. IV. Structure and function of the benthic ecosystem, its chemical and physical factors and the microfauna communities, with special reference to the ciliated protozoa, *Ophelia 6:* 1-182.

Fenchel, T., and Riedl, R. J., 1970, The sulfide system: a new biotic community underneath the oxidized layer of marine sand bottoms, *Marine Biology 7:* 255-268.

Fox, H. M., and Wingfield, C. A., 1938, A portable apparatus for the determination of oxygen dissolved in a small volume of water, *Journal of Experimental Biology 15:* 437-445.

Fry, F. E. J., 1958, Temperature compensation, *Annual Review of Physiology 20:* 207-224.

Gerlach, S. A., 1971, On the importance of marine meiofauna for benthos communities, *Oecologia* (Berlin) *6:* 176-190.

Gerlach, S. A., 1972, Die Produktionsleistung des Benthos in der Helgoländer Bucht, *Verhandlungsbericht der Deutschen Zoologischen Gesellschaft 65:* 1-13.

Glick, D., 1961, *Quantitative Chemical Techniques of Histo- and Cytochemistry*, Interscience Publications, New York, 470 pp.

Goldberg, M. L., 1973, Quantitative assay for submicrogram amounts of protein, *Analytical Biochemistry 51:* 240-246.

Grassle, F. J., Sanders, H. L., 1973, Life histories and the role of disturbances. *Deep-Sea Research 20:* 643-659.

Gray, J. S. 1965, The behaviour of *Protodrilus symbioticus* Giard in temperature gradients, *Journal of Animal Ecology, 34:* 455-461.

Gray, J. S. 1967, Substrate selection by the archiannelid *Protodrilus rubropharyngeus*, *Helgolander Wissenschaftliche Meeresuntersuchungen 15:* 253-269.

Hargrave, B. T., 1973, Coupling carbon flow through some pelagic and benthic communities, *Journal of the Fisheries Research Board of Canada 30:* 1317-1326.

Hemmingsen, A. M., 1960, Energy metabolism as related to body size and respiratory surface and its evolution, *Report of the Steno Memorial Hospital and the Nordisk Insulinlaboratorium* (Copenhagen) *9:* 7-11.

Hessler, R. P., and Jumars, P. A., 1974, Abyssal community analysis from replicate box cores in the central North Pacific, *Deep-Sea Research 21:* 185-209.

Heunert, H. H., and Uhlig G., 1966, Erfahrungen mit einer neuen Kammer zur Lebendbeobachtung beweglicher Mikroorganismen. *Research Film 5:* 642-649.

Holm-Hansen, O., and Booth, C. R., 1966, The measurement of adenosine triphosphate in the ocean and its ecological significance, *Limnology and Oceanography 11:* 510-519.

Holme, N. A., and McIntyre, A. D., 1971, *Methods for the Study of Marine Benthos* (IBP Handbook No. 16), Blackwell, Oxford, 334 pp.

Holter, H., 1943, Technique of the Cartesian diver. *Comptes-rendus des Travaux du Laboratoire de Carlsberg, Série Chimie 24:* 399-478.

Holter, H., 1945, A colorimetric method for measuring the volume of large amoebae, *Comptes-rendus des Travaux du Laboratoire de Carlsberg, Série Chimie 25:* 156-167.

Holter, H., and Zeuthen, E., 1966, Manometric techniques for single cells, in: (G. Oster, and A. W. Pollister, eds.) *Physical Techniques in Biological Research*, Vol. 3, Academic Press, New York, pp. 251-317.

Hochachka, P. W., and Somero, G. N. 1973, *Strategies of Biochemical Adaptation*, Saunders Company, Philadelphia 358 pp.

Hulings, N. D., and Gray, J. S., 1971, A manual for the study of meiofauna, *Smithsonian Contributions to Zoology 78:* 1-84.

Jansson, B. O., 1962, Salinity resistance and salinity preference of two oligochaetes., *Aktedrilus monospermathecus* Knollner and *Marionina preclitellochaeta* n. sp., from the interstitial fauna of marine sandy beaches, *Oikos 13:* 293-305.

Jansson, B. O., 1967, The availability of oxygen for the interstitial fauna of sandy beaches, *Journal of Experimental Marine Biology and Ecology 1:* 123-143.

Jansson, B. O., 1972, Ecosystem approach to the Baltic problem. *Bulletins from the Ecological Research Committee, No. 16:* 5-82, Swedish Natural Science Research Council (NFR).

Jørgensen, B. B., Fenchel, T., 1974, The sulfur cycle of a marine sediment model system, *Marine Biology*, (Berlin) *24:* 189-201.

Kanwisher, J., 1959, Polarographic oxygen electrode, *Limnology and Oceanography 4:* 210-217.

Kanwisher, J., 1962, Gas exchange of shallow marine sediments, in: *Symposium on the Environmental Chemistry of Marine Sediments*, Occasional Publication No. 1, Graduate School of Oceanography, University of Rhode Island, pp. 13-19.

Kanwisher, J. W., Lawson, K. D., and McCloskey, L. T., 1974, An improved, self-contained polarographic dissolved oxygen probe, *Limnology and Oceanography 19:* 700-704.

Klekowski, R. Z., 1971, Cartesian diver microrespirometry for aquatic animals, *Polskie Archiwum Hydrobiologii 18:* 93-114.

Larsson, S., and Løvtrup, S., 1966, An automatic diver balance, *Journal of Experimental Biology 44:* 47-58.

Lasker, R., 1966, Feeding, growth, respiration and carbon utilization of a euphausiid crustacean, *Journal of the Fisheries Research Board of Canada 23:* 1291-1317.

Lasker, R., Wells, J. B. J., and McIntyre, A. D., 1970, Growth, reproduction respiration and carbon utilization of the sand-dwelling Harpacticoid copepod, *Asellopsis intermedia*, *Journal of the Marine Biological Association of the United Kingdom 50:* 147-160.

Lasserre, P., 1969, Relations énergétiques entre le métabolisme respiratoire et la régulation ionique chez une Annélide oligochète euryhaline, *Marionina achaeta* (Hagen), *Comptes-rendus Hebdomadaires des Séances de l'Académie des Sciences, Série D. 268:* 1541-1544

Lasserre, P., 1970, Action des variations de salinité sur le métabolisme respiratoire d'oligochètes euryhalins du genre *Marionina* Michaelsen, *Journal of Experimental Marine Biology and Ecology 4:* 150–155.

Lasserre, P., 1971a, Données écologiques sur la répartition des oligochètes marins méiobenthiques. Incidence des paramètres salinité-température sur le métabolisme respiratoire de deux espèces euryhalines du genre *Marionina* Michaelsen (1889) (*Enchytraeidae, Oligochaeta*), *Vie et Milieu 22:* 523–540.

Lasserre, P., 1971b, Oligochaeta from the marine meiobenthos: taxonomy and ecology, *Smithsonian Contributions to Zoology 76:* 71–86.

Lasserre, P., 1975. Métabolisme et osmorégulation chez une annélide oligochète de la méiofaune: *Marionina achaeta* Lasserre. *Cahiers de Biologie Marine 16:* 765–799.

Lasserre, P., and Gallis, J. L., 1975, Osmoregulation and differential penetration of two grey mullets, *Chelon labrosus* (Risso) and *Liza ramada* (Risso), in esturaine fish ponds, *Aquaculture 5:* 323–344.

Lasserre, P. and Renaud-Mornant, J., 1971a, Consommation d'oxygène chez un Crustacé méiobenthi-interstitiel de la sous-classe des Mystacocarides. *Comptes-rendus hebdomadaires des séances de L'Académie des Sciences* Série D., *272:* 1011–1014.

Lasserre, P., Renaud-Mornant, J., 1971b, Interprétation écophysiologique des effets de température et de salinité sur l'intensité respiratoire de *Derocheilocaris remanei biscayensis* Delamare, 1953 (*Crustacea, Mystacocarida*), *Comptes-rendus Hebdomadaires des Séances de L'Académie des Sciences, Série D 272:* 1159–1162.

Lasserre, P., and Renaud-Mornant, J., 1973, Resistance and respiratory physiology of intertidal meiofauna to oxygen-deficiency, *Netherlands Journal of Sea Research 7:* 290–302.

Lasserre, P. and Renaud-Mornant, J. eds., 1975, Aspects of meiofauna research (Proceedings of the First International Meeting on Meiofauna Physiological Ecology, September 25–29, 1974, Arcachon, France), *Cahiers de Biologie Marine 16* (5): 593–799.

Lee, J. J., Tietjen, J. H., Stone, R. J., Muller, W. A., Rumiman, J. and McEnery, M., 1970, The cultivation and physiological ecology of members of salt marsh epiphytic communities, *Helgoländer Wissenschaftliche Meeresuntersuchugen 20:* 136–156.

Lemon, E. R., and Erickson, A. E., 1952, The measurement of oxygen diffusion in the soil with a platinum microelectrode, *Soil Science Society of American Proceedings 16:* 160–163.

Linderstrøm-Lang, K., 1943, On the theory of the Cartesian diver micro-respirometer, *Comptes-rendus des Travaux du Laboratoire Carlsberg, Série Chimie 24:* 333–398.

Løvlie, A. and Zeuthen, E., 1962, The gradient diver: a recording instrument for gasometric microanalysis, *Comptes-rendus des Travaux du Laboratoire Carlsberg 32:* 513–534.

Løvlie, A., 1964, Genetic control for division rate and morphogenesis in *Ulva mutabilis* Föyn, *Comptes-rendus des Travaux du Laboratoire Carlsberg, Série Chimie 34:* 77–168.

Løvtrup, S., 1973, The construction of a microrespirometer for the determination of respiratory rates of eggs and small embryos, *Experiments in Physiology and Biochemistry 6:* 115–152.

Low, B. W., and Richards, F. M., 1952, The use of gradient tube for the determination of crystal densities, *Journal of the American Chemical Society 74:* 1660–1666.

Lowry, O. H., 1944, A simple quartz torsion balance, *Journal of Biological Chemistry 152:* 293–294.

Lowry, O. H., and Passonneau, J. V., 1972, *A Flexible System for Enzymatic Analysis,* pp. 236–249. Academic Press, New York.

McIntyre, A. D. 1964, Meiobenthos of sub-littoral muds. *Journal of the Marine Biological Association of the United Kingdom 44:* 665–674.

McIntyre, A. D., 1968, The meiofauna and macrofauna of some tropical beaches. *Journal of Zoology* (London) *156:* 377–392.

McIntyre, A. D., 1969, Ecology of marine meiobenthos, *Biological Reviews 44:* 245-290.

McIntyre, A. D., and Murison, D. J., 1973, The meiofauna of a flatfish nursery ground, *Journal of the Marine Biological Association of the United Kingdom 53:* 93-118.

McIntyre, A. D., Munro, A. L. S., and Steele, J. H., 1970, Energy flow in a sand ecosystem, in (J. H. Steele, ed.), *Marine Food Chains*, Oliver and Boyd, Edinburg, pp. 19-31.

Maguire, C., and Boaden, P. J. S., 1975, Energy and evolution in the thiobios: An extrapolation from the marine gastrotrich *Thiodasys sterreri*, *Cahiers de Biologie Marine 16:* 635-646.

Mare, M. F., 1942, A study of a marine benthic community with special reference to the microorganisms, *Journal of the Marine Biological Association of the United Kingdom 25:* 517-554.

Marshall, N., 1970, Food transfer through the lower trophic levels of the benthic environment, in: (J. H. Steele, ed.) *Marine Food Chains*, Oliver and Boyd, Edinburg, pp. 52-66.

Muus, B., 1967, The fauna of Danish estuaries and lagoons, *Meddelelser fra Danmarks fiskeri-og Havundersøgelser (Denmark)*, *N. S. 5:* 1-316.

Nielsen, C. O., 1949, Studies on the soil microfauna. II. The soil inhabiting nematodes, *Natura Jutlantica* (Aarhus, Denmark) *2:* 1-131.

Neuhoff, V., 1971, Wet weight determination in the lower milligram range, *Analytical Biochemistry 41:* 270-271.

Neuhoff, V., 1973, *Micromethods in Molecular Biology*, Springer-Verlag, Berlin, 428 pp.

Newell, R. C. and Northcroft, H. R., 1967, A re-interpretation of the effect of temperature on the metabolism of certain marine invertebrates, *Journal of Zoology* (London) *151:* 277-298.

Nexø, B. A., Hamburger, K., and Zeuthen, E., 1972, Simplified microgasometry with gradient divers, *Comptes-rendus des Travaux du Laboratoire Carlsberg 39:* 33-63.

Noodt, W., 1957, Zur Okologie der Harpacticoidea (Crust. Cop.) des Eulittorals der deutschen Meeresküste und der angrenzenden Brackgewässer. *Zeitschrift fuer Morphologie und Oekologie der Tiere 46:* 149-242.

Ott, J. and Schiemer, R., 1973, Respiration and anaerobiosis of free living nematodes from marine and limnic sediments. *Netherlands Journal of Sea Research 7:* 233-243.

Pamatmat, M. M., 1965, A continuous-flow apparatus for measuring metabolism of benthic communities, *Limnology and Oceanography 10:* 486-489.

Pamatmat, M. M., 1968, Ecology and metabolism of a benthic community on an intertidal sandflat, *Internationale Revue der Gesamten Hydrobiologie 53:* 211-298.

Pamatmat, M. M., 1971a, Oxygen consumption by the sea bed. IV, *Limnology and Oceanography 16:* 536-550.

Pamatmat, M. M., 1971b, Oxygen consumption by the sea bed. VI, *Internationale Revue der Gesamten Hydrobiologie 56:* 675-699.

Pamatmat, M. M., and Banse, K., 1969, Oxygen consumption by the seabed. II. *In situ* measurements to a depth of 180 m, *Limnology and Oceanography, 14:* 250-259.

Pamatmat, M. M., and Bhagwat, A. M., 1973, Anaerobic metabolism in lake Washington sediments, *Limnology and Oceanography 18:* 611-627.

Pamatmat, M. M., and Fenton, D., 1968, An instrument for measuring subtidal benthic metabolism *in situ*, *Limnology and Oceanography 13:* 537-540.

Pamatmat, M. M., and Skjoldal, H. R., 1974, Dehydrogenase activity and adenosine triphosphate concentration of marine sediments in Lindåspollene, Norway, *Sarsia 56:* 1-12.

Poulson, T. L., and White, W. B., 1969, The cave environment, *Science 165:* 971-981.

Prosser, C. L., 1955, Physiological variation in animals, *Biological Reviews 30:* 229-262.

Prosser, C. L., 1958, General summary: the nature of physiological adaptation, in: (C. L.

Prosser, ed.), *Physiological Adaptation*, American Physiological Society, Washington, D. C., pp. 167–180.

Prosser, C. L., 1967, Molecular mecanisms of temperature adaptation in relation to speciation, in: (C. L. Prosser, ed.), *Molecular Mechanisms of Temperature Adaptation*, Publ. No. 84, American Association for the Advance of Science, pp. 351–376.

Rao, K. P., 1967, Some biochemical mechanisms of low temperature acclimation in tropical poiklotherms, in: (A. S. Trochin, ed.), *The Cell and Envrionmental Temperature*, Pergamon Press, New York, pp. 98–112.

Remane, A., 1933, Verteilung und Organisation der benthonischen Mikrofauna der Kieler Bucht. *Wissenschaftliche Meeresuntersuchungen* (Abt. Kiel) *21:* 161–221.

Remane, A. 1940, Einführung in die zoologische Ökologie der Nord- und Ostsee. *Tierwelt der Nord- und Ostsee*, Leipzig, 238 pp.

Renaud-Debyser, J. et Salvat, B. 1963, Eléments de prospérité des biotopes des sédiments meubles intertidaux et écologie de leurs populations en microfaune et macrofaune, *Vie et Milieu 14:* 463–550.

Richards, F. M., and Berger, J. E., 1961, Determination of the density of solids, in: (K. Lonsdale, ed.), *International Tables for X-Ray Crystallography*, Vol. 3, Kynoch, Birmingham, England.

Riedl, R. J., 1971, How much seawater passes through sandy beaches, *Internationale Revue der Gesamten Hydrobiologie u Hydrographie 56:* 923–946.

Rokop, F. J., 1974, Reproductive patterns in the deep-sea benthos, *Science 186:* 743–745.

Salvat, B., and Renaud-Mornant, J., 1969, Etude écologique du macrobenthos et du meiobenthos d'un fond sableux du Lagon de Mururoa (Tuamotu-Polynésie), *Cahiers du Pacifique 13:* 159–179.

Schiemer, F., 1973, Respiration of two species of Gnathostomulids, *Oecologia 13:* 403–406.

Schiemer, F., and Duncan, A., 1974, The oxygen consumption of a freshwater benthic nematode *Tobrilus gracilis* (Bastian), *Oecologia* (Berlin) *15:* 121–126.

Shaw, J. and Beadle, L. C., 1949, A simplified ultra-micro Kjeldahl method for the estimation of protein and total nitrogen in fluid samples of less than 1.0 µl, *Journal of Experimental Biology 26:* 15–23.

Smith, K. L., 1973, Respiration of a sublittoral community, *Ecology 54:* 1064–1075.

Smith, K. L., and Teal, J. M., 1973, Deep-sea benthic community respiration: An *in-situ* study at 1850 meters, *Science 179:* 282–283.

Smith, K. L., Burns, K. A., and Teal, J. L., 1972, *In-situ* respiration of benthic communities in Castle Harbor, Bermuda, *Marine Biology 12:* 196–199.

Swedmark, B., 1964, The interstitial fauna of marine sand, *Biological Reviews 39:* 1–42.

Teal, J. M., and Kanwisher, J., 1961, Gas exchange in a Georgia salt marsh, *Limnology and Oceanography 6:* 388–399.

Teal, J. M., and Wieser, W., 1966, The distribution and ecology of nematodes in a Georgia salt marsh, *Limnology and Oceanography 11:* 217–222.

Thane-Fenchel, A., 1970, Interstitial gastrotrichs of some South Florida beaches, *Ophelia 7:* 113–138.

Thiel, H., 1972, Meiofauna und Struktur der bentischen Lebensgemeinschaft des Iberischen Tiefseebeckens, *"Meteor" Forsch. -Ergebnisses D 12:* 36–51.

Thorson, G., 1966, Some factors influencing the recruitment and establishment of marine benthic communities, *Netherlands Journal of Sea Research 3:* 267–293.

Tietjen, J. H., 1969, The ecology of shallow water meiofauna in two New England estuaries. *Oecologia* (Berlin) *2:* 251–291.

Tietjen, J. H., and Lee J. J., 1972, Life cycles of marine nematodes. Influence of temperature and salinity on the development of *Monhystera denticulata* Timm. *Oecologia* (Berlin) *10:* 167–176.

Uhlig, G., Thiel, H., and Gray, J. S., 1973, The quantitative separation of meiofauna: A comparison of methods, *Helgoländer Wissenschaftliche Meeresuntersuchungen 25:* 173-195.

Van Doren, D. M., and Erickson, A. E., 1966, Factors influencing the platinum microelectrode method for measuring the rate of oxygen diffusion through the soil solution, *Soil Science 102:*23-28.

Vernberg, W. B., and Coull, B. C., 1974, Respiration of an interstitial ciliate and benthic energy relationships, *Oecologia* (Berlin) *16:* 259-264.

Vernberg, W. B., and Coull, B. C., 1975, Multiple factor effects of environmental parameters on the physiology, ecology, and distribution of some marine meiofauna, *Cahiers de Biologie Marine 16:* 721-732.

Wieser, W., 1960, Benthic studies in Buzzards Bay. II. The meiofauna, *Limnology and Oceanography 5:* 121-137.

Wieser, W., and Kanwisher, J., 1959, Respiration and anaerobic survival in some sea weed inhabiting invertebrates, *Biological Bulletin* (Woods Hole) *17:* 594-600.

Wieser, W., and Kanwisher, J., 1960, Growth and metabolism in a marine nematode, *Enoplus communis* Bastian, *Zeitschrift fuer Vergleichende Physiologie 43:* 29-36.

Wieser, W., and Kanwisher, J., 1961, Ecological and physiological studies on marine nematodes from a small salt marsh bear Woods Hole, Massachusetts, *Limnology and Oceanology 6:* 262-270.

Wieser, W., Ott, J., Schiemer, F., and Gnaiger, E., 1974, An ecophysiological study of some meiofauna species inhabiting a sandy beach at Bermuda, *Marine Biology 26:* 235-248.

Winberg, G. G., 1971, *Methods for the Estimation of Production of Aquatic Animals* (translation by A. Duncan), Academic Press, London and New York, 175 pp.

Zeuthen, E., 1943, A Cartesian diver micro respirometer with a gas volume of 0.1 μl, *Comptes-rendus des Travaux du Laboratoire Carlsberg*, série chimie *24:* 479-518.

Zeuthen, E. 1947a, A sensitive "Cartesian diver" balance. *Nature, 159:* 440.

Zeuthen, E. 1947b, Body size and metabolic rate in the animal kingdom. *Comptes-rendus des Travaux du Laboratoire Carlsberg*, série chimie *26:* 17-165.

Zeuthen, E. 1950, Cartesian diver respirometer, *Biological Bulletin* (Woods Hole) *98:* 303-318.

Zeuthen, E., 1953. Oxygen uptake as related to body size in organisms, *Quarterly Review of Biology, 28:* 1-12.

7
Engineering Interest in the Benthic Boundary Layer

R. B. KRONE

Engineers are interested in prediction of the effects of man's activities on the sea bed. In particular this involves knowing the controls of erosion and deposition of sediments in regard to management of navigable waterways and maintainance of water quality in nearshore areas. Geotechnical assessment of the bed is of importance for assessment of the stability of structures. Most sediments in the oceans are fine grained. In estuaries fine sediments flocculate. The physical properties of these aggregates are given and are related to the critical erosion conditions for mud beds. Deposition of these flocculent materials follows an exponential law when the fluid shear stress at the bed is below a limiting value. Increasing use is being made of these erosion and deposition relationships in numerical models. Both modeling and empirical approaches are used in water-quality studies. The increasing impact of man's activities in dumping and discharge of wastes near shore are actively being investigated by engineers. The paper concludes with suggestions for future directions of work.

Introduction

Engineering interests in the benthic boundary layer are centered primarily in the areas of transport of dissolved and suspended materials, in deposition, consolidation, and erosion of sediments that affect navigation, and in the effects of solid and liquid wastes disposed off-shore on water quality. We are particularly concerned with the development of knowledge and techniques needed to predict and monitor the effects of man's actions on shoaling, on water quality for fisheries and for aesthetic and recreational uses, and on the stability of structures. En-

R. B. KRONE • Department of Civil Engineering, University of California, Davis, California, U.S.A.

gineers are particularly concerned with near-shore and estuarial areas because these are the areas of greatest human activity. This conference is concerned with the region of the sea bed surface beyond the area of breaking waves. Much of the recent work concerned with the beds of estuaries, where both navigation and water quality are of concern, is relevant to our subject, however, and will be alluded to when appropriate.

Transport, Deposition, Consolidation, and Erosion of Sediment Materials

Beyond the region where waves continually rework the bed, hydraulic conditions favor the transport of particles having low settling velocities. Such particles include especially clay and silt particles that originate with the erosion of soils on the land as well as algae and particulate waste materials. The processes of transport of such materials are profoundly affected by their interparticle cohesion. Clay minerals are universally cohesive in sea water. Repeated interparticle collisions in estuaries are largely due to internal shearing of the suspending water (Krone, 1962). The frequency of collisions J on one spherical particle is

$$J = \frac{4}{3} \nu R_{ij}^3 \frac{du}{dz}$$

where ν is the particle concentration by number, R_{ij} is the collision radius or the sum of the radii of the colliding particles, and du/dz is the local shearing rate. The product νR_{ij}^3 is especially large for a few large particles in a suspension of many small particles: the large particles appear to gather the smaller ones. Collisions are more frequent with increasing shearing rates, but collisions result in aggregation only when du/dz does not exceed a level that breaks interparticle bonds.

Aggregates formed by fluid shearing are denser and stronger than those formed by Brownian motion or by differential settling velocities. Aggregates can be dispersed by local high shearing rates and later reaggregate, repeatedly. The settling velocities of aggregates and the probability of their sticking to the bed to which they settle are determined by their size, density, and shear strength.

Physical properties of aggregates of cohesive sediment materials taken from a number of estuaries were studied by Krone (1963) using rheological techniques and a large concentric cylinder viscometer designed for the purpose. It was found that aggregates having several different structures could be formed repeatedly in the viscometer. Each structure had its unique combination of density and shear strength. The following model was constructed: first, individual clay mineral particles would aggregate from a dispersed suspension undergoing shearing at appreciable rates. Such aggregates grew as individual particles collided with them. At lower shearing rates in a suspension of such primary aggregates these aggregates would collide with each other and bond to form aggregates of aggregates, which I will call "first-order aggregates." At still lower shearing rates first-order aggregates can collide and bond with each other to form weaker, less-dense second-order aggregates, and so on. Each higher-order aggregate would include water in the new pore volume formed, and because shear stress can be transmitted only through interaggregate contacts, the higher order is weaker. Table 1 shows the densities and shear strengths calculated from the viscometer data.

It was found that the shear strength of zero-order aggregates could be related to the cation exchange capacity of the material. The relation between shear strength of higher-order aggregates and those of zero order is not yet well defined. Kandiah (1974) has recently shown that the cation exchange capacity is a

Table 1. Properties of Sediment Aggregates

Sediment sample	Order of aggregation	Density,[a] $g\ cm^{-2}$	Shear strength, $dyn\ cm^{-2}$
Wilmington District	0	1.250	21
	1	1.132	9.4
	2	1.093	2.6
	3	1.074	1.2
Brunswick Harbor	0	1.164	34
	1	1.090	4.1
	2	1.067	1.2
	3	1.056	0.62
Gulfport Channel	0	1.205	46
	1	1.106	6.9
	2	1.078	4.7
	3	1.065	1.8
San Francisco Bay	0	1.269	22
	1	1.179	3.9
	2	1.137	1.4
	3	1.113	1.4
	4	1.098	0.82
	5	1.087	0.36
	6	1.079	0.20
White River (salt)	0	1.212	49
	1	1.109	6.8
	2	1.079	4.7
	3	1.065	1.9

[a]Density in sea water, $\rho_l = 1.025\ g\ cm^{-3}$.

useful index of interparticle cohesion of clay soils if the structure, pH, concentration, and sodium adsorption ratio of pore fluid salts are also known.

Kranck (1974) measured the sizes of aggregates suspended in marine waters using a Coulter counter, and observed changes with distance from land.

As aggregates of any order settle to a cohesive bed and stick, a bed surface is formed having the next higher order of aggregation. The shear strengths of beds formed under a variety of hydraulic conditions, using San Francisco Bay sediments in a closed recirculating flume, were measured by increasing the bed stress in steps and measuring the suspended sediment concentration (Krone, 1962). The results are shown in Fig. 1. The plots show that at the zero concentration intercept the shear strengths correspond to those measured in the viscometer. At increasing distances below the bed surface the shear strength increased linearly. The total depth eroded was only 2-3 cm, so that it is apparent that even small amounts of overburden will consolidate and stabilize deposits having such weak structures. Figure 1 also shows that the amount of material suspended at a particular applied shear stress depends on the history of the deposit, as well as on the order of aggregation of the depositing aggregates from which it was formed.

Erosion appears to occur by either of two mechanisms: A bed may fail in bulk if the applied stress exceeds the bulk shear strength of the deposit. Deposits having increasing shear strength with depth would fail to the depth where the strength equals the applied stress. Second, erosion of cohesive material also occurs particle by particle or aggregate by aggregate. A method for measuring critical shear stresses for surface erosion of core samples has been demonstrated by Sargunum et al. (1973) using a device designed by Masch et al. (1963). In this device the core is supported on a mandrel concentrically with a rotating outer cylinder. The annular space is filled with eroding water having the same composition as that of concern. The outer cylinder is rotated at the same speed for several measured short periods using clear water for each period, and the core is weighed after each eroding period to obtain an erosion rate for that applied stress. Erosion rates similarly determined at each of a number of rotating speeds are then plotted, and are found to plot as a straight line, as shown in Fig. 2. The intercept yields the critical shear stress for erosion. The sea bed at the proposed site for a floating nuclear power plant was recently sampled and tested in this manner. The advantage of this method is that core samples of the bed can be tested. It should lead to the determination of erodibilities of natural materials reworked by benthic organisms and affected by biochemical changes of the pore fluid. It cannot, however, be used to evaluate changes of bed shear strength over depths less than the height of the core.

Deposition of suspended aggregates from boundary layer flows was also investigated by Krone using the recirculating flume. Loss in concentration of suspended matter with time for concentrations so low that aggregation rates were negligibly slow (less than 300 mg liter^{-1} under the prevailing hydraulic conditions) were measured. A plot of concentrations against time of deposition is

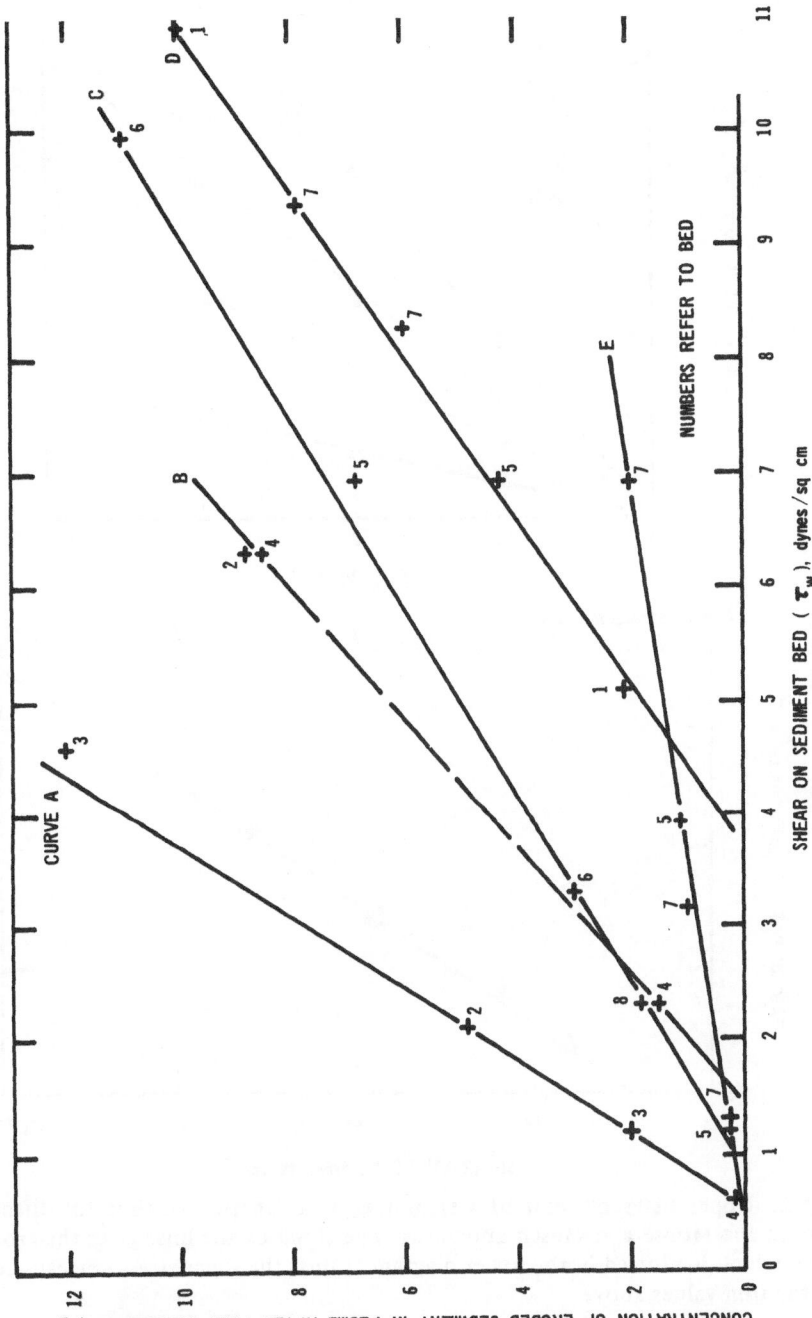

Fig. 1. Concentration of sediment in suspension measured at time of one circulation through flume loop as a function of shear stress at the bed. Extrapolation of the lines to the axis gives the erosion stress for the bed.

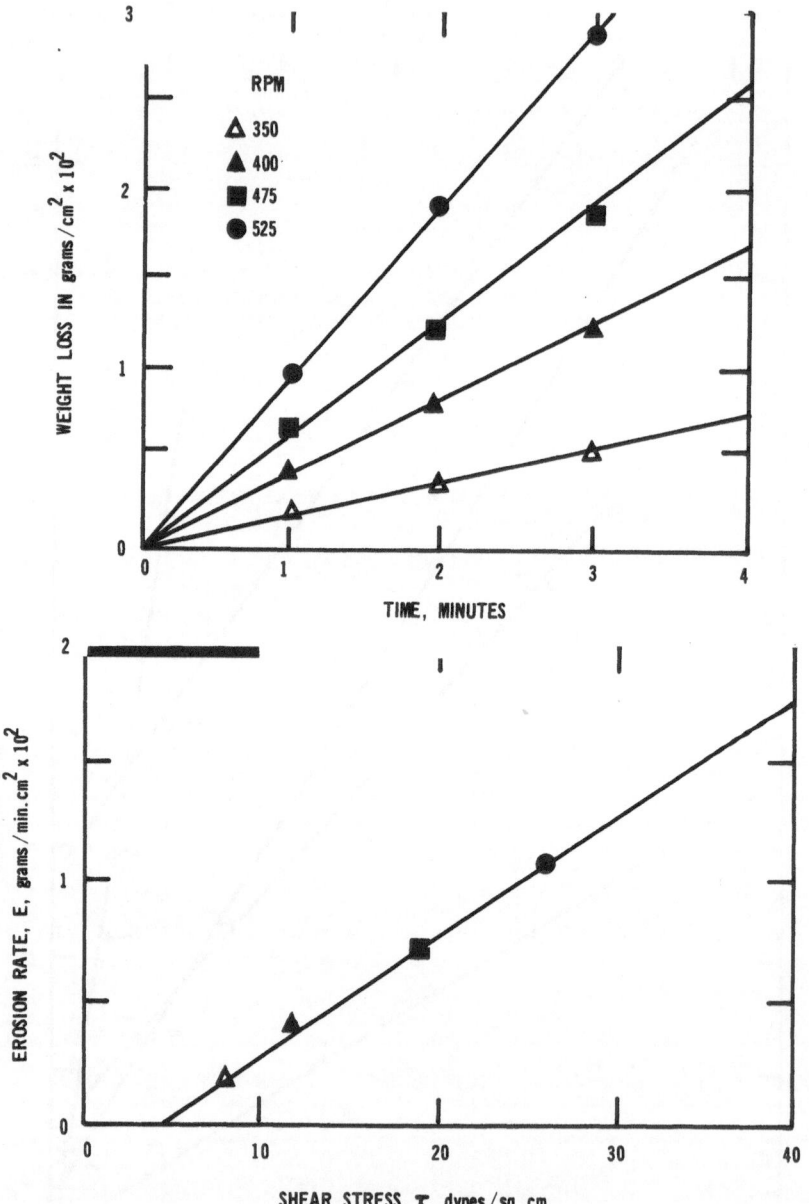

Fig. 2. Upper diagram: plot of weight loss as a function of time for different rpm in the Moore and Masch apparatus. The slope of the lines gives the erosion rate, which is plotted in the lower diagram against the shear stress corresponding to the rpm values above.

shown in Fig. 3. The loss from suspension can be described as

$$\ln \frac{C}{C_0} = -\frac{v}{\bar{y}} \left(1 - \frac{\tau}{\tau_c} \right) t \tag{1}$$

where C_0 and C are the initial concentration and the concentration at time t, respectively, v is the settling velocity, $\bar{y}$ is the average depth of settling, τ is the shear stress applied by the flow, and τ_c is the critical shear stress for deposition. The average settling velocity for an aggregating suspension that has proceeded to the point where interparticle collisions are infrequent is proportional to the concentration to the $\frac{4}{3}$ power (Krone, 1962). Most natural suspensions undergo such a variety of dispersions and dilutions, however, as well as experiencing varying shearing rates, that a direct measurement of v would be desirable for using the integrated or differentiated equation (1) in a model; τ_c would be one of the values given in Table 1.

When deposition proceeds under conditions where aggregation is actively going on, a more elaborate relation is needed:

$$\ln \frac{C}{C_0} = -\frac{v t_c}{\bar{y}} \left(\frac{v_0}{v} \right) \left(1 - \frac{\tau}{\tau_c} \right) \ln \left(1 + \frac{t}{t_c} \right) \tag{2}$$

where t_c is the average time between collisions of suspended particles and v_0/v is the average number of particles per aggregate. Expansion of the right-hand side shows that when $t \gg t_c$, equation (2) reduces to equation (1). For practical purposes equation (2) can be written as

$$\ln C = -\frac{K_2}{\bar{y}} \left(1 - \frac{\tau}{\tau_c} \right) \ln t + \text{constant}$$

when $t \ll t_c$. Data obtained in the flume under these conditions as shown in Fig. 4.

Parthenaides (1962) proposed that erosion of the bed surface contributes suspended solids to the overlying water at the rate

$$\frac{dc}{dt} = \frac{1}{d} M \left(\frac{\tau}{\tau_{ce}} - 1 \right) \tag{3}$$

where d is the average water depth, M is an erodibility constant, and τ_{ce} is the critical shear stress for erosion. M and τ_{ce} would change with depth of erosion of the bed surface, as shown in Fig. 1, which makes careful measurements essential.

Odd and Owen (1972) incorporated these relations in a one-dimensional by two-layer mathematical model of sediment transport in the Thames estuary with

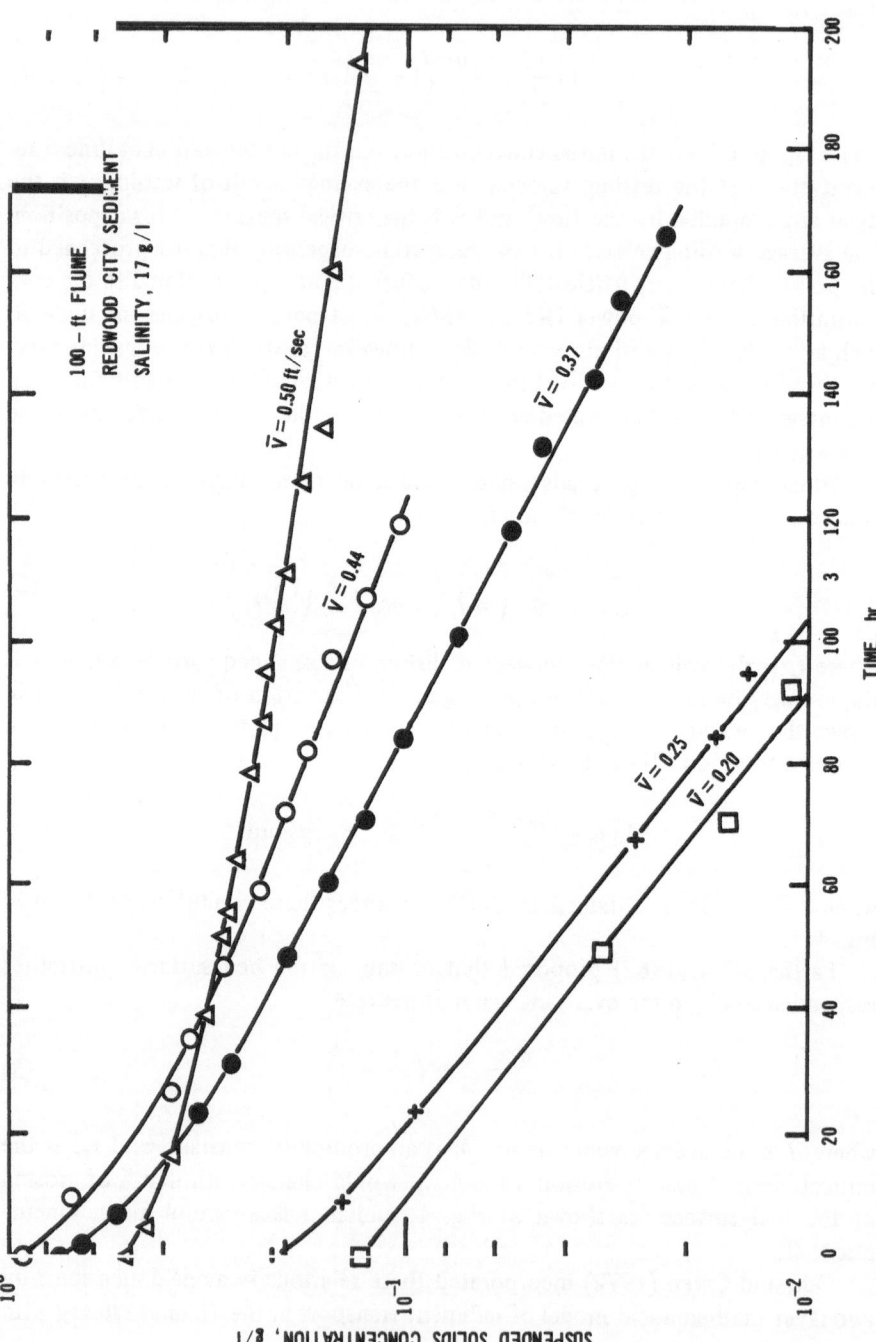

Fig. 3. Decrease in suspended material concentration with time in a 2/3 recirculation flume for initial concentrations less

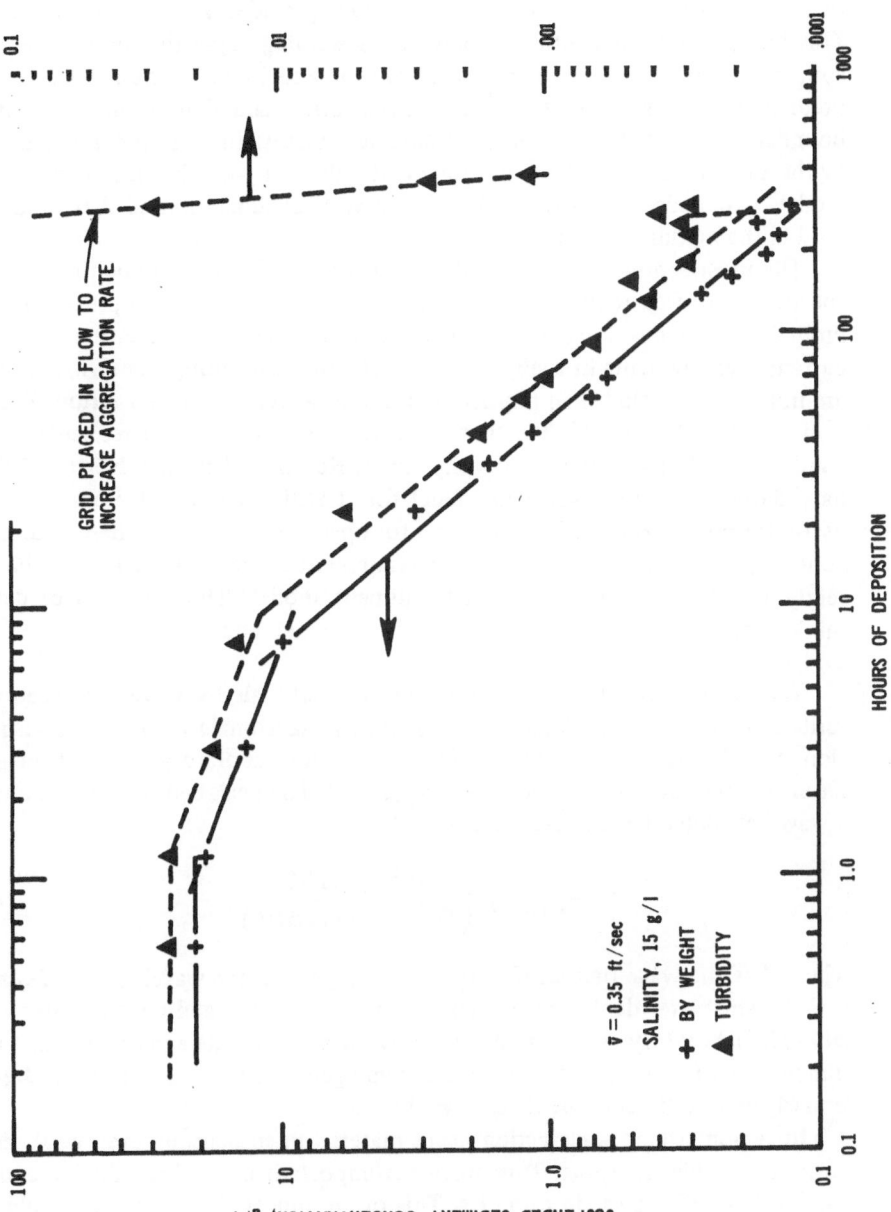

Fig. 4. Similar data to those in Fig. 3 but for initial concentration of 20 g l^{-1} .

evident success. More recently, Ariathurai (1974) developed a finite-element solution for the two-dimensional advection–diffusion equation for concentration
of suspended solids where the above relations provided source and sink terms.
This has so far been confirmed only for describing deposition patterns downstream from model pilings under steady flow in a recirculating flume. The model
predicts both suspended sediment concentration and deposition or scour if
boundary and initial conditions, currents, and diffusivities are provided, as well
as the parameters describing the sediments. Present work includes revising the
model to include one horizontal and one vertical dimension so that it can be
used in the mixing zone of an estuary.

The most serious weakness of the sediment model appears to be the need for
specification of currents and shear stresses in sufficient detail. At present we are
limited to either making many expensive prototype measurements or making
current measurements in a physical model. Currents resulting from wind stresses
are not usually included in physical models, however, which is a serious limita-
tion. Link–node models, like that of Water Resources Engineers, can include
wind stresses but not variable density flows. Recently, King and Norton (1974)
have developed a finite-element model for stratified lakes that promises to be
useful for mixing zones of estuaries and for open off-shore areas where boundary
conditions can be prescribed from measurements. Such a model could be directly coupled with the suspended sediment model. The problems of determining diffusivities for both the hydrodynamic and sediment models are
evident.

The suspension of noncohesive sediment materials by waves has been described by Einstein and Wiegel (1970) and by MacDonald (1973). The suspension of cohesive materials was investigated in the laboratory by Alishahi and
Krone (1964) and in the field by Krone (1961). The peak bed shear stress under
a wave calculated from linear wave theory is

$$\tau_{max} = \mu \left(\frac{\rho\pi}{T\mu}\right)^{1/2} \frac{\pi H/T}{\sinh\left(2\pi d/L\right)} \tag{4}$$

where T is the wave period, H is the wave height, L is the wavelength ($=T\sqrt{gd}$),
d is the still-water depth, μ is the coefficient of viscosity, and ρ is the water density. The laboratory studies indicated only an approximate agreement when the
dimensions of the highest one-third of wind-generated waves were used. Equation (4) underestimated the shear stress by 20%.

In regions where wave action keeps material in suspension, even weak tidal
currents provide transport. This process is important in the littoral zone and in
the large mudflat areas in estuaries. This or an improved description should be
useful in mathematical models of sediment transportation, and will be essential
for descriptions of transport patterns in many estuaries.

Water Quality

In a number of areas contaminated dredged spoil and industrial wastes are barged to the open ocean for disposal. Wastes released in the New York Bight area and outside of the Golden Gate at San Francisco are typical. We are faced with the need to predict the effects of such practices on water quality.

The effects of solids disposal are of two kinds. The turbidity resulting from suspended particulates and the effects of particles in smothering or physically harming aquatic biota are direct effects which might be predictable from the sediment model output plus knowledge of effects on organisms. Unfortunately, such effects are not well known. Indirect effects include those resulting from sorption of toxic materials, such as PCB, pesticides, and heavy metals. The effects may be either beneficial or harmful. Sorption of dissolved toxic materials by suspended sediment, for example, followed by deposition of the sediment, can scavenge significant amounts of toxic material and provide effective assimilation capacity for such materials, thereby enhancing water quality. Sorbed contaminants, on the other hand, may directly harm filter feeders. Turbidity may be beneficial if nutrient levels are such that undesirable concentrations of algae would occur, but would otherwise be undesirable for many fish or from an aesthetic point of view.

The U.S. Army Corps of Engineers has been conducting investigations for years on effects and efficiencies of existing and new technologies for maintenance dredging. A $30 million, five-year investigation, with headquarters at the Waterways Experiment Station, Vicksburg, Mississippi, U.S.A., is in its second year (Boyd, et al., 1972). The study will include environmental impacts of dredging, open water disposal, and land disposal, as well as productive uses of spoil, spoil treatment, and the development of new dredging, transporting, and disposal techniques. Knowledge leading to significant improvement of maintenance-dredging equipment, techniques, procedures, and utilization of dredged material can be expected from the results of this study.

Underwater mining for shell and for sand is practiced widely. The effects of mining and on-site washing of such materials on water quality is currently under scrutiny. The factors above for dredging and for disposal of dredged spoils apply.

The increasing interest of engineers in chemical and biological interactions result from declining catches of commercial fish and concern for estuarial water quality. It appears that fine sediment materials sorb dissolved toxic compounds while being transported in suspension; anoxic conditions or the presence of benthic organisms after deposition then alter the sorbed materials to a nonsorbable form or to a more toxic form. Ongoing concern for bacteriological conversion of mercury to ingestible organic compounds, believed to proceed in sediment beds, has encouraged the investigation of subsurface conditions and biological processes (Krenkle *et al.*, 1974). Other heavy metals and toxic materials sorbed by sediments include cadmium, zinc, lead, and pesticides. There is very little information available on sorption–desorption rates and equilibrium coefficients, or on biosynthesis and biological uptake rates for these materials. How much undesirable material sorbed during transport in suspension is released after such sediments are deposited? What are the overall effects on aquatic biota of releases of such materials to estuary or ocean waters? It seems evident that dissolved toxic materials should not be released to surface waters, but the will and the resources to control them have been limited. We need knowledge of rates of sorption and desorption of toxic materials to establish priorities for control action and the development of analytical methods that will relate analysis of sediments and effects of desorption on water quality.

Conclusion

Engineers' needs for knowledge of the benthic boundary layer originate in the needs for predicting deposition rates, for developing dredge spoil disposal methods that have minimum impact on water quality, and for management of water quality in the face of diminishing freshwater inflows and increasing production. Specifically, we need quantitative descriptions of the following:

1. Descriptions of unsteady, near-bed flows of water having an upward density gradient are needed to calculate the transport of suspended and dissolved material and the shear stresses applied to the sediment bed.

2. Mechanical and surface properties of cohesive sediment materials need to be further characterized in:

   a. aerobic suspensions of waters having a range of dissolved salts and internal shearing,
   b. freshly deposited beds of various depths, and
   c. aging, anaerobic deposits.

3. Sorption and desorption characteristics of suspended particulates in waters containing commonly occurring toxic materials are needed both for their direct interaction with surrounding waters and to determine the initial composition of pore fluids of deposits.

4. Biological alterations of salts and organic matter in the pores of estuarine and oceanic deposits need to be described to evaluate the contributions of dissolved materials and the consumption of oxygen by the bed surface, to select treatment facilities for waste waters, and to estimate effects of open water disposal of wastes on water quality.

This information will make possible the development of mathematical models that will facilitate prediction of transport of dissolved, sorbed, and particulate materials under altered hydraulic conditions and waste discharges. Such predictive capability will make possible the management of estuarial and coastal waters for the maximum benefit of man.

References

Alishahi, M. R., and Krone, R. B., 1964, *Suspension of cohesive sediments by wind-generated waves*, Hydraulic Engineering Laboratory, University of California, Berkeley.

Ariathurai, C. R., 1974, A finite element model for sediment transport in estuaries. Ph.D. Thesis, University of California, Davis.

Boyd, M. B., Saucier, R. T., Keeley, J. M., Montgomery, R. L., Brown, R. D., Mathis, D. B., and Bruice, C. J., 1972, *Disposal of dredge spoil*, Technical Report H-72-8, U.S. Army Engineer Waterway Experiment Station, Vicksburg, Mississippi.

Einstein, H. A., and Wiegel, R. L., 1970, A literature review on erosion and deposition of sediments near structures in the ocean, *U.S. Navy Civil Engineering Laboratory Report CR-70.008.*

Gilbert, G. K., 1917, Hydraulic mining debris in the Sierra Nevada. *U.S. Geological Survey Professional Paper, 105.*

Kandiah, A., 1974, Fundamental aspects of surface eriosion of cohesive soils. Ph.D. Thesis, University of California, Davis.

King, I., and Norton, W., 1974, *Report on modelling mixing in reservoirs*, Water Resources Engineers for the U.S. Army Corps of Engineers, Walnut Creek, California.

Kranck, K., 1974, The role of flocculation in the transport of particulate pollutants in the marine environment, Presented at the International Conference on the Transport of Persistent Chemicals in Aquatic Ecosystems, Ottawa, Ontario, Canada, May, 1974.

Krenkel, P. A., Burrows, W. D., Shin, E. B., Taimi, K. I., and McMullen, E. D., 1974, *Mechanisms of mercury transformation in bottom sediments*, Part II, Environmental and Water Resources Engineering, Vanderbilt University, Nashville, Tenn.

Krenkel, P. A., Reimers, R. S., and Burrows, W. D., 1974, *Mechanisms of mercury transformation in bottom sediments*, Final Report, Part I, Environmental and Water Resources Engineering, Vanderbilt University, Nashville, Tenn.

Krone, R. B., 1961, *Third annual progress report on the silt transport studies utilizing radio isotopes*, Hydraulic Engineering Laboratory and Sanitary Engineering Research Laboratory, University of California, Berkeley, 52 pp.

Krone, R. B., 1962. *Flume studies of the transport of sediment in estuarial shoaling processes*, Hydraulic Engineering Laboratory and Sanitary Engineering Research Laboratory, University of California, Berkeley, 110 pp.

Krone, R. B., 1963, *A study of rheologic properties of estuarial sediments*, Hydraulic Engineering Laboratory and Sanitary Engineering Research Laboratory, University of California, Berkeley, 91 pp.

Krone, R. B., 1966, *Predicted suspended sediment inflows to the San Francisco Bay system*, prepared for the Central Pacific River Basins Comprehensive Water Pollution Control Project, Federal Water Pollution Control Adm., U.S.A., 1966.

Krone, R. B. 1972, *A field study of flocculation as a factor in estuarial shoaling processes*, U.S. Army Corps of Engineers Committee on Tidal Hydraulics, Technical Bulletin 19.

MacDonald, T. C., 1973, Sediment transport due to oscillatory waves, Ph.D. Thesis, University of California, Berkeley.

Masch, F. D., Espey, W. H. Jr., and Moore, W. L., 1963, Measurement of the shear resistance of cohesive sediments, *Proceedings of the Federal Interagency Conference on Sedimentation*, Paper 23.

Meade, R., 1972. Transport and deposition of sediments in estuaries, *Geological Society of America, Memoir 133: 91-120*.

Odd, N. V. M., and Owen, M. W., 1972, A two-layer model of mud transport in an estuary, *Proceedings of the Institute of Civil Engineers*, Paper 7517 S.

Parthenaides, E., 1962, A study of erosion and deposition of cohesive soils in salt water, Ph.D. Thesis, University of California, Berkeley.

Sargunum, A., Riley, P., Arulanandan, K., and Krone, R., 1973, A method for measuring the erodibility of soils, *Proceedings of the American Society of Civil Engineers, Journal of the Hydraulics Division 99: 555-558*.

8
Marine Geotechnology
Average Sediment Properties,
Selected Literature and Review of
Consolidation, Stability, and
Bioturbation–Geotechnical
Interactions in the Benthic
Boundary Layer

ADRIAN F. RICHARDS and
JAMES M. PARKS

The basic geotechnical literature relevant to nongeotechnical benthic boundary layer scientists is briefly reviewed. The range of Atterberg limits, water content, bulk density, shear strength, and sensitivity for the common deep-sea cohesive sediments is summarized; minimum and maximum geotechnical properties are given for deep-sea cohesive sediments without regard to sediment type; and the range of grain size, bulk density, and sound velocity is reported for continental terrace cohesionless sediments. Geotechnical data for cohesive sediments are plotted with respect to depth below the seafloor to demonstrate the range of average values within selected continental shelf and deep-sea environments. The method of evaluating consolidation within the benthic boundary layer using sedimentation–compression diagrams relating water content to effective overburden pressure is illustrated by the use of examples of cohesive silty clay and carbonate sediments. Three principal processes affecting sea-floor sediment stability are micromovements caused by organisms, gas, and small environmental loads; macromovements or sea-slides and sea-slumps caused by various processes; and

ADRIAN F. RICHARDS ● Marine Geotechnical Laboratory, and JAMES M. PARKS ● Center for Marine and Environmental Studies, Lehigh University, Bethlehem, Pennsylvania 18015, U.S.A.

movements associated with turbidity currents. Reference is made to the literature describing these processes. In a discussion of bioturbation-geotechnical relationships, the use of geotechnical data for determining the bearing capacity of the epifauna and quantifying the mobility of the infauna is cited. The laboratory fall-cone and the in situ static-cone penetrometer tests are recommended for marine biological investigations in cohesive sediments. The use of the minimum-maximum shear strength concept is recommended to quantify the magnitude of strength heterogeneity produced by bioturbation of cohesive sediments.

Introduction

Marine geotechnology may be defined as the application of scientific methods and engineering principles to the acquisition, interpretation, and use of knowledge of sea-floor sediments and rocks with the goal of solving marine problems in civil engineering and the earth sciences. This definition is modified from that given for the word *geotechnics* in the *Glossary of Geology* (AGI, 1972).

The objectives of this paper are (1) to briefly summarize the range of geotechnical properties occurring in fine-grained cohesive and coarse-grained non-cohesive materials on the continental shelf and on the deep-sea floor, and (2) to briefly review the marine geotechnical literature relevant to benthic boundary layer investigations of sea-floor deposition, stability, and bioturbation caused by organisms. Emphasis has been placed on acquainting nongeotechnologists with the marine geotechnical literature considered to be particularly applicable to their interests, and on phenomena occurring at very shallow burial depths within the benthic boundary layer.

Units in this paper are in SI, or modern metric, and the symbols conform to the international standards that have been summarized by Richards (1974); a pascal (Pa), the SI unit for stress or pressure, is a newton per square meter (Nm^{-2}).

Basic Literature on Marine Geotechnics

Noorany and Gizienski (1970) and Fukuoka and Nakase (1973) have surveyed the state-of-the-art on the engineering properties of marine sediments. Continental shelf geotechnics have been reviewed in general by Richards *et al.* (1975) and by Bara (1974) specifically for the California shelf and continental borderland. Two regional geotechnical studies of bay sediments are by Mitchell (1963) on San Francisco Bay mud and Harrison *et al.* (1964) on the Chesapeake Bay. Noorany (1972), Tirey (1972), and Ling (1972) have surveyed marine soil instrumentation. Six symposium volumes (Richards, 1967; Proceedings, 1971;

ASTM, 1972; Inderbitzen, 1974; Kaplan, 1974; Hampton, 1974) cover a variety of marine geotechnical subjects. Garcia (1971) has made a partial literature survey on engineering properties of marine soils.

In the 1960s, the U.S. Naval Civil Engineering Laboratory sponsored detailed state-of-the-art reports: on breakout resistance (Vesic, 1969); on earthquake occurrence and effects (Wilson, 1969); on penetration of objects into the bottom (Schmid, 1969); on slope instability (Scott and Zuckerman, 1970); on erosion and deposition of soils near structures (Einstein and Wiegel, 1970); and on sea-floor trafficability (Wiendieck, 1970). A wealth of marine geotechnical material is contained in the technical reports and other publications of this laboratory, which recently changed its name to the Civil Engineering Laboratory of the U.S. Naval Construction Battalion Center, Port Hueneme, California.

Range of Geotechnical Properties

Keller and Bennett (1968) and Horn et al. (1974) have mapped selected geotechnical properties over entire ocean basins. Herrmann et al. (1972) recast the common range of deep-sea geotechnical data given by Keller (1969) into a table summarizing the principal types of ocean basin sediments occurring at burial depths of a few meters. This table has been modified rather extensively to include a few data from Horn et al. (1974) and from the Marine Geotechnical Laboratory (Table 1). The shear strength and sensitivity data are at variance

Table 1. Range of Common Deep-Sea Cohesive Sediment
Geotechnical Properties[a]

Sediment	Liquid limit	Plastic limit	Plasticity index	Water content, % dry wt.	Bulk density, Mg m^{-3}	Shear strength, kPa	Sensitivity[b]
Terrigenous	Data not synthesized			50–100	1.5–1.7	0.3–20	Variable
Pelagic clay	150–70	70–35	80–35	50–200	1.2–1.7	2–7	5–6
Calcareous ooze	Probably lower than above			50–200	1.2–1.6	5–10	6–11
Siliceous ooze	?	?	?	50–200	1.2–1.7	10–19	5–10 (?)

[a]Primarily for a depth of about one meter below the water-sediment interface
[b]Ratio of natural shear strength to remolded shear strength

Table 2. Minimum and Maximum Deep-Sea Cohesive Sediment
Geotechnical Properties

Parameter	Minimum	Maximum
Liquid limit	25	150
Plastic limit	15	70
Plasticity index	<2	92
Liquidity index	0.8	33
Water content, % dry weight	15	673
Void ratio (from water content)	0.4	18
Bulk density, Mg/m^3	1.1	2.6
Shear strength, kPa		
natural	0.02	90
remolded	0.02	>0.5
Sensitivity	1	88

with some published data. The numbers represent the common range of cohesive
sediments, primarily in the North Pacific; they should be considered only approximate. Richards and Parker (1968) tabulated deep-sea minima and maxima
values. In Table 2 these have been updated in part from Keller (1974), Horn *et al.*
(1974), and the Marine Geotechnical Laboratory, without regard to sediment
type or geological environment.

Table 3. Geotechnical Properties of North Pacific Continental Shelf and
Slope Soils[a]

Soil type	Number of samples	Sand, %	Silt, %	Clay, %	Mean grain diameter,[b] μm	Bulk density,[c] Mg m^{-3}	Sound velocity, m s^{-1}
Coarse sand	2	100	0	0	528	2.03	1836
Fine sand	18	92	4	3	164	1.96	1753
Very fine sand	6	84	10	6	91.5	1.87	1697
Silty sand	14	64	23	13	67.9	1.81	1668
Sandy silt	17	26	61	13	30.8	1.79	1664
Silt	12	6	81	13	21.3	1.77	1623
Sand–silt–clay	18	33	40	27	18.3	1.58	1580
Clayey silt	54	6	61	33	7.4	1.47	1546
Silty clay	19	5	41	54	2.7	1.42	1520

[a]After Hamilton (1974).
[b]Computed salt free.
[c]Water-saturated samples.

The geotechnical properties of the predominantly cohesionless materials of the continental shelf are not well known. Data from Hamilton (1974) for the Pacific continental terrace are summarized in Table 3. These properties should not be greatly different for continental shelves in other parts of the world.

A series of computer-drawn graphs of selected geotechnical properties related to depth below the bottom were prepared for two continental shelf areas having cohesive sediments considered to be the shallow-water analogs of deep-sea sediments. One of these areas is the Wilkinson Basin Geotechnical Test Area, which is located 120 km east of Boston, Massachusetts (Fig. 1). The other is the San Diego Trough Geotechnical Test Area, which is located 24 km southwest of San Diego, California (Fig. 5). In each test area, geotechnical measurements were made *in situ* and on cores; data taken within 10-cm depth intervals were averaged; the resulting values were plotted against depth; and a linear least-squares regression line was computer-fitted to all of the plotted values shown by an X. Values not used in the regressions are shown by open circles. The least-squares predictor equation, the standard error of estimate, and the correlation coefficient are also given. Because the initial terms of reference established by the

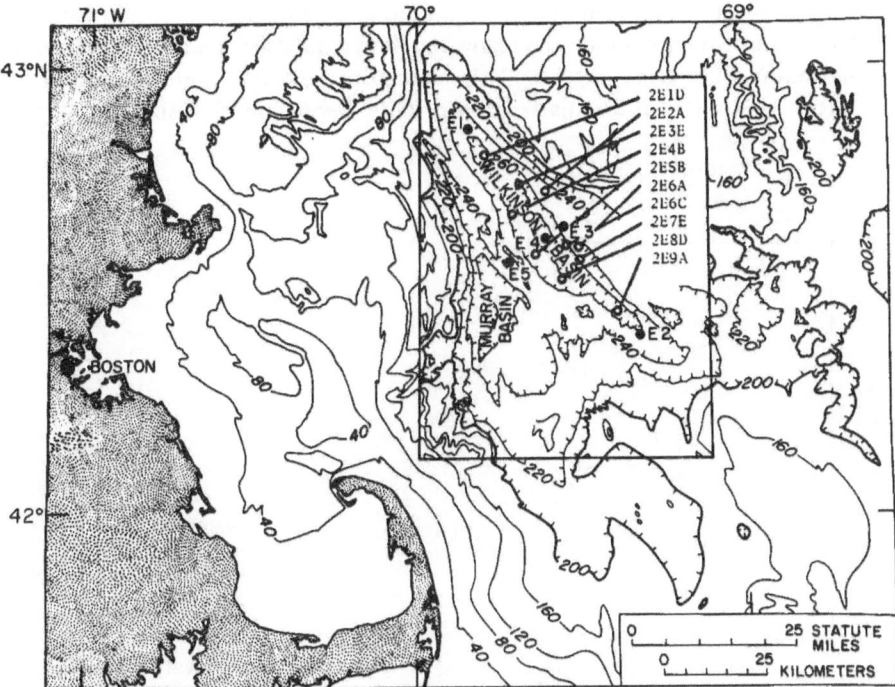

Fig. 1. Location map of the Wilkinson Basin Geotechnical Test Area showing density of cores and in-place measurements. Depths in meters.

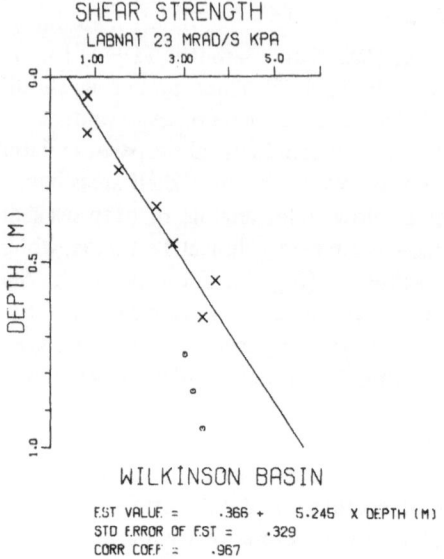

Fig. 2. Average shear strength measured on cores in the laboratory using a 12 × 25 mm vane rotated at 23 mrad s^{-1}. See text for averaging details.

Conference Steering Committee emphasized a consideration of only the top 1 m of the sedimentary column, the regression lines emphasize data occurring nearest the sea-floor surface. Figures 2–4 summarize selected averaged parameters plotted against depth for the Wilkinson Basin Geotechnical Test Area and Figs.

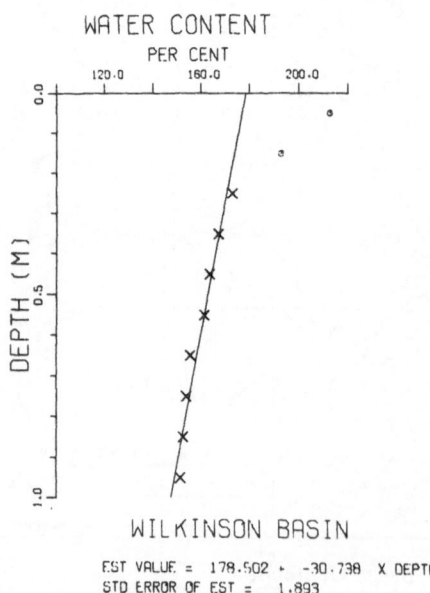

Fig. 3. Average water content, in percent dry weight, measured in the laboratory. See text for averaging details.

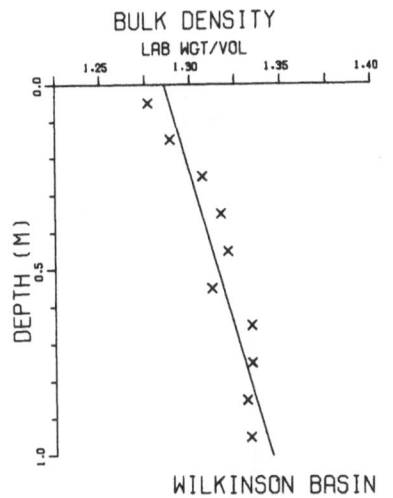

BULK DENSITY
LAB WGT/VOL

Fig. 4. Average bulk density, in mg m^{-3}, measured in the laboratory. See text for averaging details.

WILKINSON BASIN

EST VALUE = 1.286 + .059 X DEPTH (M)
STD ERROR OF EST = .009
CORR COEF = .903

6-9 summarize selected average parameters for the San Diego Trough Geotechnical Test Area (Fig. 5).

Geotechnical data from seven areas in the Atlantic and Pacific Oceans and one in the Mediterranean Sea (Fig. 10) exist in the Marine Geotechnical Laboratory's data bank from continental slope and deep-sea sediment investigations reported by Richards (1961, 1962) and Richards and Hamilton (1967). For all areas all of the data for each 10-cm interval were averaged and are shown in Figs. 11-16. Because different environments were combined, it would be inappropriate to fit regression lines to these data.

The data presented in the graphs must be used with caution. While the primary purpose of this method of analysis was to portray to nongeotechnologists average properties from several environments, the low standard error of estimates and the high correlation coefficients indicate that there is validity to the predictor equations presented for the two well-defined Geotechnical Test Areas. Even for the data that were averaged together from the eight other worldwide areas, the values show less variation with respect to depth below the sea-floor than might be expected.

Consolidation

The principal process of consolidation, or gravitational compaction, is the dewatering of sediments under their own weight. The consolidation of cohesive

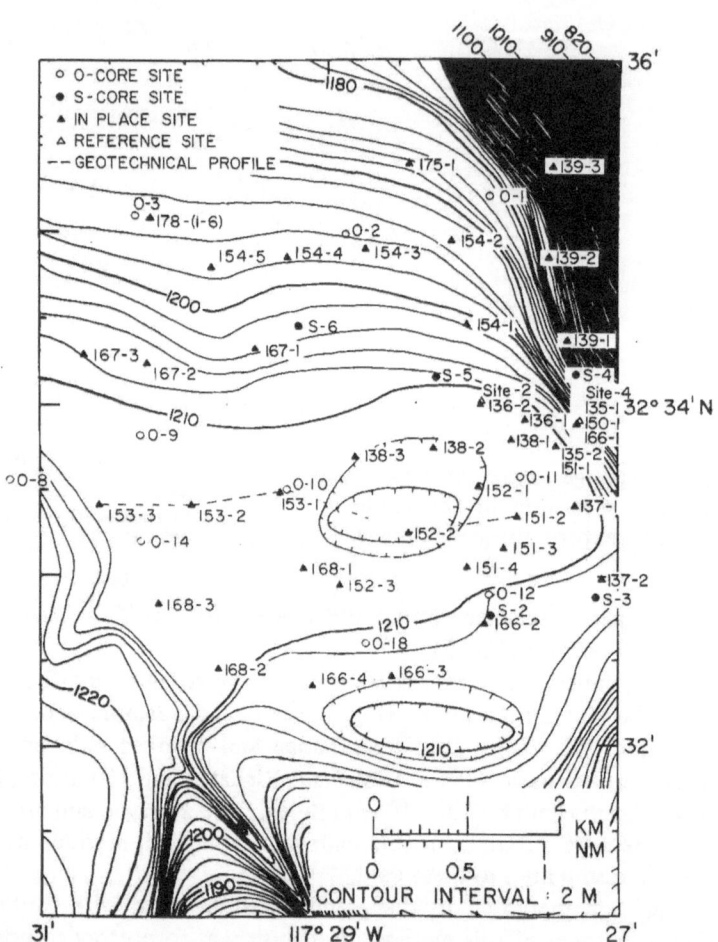

Fig. 5. Map of San Diego Trough Geotechnical Test Area showing sampling density of cores and in-place measurements. (After Carius and Richards, unpublished.)

deep-sea sediments has been summarized by Richards and Hamilton (1967), Skempton (1970), and Bryant *et al.* (1974), among others. The recent review by Riecke and Chilingarian (1973) contains relatively little information on marine sediment consolidation.

Consolidation of cohesive sediments over the first few meters of burial depth can best be studied using sedimentation–compression diagrams. Miller and Richards (1969) described how such diagrams are constructed. Richards and Dzwilewski (1974) interpreted the sedimentary history of the Golfo San Matías

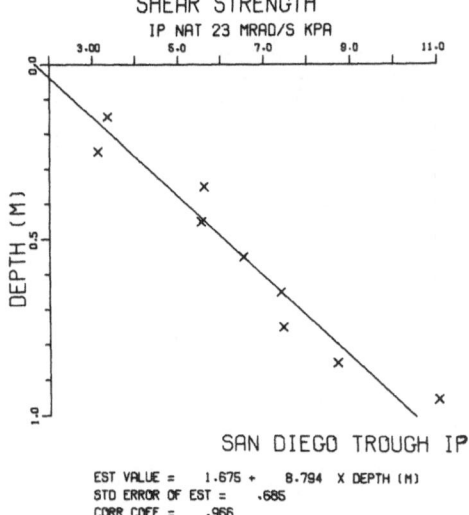

Fig. 6. Average shear strength measured in place using vanes of different sizes rotated at 23 mrad s^{-1}. See text for averaging details.

using these diagrams. For silty clay sediments, Chough and Richards (1974) show how a laboratory nuclear transmission densitometer can be used to obtain both the water content (expressed as void ratio, or *e*) and bulk density, from which the effective overburden pressure, $\bar{p}$, can be calculated. Richards *et al.*

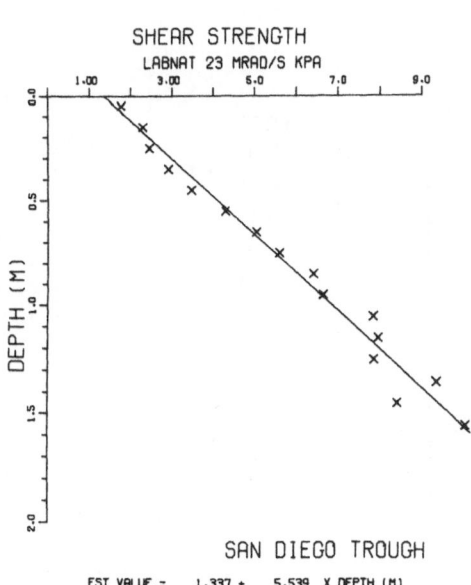

Fig. 7. Average shear strength measured in the laboratory using 12 X 25 mm vane rotated at 23 mrad s^{-1}. See text for averaging details.

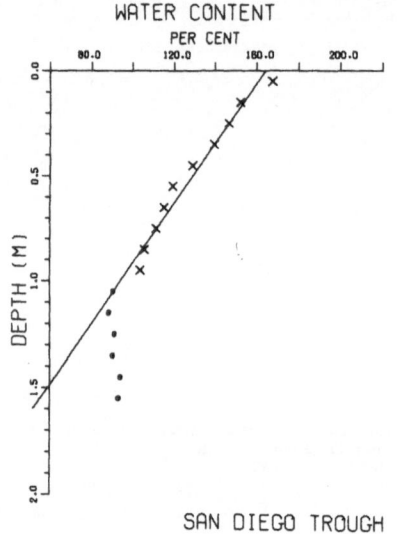

Fig. 8. Average water content, in percent dry weight, measured in the laboratory. See text for averaging details.

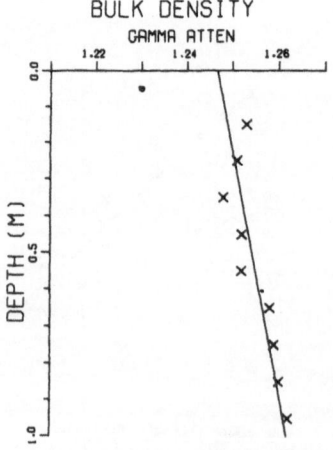

Fig. 9. Average bulk density, in Mg m^{-3}, measured by a nuclear transmission densitometer. See text for averaging details.

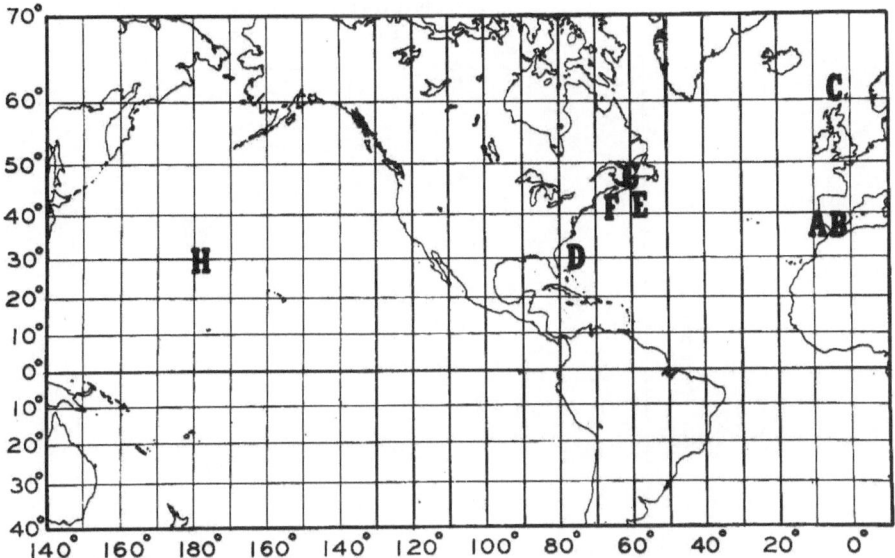

Fig. 10. Map showing where a number of Hydrographic Office cores were
collected from each lettered area. See text for details.

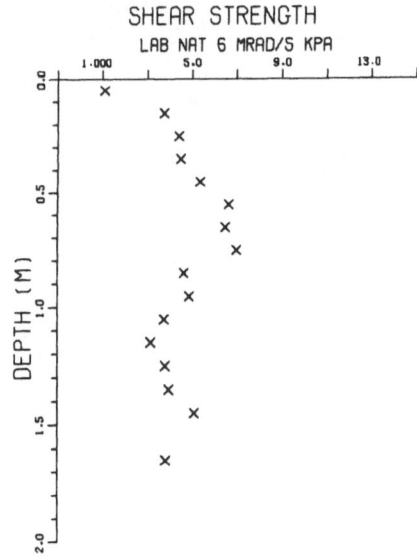

Fig. 11. Average shear strength measured on some cores using a 12 × 25
mm vane rotated at 6 mrad s^{-1}. See
text for averaging details.

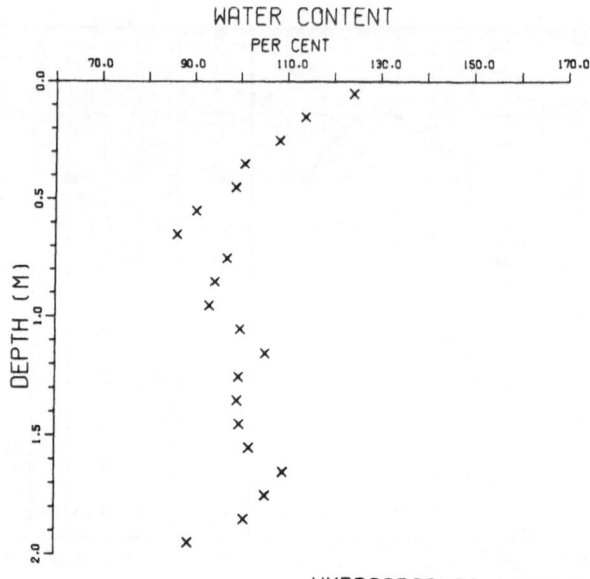

Fig. 12. Average water content, in percent by dry weight. See text for averaging details.

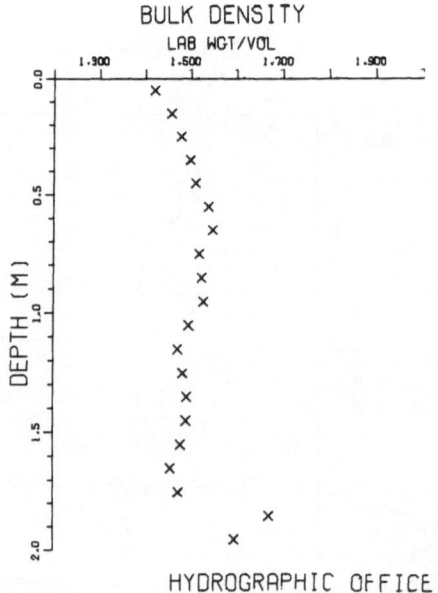

Fig. 13. Average bulk density, in Mg m^{-3}. See text for averaging details.

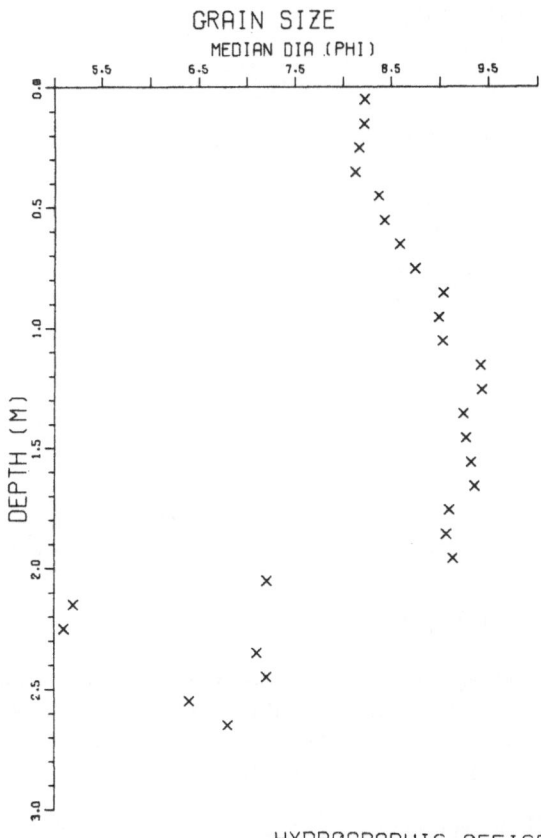

Fig. 14. Average grain median diameter in *phi* units. See text for averaging details.

(1974) present polynomial regression equations that simplify the calculation of e and $\bar{p}$. These parameters can be measured *in situ* rather precisely using nuclear transmission densitometers (Preiss, 1968; Hirst *et al.*, 1975) or less accurately using nuclear backscatter densitometers (Keller, 1965; Rose and Roney, 1971). Examples of a silty clay and a carbonate $e \log \bar{p}$ sedimentation–compression diagram are shown in Figs. 17–18.

In cohensionless sandy sediments, the influence of gravity causing dewatering may be less important than other processes. The change in the arrangement of clastic grains, or packing, toward a more dense state appears to be more commonly controlled by the action of waves (Webb and Theodor, 1972) or organisms (Webb, 1969) than by gravity.

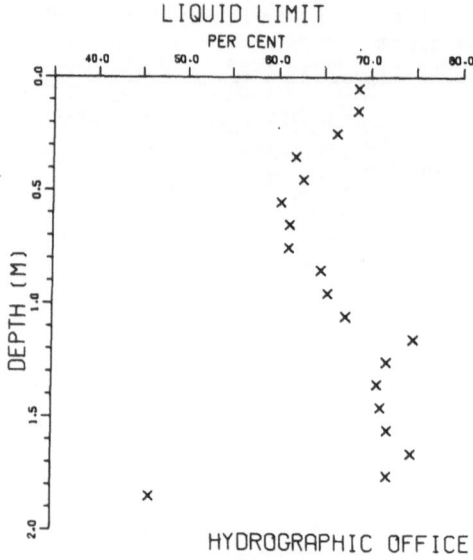

Fig. 15. Average liquid limits.
See text for averaging details.

Erosion and Transportation

Krone (this volume) has reviewed estuarine erosion and transportation. Three excellent general compendia for the continental shelf are edited by Swift *et al.* (1972), the U.S. Army Coastal Engineering Research Center (1973), and Stanley and Swift (1976). A brief summary of the effects of wave-induced scour and

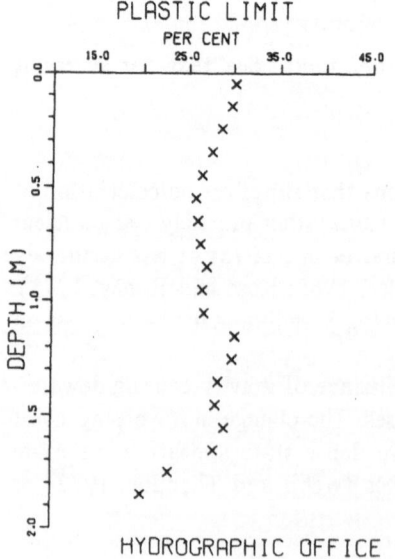

Fig. 16. Average plastic limits. See text for averaging details.

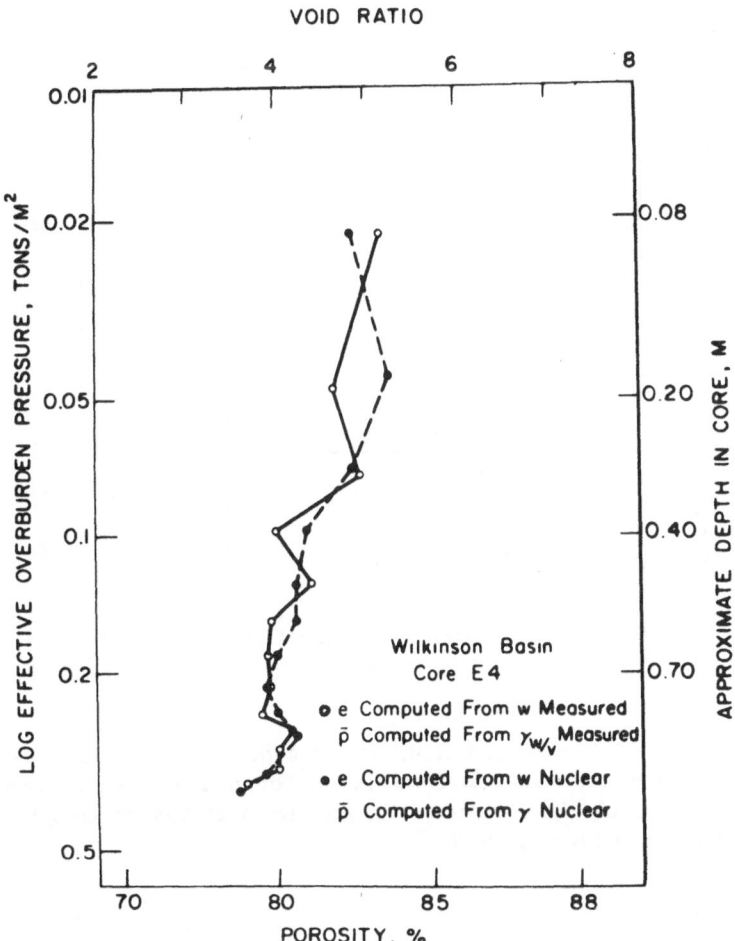

Fig. 17. Sedimentation–compression, or $e \log \bar{p}$, diagram for silty clay, cohesive sediment in the Wilkinson Basin Geotechnical Test Area (Chough and Richards, 1974).

bottom currents affecting off-shore structures has been made by Richards *et al.* (1975). A more detailed discussion of erosion and transportation has been made by Hollister and in the report of Group F in this volume.

Unfortunately, quantitative studies of bottom erosion and transportation of both cohesive and cohesionless sediments are in their infancy for the continental shelf and deeper water depths. A particularly fruitful line of investigation would be to use upside-down flumes on the sea-floor coupled with the *in situ* measurement of relevant geotechnical properties of the sediments immediately adjacent to the flume.

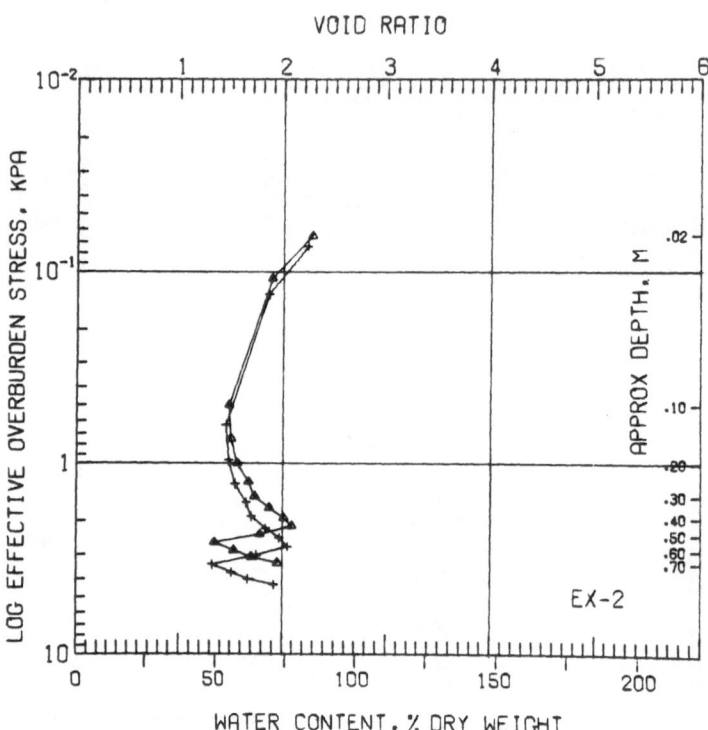

Fig. 18. Sedimentation–compression, or *e* log $\bar{p}$, diagram for a fine-grained, cohesive calcium carbonate core from Exuma Sound, Bahamas (Richards *et al.*, 1974). The double lines represent alternative methods of computing the sedimentation–compression curve.

Stability

Sea-floor stability may be considered to be affected by three types of processes: (1) microbathymetric movements caused by organisms, by gas, and probably by small environmental loads (waves, currents, etc.); (2) macrobathymetric movements or sea-slides and sea-slumps; and (3) movements associated with turbidity currents. Instability results from: (1) a change in the in-place sedimentary stress caused by artificial or sudden natural loading, by erosion at the base of a slope, and by sedimentation—particularly rapid sedimentation in the vicinity of deltas; and (2) a decrease in the shear strength of the sediment caused by disturbance or by an increase of pore-water pressure.

Selected type-1 movements are briefly summarized in the next section. Type-3 movements are not considered here because they are generally well known (e.g., Morgenstern, 1967; Kuenen, 1967). Type-2 movements appear to be further divisible into two subcategories: 2a) the slides and slumps on the upper continental slope and the continental shelf that are caused by large environmental loads (such as storm waves, perhaps the breaking of internal waves, earthquakes, etc.), with or without the presence of free gas in the sediments; and 2b) the small- to large-scale slides and slumps occurring at the base of the continental slope, on the continental rise, and elsewhere on the deep-sea floor, which are caused by various processes.

A great deal of research on type-2a processes affecting sea-floor stability is underway in the vicinity of the Mississippi River Delta, because of the instability of gas-rich sediments. Papers by Morgenstern (1967), Henkel (1970), Wright and Dunham (1972), Mitchell *et al.* (1973), Bea and Arnold (1973), Bea and Bernard (1973), Garrison (1974), and Singh (1974) review the problem of sea-sliding, particularly as the result of cyclical storm-wave-induced loading. These loads are believed to increase the excess pore-water pressure, above hydrostatic, until a condition of sedimentary instability results. Elsewhere, Bjerrum (1971) has reviewed submarine slope failures in Norwegian fjords and Monney (1967) has evaluated slope stability factors.

The geological bibliography of type-2b slides and slumping is voluminous; representative papers are by Morelock and Bryant (1966), Uchupi (1967), Stanley and Silverberg (1969), and Lewis (1971). The classic paper by Moore (1961) is now dated because it is believed that the conditions of drainage during his tests were not adequately controlled.

Bioturbation–Geotechnical Relationships

Activity of benthic invertebrates and vertebrates affects sediment in two principal ways: (1) mechanical disturbances are caused by the movement of organisms, and (2) biogeochemical changes occur as a result of sediment ingestion by organisms. Rowe (1974) has summarized the general effects of the benthic fauna on the physical properties of deep-sea sediments. Representative papers relating sedimentary mass properties and organism activity are by Harrison and Wass (1965), Rhoads (1970), and Stanley (1971). In continental shelf water depths, Muraoka (1970) has described the organism undermining of structures emplaced on the sea floor, and Cory and Pierce (1967) reported on the activity of lancelets (small fishlike chordates) that move through the top few centimeters of sandy sediments.

Benthic biologists are concerned with the "hardness" of the bottom to quantify the burrowing constraints on the infauna and the sedimentary bearing capacity affecting the mobility or trafficability of the epifauna. The latter problem is relatively simpler to analyze than the former. A simplified up-to-date summary of bearing capacity evaluations for sand and for clay that may assist marine biologists concerned with the epifauna has been made by Valent (1974). The geotechnical quantification of the constraints to infauna mobility are less easily definable. For fine-grained, cohesive, silty, and clayey sediments, it would be desirable to know the grain size (% sand >60 μm, <2 mm; % silt >2 μm, <60 μm; and % clay <2 μm); the plasticity index or range of water content over which the sediment is plastic (liquid limit–plastic limit); and the shear strength. For coarse-grained, cohesionless, sandy sediments, it would be desirable to know the true *in situ* bulk density—which is not yet obtainable by state-of-the-art equipment—and the relative density. The relative density test in the United States is designated D2049-69 (ASTM, 1974), and relative density measurements are evaluated in ASTM (1973).

Shear strength can most easily be measured on cores or *in situ* by the vane shear strength test, the fall-cone test, or by using a static cone penetrometer; references to test methods and equipment have been given by Noorany (1972), Hirst *et al.* (1972), and Richards (1974). Comparison of vane and fall-cone tests are given by Flaate (1965) and Kessler and Stiles (1968). Standard soil mechanics tests to directly measure the parameter of concern are strongly recommended. The difficulties of interpreting penetrator-type tests, such as described by Gordon (1972), tend to lead to simplified approaches that negate the conclusions drawn from the research.

An example of how the bioturbation in sedimentary basins affects the shear strength was recently presented (Richards, 1973). It was reported that the fall-cone shear strength measurements on some cores from the Oslofjorden, Norway, exhibited an unusually wide scatter of data points. Considerable heterogeneity existed in such samples when they were remolded with a finger. The nature of the heterogeneity was found to be soft soil filling tubelike structures in the relatively stiffer soil adjacent to the structures. It was hypothesized that many of the soils were highly bioturbated and that the tubes probably were the remnants of polychaete worm tubes or burrows. The subsequent discovery and identification of polychaete tubes on the surface of a core corroborated this hypothesis. Many of the tubes or burrows could be clearly observed in X-radiographs. The biogenic origin of the heterogeneity having been explained, it appeared unreasonable to average the individual fall-cone measurements made at a given depth. A preferable method was developed, which was called the concept of minimum and maximum natural shear strength. This hypothesis is that the greater the amount of bioturbation, such as worm burrows or the like, the greater will be the variation in shear strength measurements made on a volume of soil of sufficient size to show biogenically produced heterogeneity. The burrows or tubes filled with soft clay that contains a higher water content than the surrounding soil will have a small shear strength, while the soil adjacent to the tube or burrow will have a relatively larger shear strength. The greater the degree of bioturbation, the greater the range of natural shear strength values. Soils that exhibit little or no bioturbation show a narrow range of natural shear strength. The minimum measured natural shear strength, in a profile of shear strength related to depth below the soil-water interface, can be defined by drawing a line connecting the smallest values of natural shear strength. A similar line connecting the maximum natural shear strength values at each depth interval defines the profile of maximum measured natural shear strength. The magnitude of bioturbation may be noted easily by observing the range of shear strength between the minimum and maximum profiles. The measured minimum and maximum shear strength does not necessarily represent the true minimum and maximum strength of the soil. This is because the position of the three fall-cone tests at a given stratum does not necessarily coincide with the location of weakest or strongest soil in the core tube.

From this example, it is concluded that the laboratory fall-cone test for the measurement of the shear strength of cohesive sediments may be particularly useful for marine biologists. Only a small amount of material is sampled by the cone, because the recommended cone penetration is not less than 5 mm or greater than 15 mm (e.g., Richards, 1973). A suggested calibration for deep-sea sediments has been given by Moore and Richards (1962). The static-cone penetrometer would appear to be well suited for *in situ* strength determinations because it can take continuous measurements of cone resistance, which can be interpreted in terms of shear strength (e.g., Hirst *et al.*, 1972).

Summary

1. Reference is made to the basic geotechnical literature of relevance to marine biologists, chemists, geologists, and physicists concerned with the benthic boundary layer.

2. For the common types of deep-sea cohesive sediments, the approximate range of the Atterberg limits, water content, bulk density, shear strength, and sensitivity is summarized.

3. For deep-sea cohesive sediments without regard to sediment type, the approximate minimum and maximum range of geotechnical properties is tabulated.

4. The range of grain size, bulk density, and sound velocity in predominantly cohesionless sediments of the continental terrace is reported.

5. Sedimentation-compression water content-effective overburden pressure diagrams are considered to be the best method of presenting consolidation data and interpreting the effects of sedimentary dewatering with respect to burial

depth for cohesive sediments within the benthic boundary layer, as well as at greater burial depths.

6. Wave and organism change of clastic grain packing to a more dense state appears to be of greater importance than gravity alone for cohesionless sediments.

7. Average geotechnical data in the two well-known Geotechnical Test Areas off the West and East coasts of the United States are graphed with respect to burial depth and presented with linear least-squares predictor equations and other statistical data.

8. Geotechnical data for eight predominantly continental slope and deep-sea areas are averaged and plotted with respect to depth to indicate the approximate range of values in these selected environments.

9. For the shallow- and deep-ocean floor, quantitative studies of erosion and transportation have yet to be coupled with *in situ* measurements of relevant geotechnical properties for cohesive or cohesionless sediments.

10. Three types of processes appear to be dominant in controlling sea-floor stability or instability: (1) microbathymetric movements caused by organisms, gas, and small environmental loads; (2) macrobathymetric movements or sea-slides and sea-slumps; and (3) movements associated with turbidity currents. Much current geotechnical research is focused on sea-slides that are believed to be caused by cyclical storm-wave loading, particularly in gas-rich cohesive sediments that have little natural shear strength.

11. Studies of bioturbation–geotechnical relationships appear to warrant greater attention by both marine biologists and geotechnologists. Reference is made to literature that summarizes how to determine the epifauna bearing capacity on sandy and clayey sediments, and an assessment of the geotechnical quantification of constraints to infauna mobility is made for both cohesive and cohesionless sediments.

12. The laboratory fall-cone test and the *in situ* static-cone penetrometer test for the indirect measurement of shear strength of cohesive sediments may be the most useful tests for many marine biological investigations. The penetrometer has the added advantage of being able to make continuous measurements.

13. The concept of minimum and maximum shear strength has been introduced in the literature to quantify the magnitude of strength–heterogeneity in cohesive sediments produced by bioturbation. Its use is recommended.

Acknowledgments

This paper utilizes data and computer programs compiled and developed under the sponsorship of the National Oceanic and Atmospheric Administration,

Office of Sea Grant, Department of Commerce, Grants 1-35357 and NG-2-72. We thank Dr. T. J. Hirst for his critical review of this manuscript. Mr. Homa Lee kindly provided information modifying a preliminary copy of Table 1; we appreciate his helpful comments and the use of his unpublished data.

References

AGI, 1972 *Glossary of Geology*, American Geological Institute, Washington, p. 295.

ASTM, 1972, Underwater soil sampling, testing, and construction control, *American Society for Testing and Materials, Special Technical Publication 501*, 241 pp.

ASTM, 1973, Evaluation of relative density and its role in geotechnical projects involving cohesionless soils, *American Society for Testing and Materials, Special Technical Publication 523*, 510 pp.

ASTM, 1974, *1974 Annual Book of ASTM Standards. Part 19, Natural Building Stones; Soil and Rock, Peats, Mosses and Humus*, American Society for Testing and Materials, Philadelphia, 462 pp.

Bara, J. P., 1974, Geotechnical properties of marine sediments on the California continental shelf, appendix 4 of *California Undersea Aqueduct (CUA) Oceanographic Reconnaissance Study Program, Marine Soils Engineering and Foundations Study*, U.S. Bureau of Reclamation, Denver, 126 pp.

Bea, R. G., and Arnold, P., 1973, Movements and forces developed by wave-induced slides in soft clays, *Offshore Technology Conference Preprints 2:* 731–742.

Bea, R. G., and Bernard, H. A., 1973, Movements of bottom soils in the Mississippi delta offshore, *in: Offshore Louisiana Oil and Gas Fields*, Lafayette Geological Society and New Orleans Geological Society, pp. 13–28.

Bjerrum, L., 1971, Subaqueous slope failures in Norwegian fjords, *Proceedings of the 1st International Conference on Port and Ocean Engineering Under Arctic Conditions, Trondheim 1:* 1–24 (essentially the same paper in *Norwegian Geotechnical Institute, Publication 88*, 1971).

Bryant, W. R., Deflache, A. P., and Trabant, P. K., 1974, Consolidation of marine clays and carbonates, in (A. L. Inderbitzen, ed.), *Deep-Sea Sediments: Physical and Mechanical Properties*, New York, Plenum Press, pp. 209–244.

Chough, S. K., and Richards, A. F., 1974, Comparison of sedimentation–compression curves obtained by nuclear and gravimetric methods for Wilkinson Basin, Gulf of Maine, sediments, *Marine Geology 17:* 249–261.

Cory, R. L., and Pierce, E. L., 1967, Distribution and ecology of lancelets (Order Amphioxi) over the continental shelf of the Southeastern United States, *Limnology and Oceanography 12:* 650–656.

Einstein, H. A., and Wiegel, R. L., 1970, A literature review on erosion and deposition of sediments near structures in the ocean, *U.S. Navy Civil Engineering Laboratory Report, CR-70.008*, 183 pp.

Flaate, K., 1965, A statistical analysis of some methods for shear strength determination in soil mechanics, *Statens Vegvesen, Veglaboratoriet, Meddelelse 24* (reprinted in *Norwegian Geotechnical Institute, Publication 62*, 8pp.).

Fukuoka, M., and Nakase, A., 1973, Problems of soil mechanics of the ocean floor, *8th International Conference on Soil Mechanics and Foundation Engineering*, Moscow, 18 pp. (preprint).

Garcia, W. J., Jr., 1971, Literature survey and bibliography of engineering properties of marine sediments, *University of California, Hydraulic Engineering Laboratory, Technical Report HEL-2-27*, 31 pp.

Garrison, L. E., 1974, The instability of surface sediments on parts of the Mississippi delta front, *U.S. Geological Survey, Corpus Christi, Texas, Open-File Report.*

Gordon, R. B., 1972, Hardness of the sea floor in near-shore waters, *Journal of Geophysical Research 77:* 3787–4293.

Hamilton, E. L., 1974, Prediction of deep-sea sediment properties: state-of-the-art, in: (A. L. Inderbitzen, ed.), *Deep-Sea Sediments: Physical and Mechanical Properties*, Plenum Press, New York, pp. 1–43.

Hampton, L., ed., 1974, *Physics of Sound in Marine Sediments*, Plenum Press, New York, 569 pp.

Harrison, W., and Wass, M. L., 1965, Frequencies of infaunal invertebrates related to water content of Chesapeake Bay sediments, *Southeastern Geology 6:* 177–187.

Harrison, W. M., Lynch, M. P., and Altschaeffl, A. G., 1964, Sediments of lower Chesapeake Bay, with emphasis on mass properties, *Journal of Sedimentary Petrology 34:* 727–755.

Henkel, D. J., 1970, The role of waves in causing submarine landslides, *Géotechnique 20:* 75–80.

Herrmann, H. G., Raecke, D. A., and Albertson, N. D., 1972, Selection of practical seafloor foundation systems, *U.S. Naval Civil Engineering Laboratory, Technical Report R. 761*, 113 pp.

Hirst, T. J., Richards, A. F., and Inderbitzen, A. L., 1972, A static cone penetrometer for ocean sediments, *in: Underwater Soil Sampling, Testing, and Construction Control, American Society for Testing and Materials, Special Technical Publication 501*, pp. 69–80.

Hirst, T. J., Burton, B. S., Perlow, M., Jr., Richards, A. F., and Van Sciver, W. J., 1975, Improved *in situ* gamma-ray transmission densitometer for marine sediments, *Ocean Engineering 3:* 17–27.

Horn, D. R., Delach, M. N., and Horn, B. M., 1974, Physical properties of sedimentary provinces, North Pacific and North Atlantic Oceans, *in:* (A. L. Inderbitzen, ed.), *Deep-Sea Sediments: Physical and Mechanical Properties*, Plenum Press, New York, pp. 417–441.

Inderbitzen, A. L., ed., 1974, *Deep-Sea Sediments: Physical and Mechanical Properties*, Plenum Press, New York, 497 pp.

Kaplan, I. R., ed., 1974, *Natural Gases in Marine Sediments*, Plenum Press, New York, 324 pp.

Keller, G. H., 1965, Nuclear density probe for in place measurement in deep-sea sediments, *Transactions of the Joint Conference and Exhibit, Marine Technical Society and the American Society of Limnology and Oceanography 1:* 363–372.

Keller, G. H., 1969, Engineering properties of some sea-floor deposits, *Journal of the Soil Mechanics and Foundations Division Proceedings of the American Soceity of Civil Engineers, 95* (SM6): 1379–1392.

Keller, G. H., 1974, Marine Geotechnical properties: interrelationships and relationships to depth of burial, *in:* (A. L. Inderbitzen, ed.), *Deep-Sea Sediments: Physical and Mechanical Properties*, Plenum Press, New York, pp. 77–100.

Keller, G. H., and Bennett, R. H., 1968, Mass physical properties of submarine sediments in the Atlantic and Pacific basins, *Proc. 23rd International Geological Congress, Prague 8:* 33–50.

Kessler, R. S., and Stiles, N. T., 1968, Comparison of shear strength measurements with the laboratory vane shear and fall-cone devices, *U.S. Naval Oceanographic Office, Informal Report 68-75* (unpublished), 18 pp.

Kuenen, P. H., 1967, Emplacement of flysch-type and sand beds, *Sedimentology 9:* 203–243.

Lewis, K. B., 1971, Slumping on a continental slope inclined at 1°–4°, *Sedimentology 16:* 97–110.

Ling, S. C., 1972, State-of-the-art of marine soil mechanics and foundation engineering, *U.S. Army Engineers Waterways Experiment Station, Technical Report S-72-11*, 176 pp.

Miller, D. G., Jr., and Richards, A. F., 1969, Consolidation and sedimentation–compression studies on a calcareous core, Exuma Sound, Bahamas, *Sedimentology 12:* 301–316.

Mitchell, J. K., 1963, Engineering properties and problems of the San Francisco Bay mud, in: Short contributions to California geology, *California Division of Mines and Geology, Special Report 82*, pp. 25–32.

Mitchell, R. J., Tsui, K. K., and Sangrey, D. A., 1973, Failure of submarine slopes under wave action, *Proceedings of the 13th Coastal Engineering Conference*, Vol. 2, Americal Society of Civil Engineers, New York, pp. 1515–1541.

Monney, N. T., 1967, Slope stability factors to consider in offshore drilling operations, *American Society of Mechanical Engineers, Paper 67-UNT-3*, 12 pp.

Moore, D. G., 1961, Submarine slumps, *Journal of Sedimentary Petrology 31:* 343–357.

Moore, D. G., and Richards, A. F., 1962, Conversion of "relative shear strength" measurements by Arrhenius on East Pacific deep-sea cores to conventional units of shear strength, *Géotechnique 12:* 55–59.

Morelock, J., and Bryant, W. R., 1966, Physical properties and stability of continental slope deposits, northwest Gulf of Mexico, *Transactions of the Gulf Coast Association of Geological Societies 16:* 279–295.

Morgenstern, N. R., 1967, Submarine slumping and the initiation of turbidity current, *in:* (A. F. Richards, ed.), *Marine Geotechnique*, University of Illinois Press, Urbana, pp. 189–220.

Muraoka, J. S., 1970, Animal undermining of naval seafloor installations, *U.S. Naval Civil Engineering Laboratory, Technical Note N-1124*, 18 pp.

Noorany, I., 1972, Underwater soil sampling and testing–a state-of-the-art review, in: *Underwater Soil Sampling, Testing, and Construction Control, American Society for Testing and Materials, Special Technical Publication 501:* 3–41.

Noorany, I., and Gizienski, S. F., 1970, Engineering properties of submarine soils: state-of-the-art review, *Journal of the Soil Mechanics and Foundations Division, Proceedings of the American Society of Civil Engineers 96* (SM5): 1735–1762.

Preiss, K., 1968, *In situ* measurement of marine sediment bulk density by gamma radiation, *Deep-sea Research 15:* 637–641.

Proceedings, 1971, *Proceedings of the International Symposium on the Engineering Properties of Sea-floor Soils and Their Geophysical Identification*, Seattle, Washington, July 25, 1971 (Sponsored by UNESCO, National Science Foundation, University of Washington), 374 pp.

Rhoads, D. C., 1970, Mass properties, stability, and ecology of marine muds related to burrowing activity, in: (T. P. Crimes, and J. C. Harper, eds.), *Trace Fossils, Geological Journal, Special Publication No. 3*, pp. 391–406.

Richards, A. F., 1961, Investigations of deep-sea sediment cores. 1. Shear strength, bearing capacity, and consolidation, *U.S. Navy Hydrographic Office, Technical Report 63*, 70 pp.

Richards, A. F., 1962, Investigations of deep-sea sediment cores. II. Mass physical properties, *U. S. Navy Hydrographic Office, Technical Report 106*, 146 pp.

Richards, A. F., ed., 1967, *Marine Geotechnique*, University of Illinois Press, Urbana, 327 pp.

Richards, A. F., 1973, Geotechnical Properties of submarine soils, Oslofjorden and vicinity, Norway, *Norwegian Geotechnical Institute Technical Report*, No. *13*, 107 pp.

Richards, A. F., 1974, Standardization of marine geotechnics symbols, definitions, units, and test procedures, *in:* (A. L. Inderbitzen, ed.), *Deep-Sea Sediments: Physical and Mechanical Properties*, Plenum Press, New York, pp. 271–292.

Richards, A. F., and Dzwilewski, P. T., 1974, Consolidation properties of Wilkinson Basin soils, *Journal of the Geotechnical Engineering Division, Proceedings of the American Society of Civil Engineers 100* (GT10): 1175–1179.

Richards, A. F., and Hamilton, E. L, 1967, Investigations of deep-sea sediment cores. III. Consolidation, in: (A. F. Richards, ed.), *Marine Geotechnique*, University of Illinois Press, Urbana, pp. 93–117.

Richards, A. F., and Parker, H. W., 1968, Surface coring for shear strength measurements, in: *Civil Engineering in the Oceans*, American Society of Civil Engineers, New York, pp. 445–489.

Richards, A. F., Hirst, T. J., and Parks, J. M., 1974, Bulk density–water content relationships in marine silts and clays, *Journal of Sedimentary Petrology 44:* 1004–1009.

Richards, A. F., Palmer, H. D., and Perlow, M., Jr., 1975, Review of continental shelf marine geotechnics: distribution of soils, measurement of properties, and environmental hazards, *Marine Geotechnology 1:* 33–67.

Rieke, H. H., and Chilingarian, G. V., 1973, *Compaction of Argillaceous Sediments*, Elsevier Scientific Publishing Company, Amsterdam, 424 pp.

Rose, V. C., and Roney, J. R., 1971, A nuclear gage for in-place measurement of sediment density, *Offshore Technology Conference Preprints 1:* 43–52.

Rowe, G. T., 1974, The effects of the benthic fauna on the physical properties of deep-sea sediments, *in:* (A. L. Inderbitzen, ed.), *Deep-Sea Sediments: Physical and Mechanical Properties*, Plenum Press, New York, pp. 381–400.

Schmid, W. E., 1969, Penetration of objects into the ocean bottom (the state-of-the-art), *U. S. Naval Civil Engineering Laboratory*, *Report CR-69.030*, 179 pp.

Scott, R. F., and Zuckerman, K. A., 1970, Study of slope instability in the ocean floor, *U.S. Naval Civil Engineering Laboratory*, *Report CR-70.007*, 72 pp.

Singh, H., 1974, The effects of waves on ocean sediments, *Danes and Moore Engineering Bulletin 44:* 11–21.

Skempton, A. W., 1970, The consolidation of clays by gravitational compaction, *Quarterly Journal of the Geological Society of London 125:* 373–411.

Stanley, D. J., 1971, Bioturbation and sediment failure in some submarine canyons, *Vie et Milieu, 3rd European Symposium of Marine Biology*, Supplement 22: 541–555.

Stanley, D. J., and Silverberg, N., 1969, Recent slumping on the continental slope off Stable Island Bank, southeast Canada, *Earth and Planetary Science Letters 6:* 123–133.

Stanley, D. J., and Swift, D. J. P., eds., 1976, *The New Concepts of Continental Margin Sedimentation*, Wiley-Interscience, New York.

Swift, D. J. P., Duane, D. B., and Pilkey, O. H., eds., 1972, *Shelf Sediment Transport: Process and Pattern*, Dowden, Hutchinson & Ross, Stroudsburg, Pennsylvania, 656 pp.

Tirey, G. B., 1972, Recent trends in underwater soil sampling methods, *in: Underwater Soil Sampling, Testing, and Construction Control*, American Society for Testing and Materials, Special Technical Publication 501, pp. 42–54.

Uchupi, E., 1967, Slumping on the continental margin southeast of Long Island, New York, *Deep-Sea Research 14:* 635–639.

U.S. Army Coastal Engineering Research Center, 1973, *Shore Protection Manual*, Vol. 1–3, U.S. Government Printing Office, Washington (variant paging).

Valent, P. J., 1974, Deep-sea foundation and anchor engineering, in: (A. L. Inderbitzen, ed.), *Deep-sea Sediments Physical and Mechanical Properties*, Plenum Press, New York, pp. 245–269.

Vesic, A. S., 1969, Breakout resistance of objects embedded in ocean bottom, *U.S. Naval Civil Engineering Laboratory, Report CR 69.031*, 44 pp. (also, *Journal of the Soil Mechanics and Foundations Division, Proceedings of the American Society of Civil Engineers 97*, (SM10): 1415–1429).

Webb, J. E., 1969, Biologically significant properties of submerged marine sands, *Proceedings of the Royal Society of London B174:* 355–402.

Webb, J. E., and Theodor, J. L., 1972, Wave-induced circulation in submerged sands, *Journal of the Marine Biological Association of the United Kingdom 52:* 903–914.

Wiendieck, K. W., 1970, A Preliminary study of seafloor trafficability and its prediction, *U.S. Army Engineers Waterways Experiment Station, Technical Report M-70-8*, 113 pp.

Wilson, B. W., 1969, Earthquake occurrence and effects in ocean areas, *U.S. Naval Civil Engineering Laboratory, Technical Report CR-69.027*, 188 pp.

Wright, S. G., and Dunham, R. S., 1972, Bottom stability under wave-induced loading, *Offshore Technology Conference Preprints 1:* 853–862.

9
Flow Phenomena in the Benthic Boundary Layer and Bed Forms beneath Deep-Current Systems

C. D. HOLLISTER, J. B. SOUTHARD, R. D. FLOOD, and P. F. LONSDALE

The discovery of erosional and depositional bed forms in the deep ocean provides a means for estimating some properties of the benthic boundary layer and also raises questions about secondary circulation near the bed. Previously unexplained hyperbolic echoes are now seen (using a deeply towed instrument package) to be caused by furrows on the sea bed. Small furrows are 1-4 m wide, ¾-2 m deep, and spaced 20-125 m apart; large furrows are 50-150 m wide, 20 m deep, and spaced 50-200 m apart. Analogy with shallow-water furrows and bed forms produced in deserts and in laboratory experiments suggest an origin for the furrows through interaction of large-scale helical vortices developed in the boundary layer with the sediments. The thickness of the bottom-mixed water layer should be about one-half the furrow spacing. Measurements show that mixed layer thickness (20-100 m) are of the correct order of magnitude.

Introduction

Although the interrelationships of current systems and bed forms in shallow seas have been quite well known for some years, it is only recently that similar

C. D. HOLLISTER • Woods Hole Oceanographic Institution, Woods Hole, Massachusetts 02543. J. B. SOUTHARD • Department of Earth and Planetary Sciences, Massachusetts Institute of Technology, Cambridge, Massachusetts 02139. R. D. FLOOD • Woods Hole Oceanographic Institution and Department of Earth and Planetary Sciences, M.I.T. P. F. LONSDALE • Marine Physical Laboratory, Scripps Institution of Oceanography, La Jolla, California 92037.

bed forms have been discovered in the deep oceans. The light shed on boundary
layer processes by shallow-water studies makes the discovery of deep-ocean bed
forms of importance because the current systems are much less well known there.
By analogy with forms and flow patterns found in the atmosphere and in shallow
water, as well as in laboratory studies, we infer some properties of deep-sea
boundary layers in this paper.

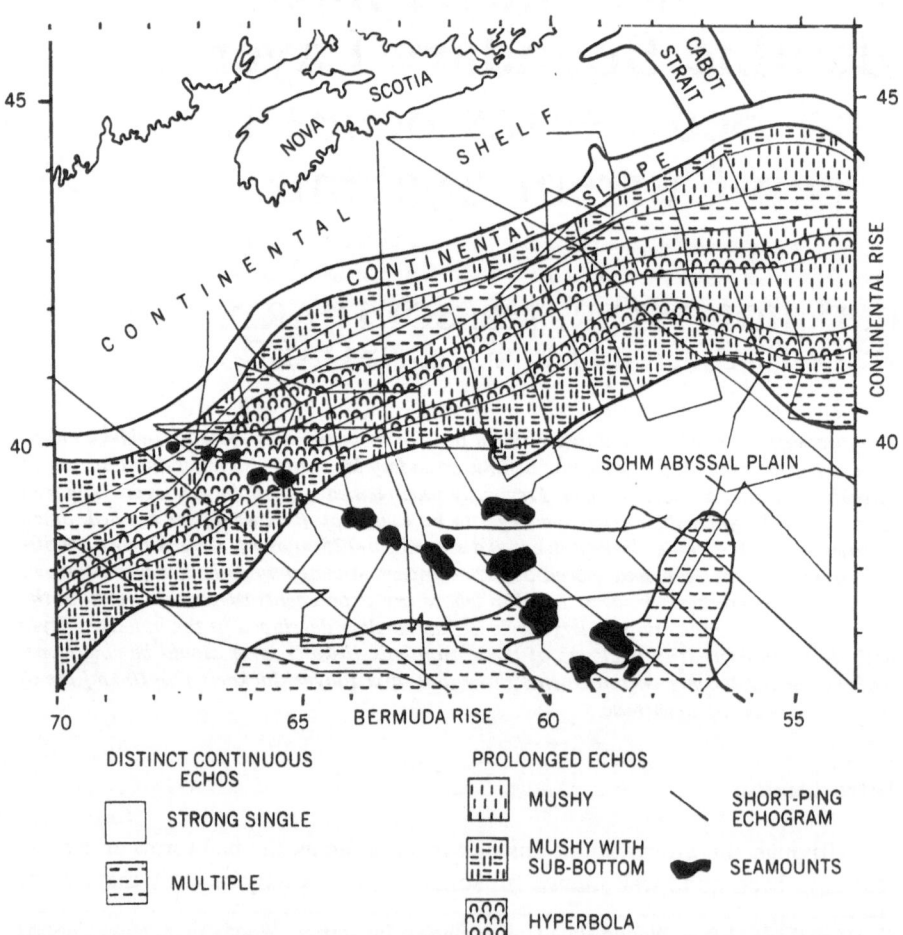

Fig. 1. Echogram character on the continental margin and Sohm Abyssal Plain
off Nova Scotia. Zones of similar echo reflectivity generally follow bathymetric
contours and can easily be correlated from profile to profile. The zones of
hyperbolae and "mushy" echoes also run parallel to near-bottom isotherms,
suggesting a relationship between echo character and water masses (From
Hollister and Heezen, 1972).

Echograms taken from research vessels have been used to provide semiquantitative information on the nature of the microtopography and the sediment stratification present on the sea floor and have been used in various attempts at understanding erosional and depositional processes acting within the benthic boundary layer of the deep sea (Heezen *et al.*, 1959; Heezen *et al.*, 1966; Schneider *et al.*, 1967, Heezen and Johnson, 1969; Hollister and Heezen, 1972; Damuth, 1975).

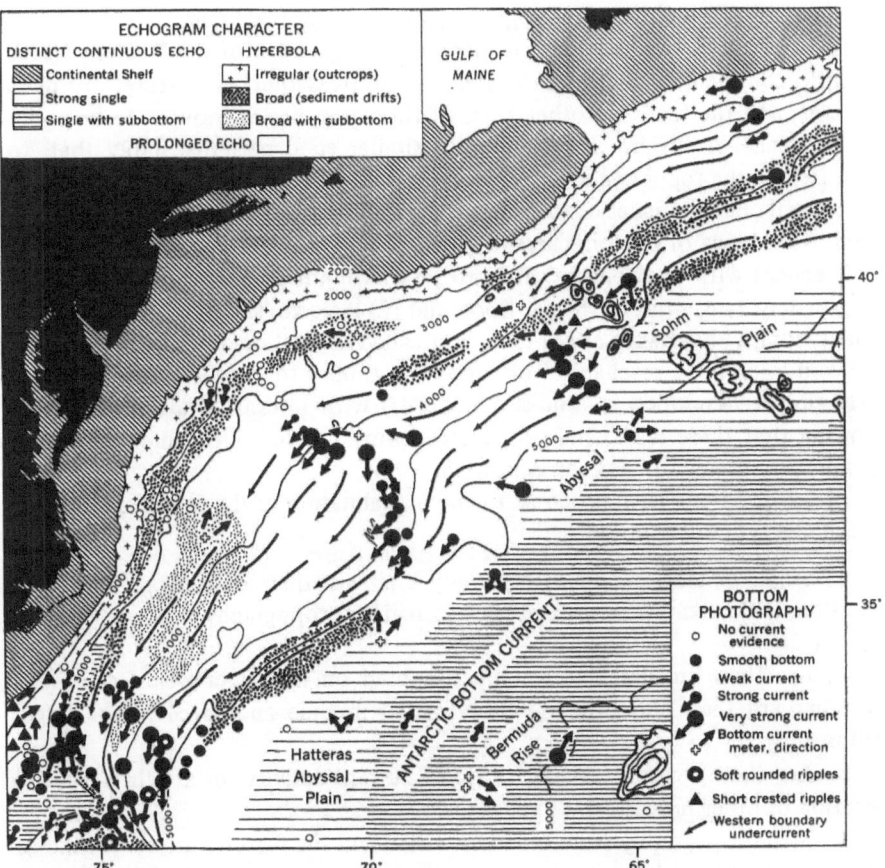

Fig. 2. The echogram character off the East Coast of North America can be related to the inferred and measured bottom currents of the Western Boundary Undercurrent (Schneider *et al.*, 1967). Zones of prolonged and hyperbolic echoes on the continental rise parallel the regional contours. Above 3500 m a tranquil current-free bottom is usually seen, but below 3500 m swift contour-following currents appear to be transporting sediment towards the South (contours in meters).

The deeper portion of the continental margin—especially the continental rise—is one region of the sea floor where such a mapping technique has proved especially useful in defining areas that are active sedimentologically. The classification of the echo types recorded on the continental rise is generally based on three criteria: (1) the coherence of the echo return, i.e., whether the echo is made up of one or several "mushy" (indistinct) returns versus one or more sharp returns; (2) the presence of topography below the limit of resolution of surface-ship echo sounding, as evidenced by the presence of hyperbolic echoes; and (3) the wavelength, amplitude, and regularity of the larger-scale relief.

The fact that the echogram character of large areas of the Western North Atlantic continental margin corresponds with areas that are affected by deep currents, and the fact that other parts of the ocean basins thought to be under the influence of current activity exhibit similar echo patterns, imply that, to some extent, the topographic forms developed may be related to current activity.

Several characteristics of the microtopography, in particular, have often been cited as evidence of bottom-current activity. These characteristics are (1) hyperbolic echoes with wavelengths of up to several hundred meters and vertices approximately tangential to the sea floor, and (2) large abyssal mud waves (termed "giant ripples" by Ewing *et al.*, 1971, but called abyssal mud waves in this paper to avoid implying a mechanism of formation similar to the more familiar transverse ripples developed in sand-size material) with internal stratification often showing that these waves have migrated normal to their crest direction as the sedimentary sequence has been deposited.

Many questions have arisen as to the actual nature of the topography which gives these types of echo traces:

1. What, in fact, are these features? What are their dimensions, degree of lineation, and orientation with respect to regional topography and to measured currents?

2. Are there smaller features superimposed on the larger-scale features? If so, what are they and how are they related in time and space to the larger features?

3. Are these features related at all to current activity, or are they a manifestation of other processes?

4. If the topography is current controlled and the features are some kind of current-generated bed form, are these bed forms presently active? If they are, what are the dynamics of their formation? If not, how recently have they been active and what can they tell us about the flow conditions at the time they were active?

Previous investigations into the nature of the topographic forms present on the continental rise and on outer ridges have been severely hampered by the sampling techniques employed. These difficulties have been twofold. First, the features to be studied are larger than the field of view of a normal bottom photograph (several square meters) and the relationship between successive bottom photographs is not usually known. Second, these forms are at or below the lower limit of detection by surface 3.5- and 12-kHz profilers. Often only a hyperbolic echo is recorded at the surface, or, if the features are larger, side-echoes tend to obscure the true form of the features. For the larger features (greater than several hundred meters) the navigational accuracy has not usually been good enough and the sounding lines have not been close enough to establish an unambiguous trend.

Other problems have been encountered in the investigation of these features. First, since even the general form of the feature is not known, the sampling strategy has not been adequate and has apparently missed the important features altogether, or, if something peculiar is seen, it is not understood and not reported. Second, an investigation into the formation of these features, if they are current controlled, requires a study of the nature and structure of the benthic boundary layer as well as a study of the sediments which make up these bed forms.

Most of the observational difficulties listed above have been overcome by the development of a deeply towed instrument package by the Marine Physical Laboratory of the Scripps Institution of Oceanography (Spiess and Mudie, 1970; Spiess and Tyce, 1974). This instrument package is towed at a height of 10–100 m above the sea floor, and with its side-scan sonar, narrow-beam echo sounder, 4-kHz subbottom profiler, stereo photography, continuous temperature measurements, and transponder navigation it can resolve sea-floor features, such as bed forms, with relatively small dimensions and can give information as to the temperature structure of the overlying water column.

Near-Bottom Investigations of Microtopography on the Blake–Bahama Outer Ridge

Previous investigations of the microtopography of the Blake–Bahama Outer Ridge have been conducted by Clay and Rona (1964) and Bryan and Markl (1966). Clay and Rona (1964) noted the presence of hyperbolic echo returns in the Blake–Bahama Basin and related them to north–south corrugations which they thought could be sand waves. Later, Bryan and Markl (1966) mapped the distribution of different echo types and determined, on the basis of the shape of the hyperbolic echoes, that these corrugations were generally parallel to the regional contours.

The Scripps deeply towed instrument package (Spiess and Mudie, 1970) was used in 1973 (Hollister et al., 1974a) to determine the nature of intermediate-

scale bed forms responsible for the hyperbolic echoes on the Bahama Oute
Ridge, the secondary ridge of the Blake–Bahama Outer Ridge complex, and to
study the abrupt acoustic–sedimentologic contact between the Blake–Bahama
Abyssal Plain and the Bahama Outer Ridge. Four giant piston cores (Holliste
et al., 1973), 25 bottom-camera stations, 3 boomerang cores, and 7 current
meter records of up to 153 hours duration were obtained in the two areas studied
on the Bahama Outer Ridge (Fig. 3). Area 1 is in a region of large (2-km wave
length) mud waves near the crest of the secondary ridge, while area 2 is a
the contact between the Bahama Outer Ridge and the Blake–Bahama Abyssa
Plain.

Previous studies of the Blake–Bahama Outer Ridge region have generally
supported the concept that bottom currents have transported mud into the
region and deposited it in the form of large lens-shaped ridges and swells (Heezer

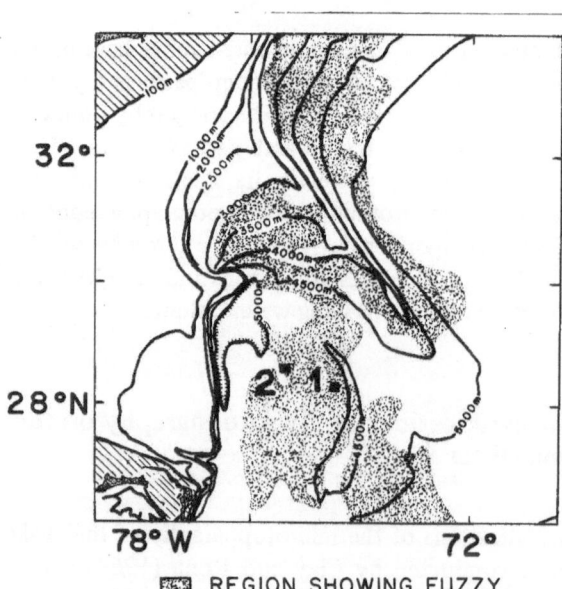

■■■ REGION SHOWING FUZZY
AND HYPERBOLIC ECHOES

Fig. 3. Index map of areas surveyed with the MPL deep tow on the Blake-
Bahama Outer Ridge. Area 1 is in a region of large (2 km wavelength) mud waves
near the crest of the secondary ridge. Area 2 is at the contact between the outer
ridge and the Blake–Bahama abyssal plain. Distribution of fuzzy and hyperbolics
from Bryan and Markl (1966), and echo-sounding records of the Woods Hole
Oceanographic Institution.

and Hollister, 1964; Heezen *et al.*, 1966). Deep-sea drilling results have confirmed that at least the upper kilometer of the Blake–Bahama Outer Ridge is composed of hemipelagic silty clay transported south from sources north of Cape Hatteras (Hollister, Ewing *et al.*, 1971). The entire suite of ridges and rises in the Blake–Bahama Outer Ridge area appears to have been formed since approximately the middle Tertiary, when the opening of the North Atlantic to sources of dense bottom water from the Norwegian Sea allowed the initiation of a vigorous bottom circulation, thus causing wholesale migration and construction of sediment drifts along the entire western side of the North Atlantic (Hollister and Heezen, 1972; Berggren and Hollister, 1974).

The preliminary results of our investigation have been reported (Flood *et al.*, 1974; Hollister *et al.*, 1974a) and are summarized below.

Small Furrows on Large Mud Waves (Area 1)

Transponder-navigated bathymetric profiles were collected in the area of large mud waves with the deeply towed instrument package. The large waves trend N10°E, and have amplitudes of 20–60 meters and wavelengths of about 2 km (Fig. 4). Available bathymetric data (Bryan and Markl, 1966; Markl *et al.*, 1970) indicate that the large waves are not parallel to the regional contours, which trend N25°W in this area, but are at an angle of approximately 35° to this trend. High-resolution subbottom profiles indicate that the mud waves have migrated to the east (Fig. 5A).

Superimposed on the large mud waves are long and remarkably straight furrows (Figs. 4, 6, 7A and 7B, and Fig. 8). Measurements on a few stereo photographs indicate that the furrows are from 1 to 4 m wide and from $\frac{3}{4}$ to 2 m deep (Fig. 8). The spacing of the furrows is 20–125 m. These furrows appear to be responsible for the very fine scale surface hyperbolic echoes seen on 3.5-kHz and 12-kHz records (Fig. 5A). The furrows are steep-sided and have flat floors (Fig. 7A; Fig. 8) with an occasional median ridge present.

The furrows trend N25°W, parallel to the trend of the regional contours but at a 35° angle to the strike of the large waves. The furrows are parallel to the measured current directions and generally join in tuning-fork junctions which open into the measured current directions (Figs. 4 and 6). Some of the furrows can be traced for at least 5 km.

Current meters located 20 m off the bottom recorded maximum velocities of 9–10 cm sec^{-1} toward 325–335° for approximately three quarters of the 102–108 h of record. This is in general agreement with the currents calculated by Amos *et al.* (1971). Most of the furrows photographed seem to be presently active or possibly only maintained; their walls are sharp and are sparsely tracked

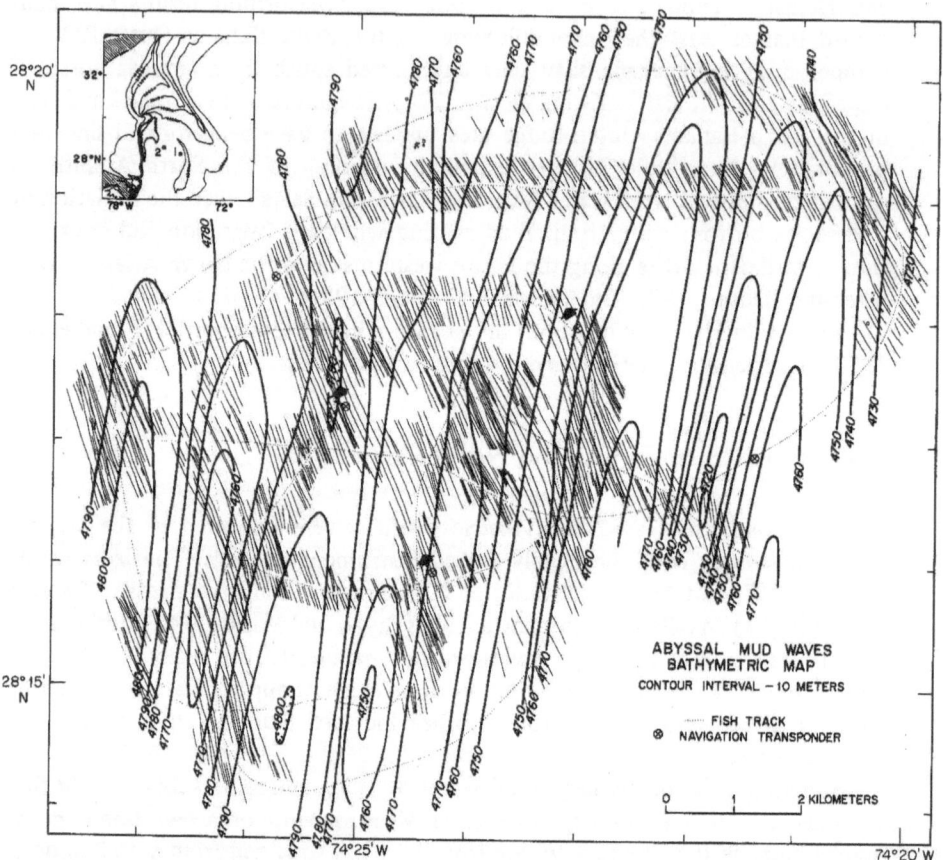

Fig. 4. Bathymetric map of abyssal mud waves area (area 1) with furrows plotted from side-scan sonar records. Arrows indicate direction of maximum current velocities measured by current meters located at the navigation transponders.

by bottom animals, and the associated ripples are well developed (Figs. 7A and 7B). However, sedimentary horizons cored at depths of 36-37 cm containing iron-rich layers, dated by [14]C- dates at 11,000 ybp and similar to those reported by McGeary and Damuth (1973), are not seen outcropping on furrow walls. This implies that these furrows are either depositional in origin, or, if erosional, they have been partially filled in. Only detailed sediment sampling can hope to solve this problem. Occasional clumps of seaweed (possibly *Sargassum*) can be seen lying in the furrows.

Large Furrows near the Contact between Abyssal Plain and Outer Ridge (Area 2)

Surface-ship profiles across this important boundary show an abrupt transition between very pronounced hyperbolic echoes on the outer ridge and a strong reverberant coherent echo on the abyssal plain (Fig. 5B). The deep-tow profiles show that this contact is abrupt on a horizontal scale of several meters (Fig. 5C). The surface hyperbolic echoes are caused by a system of steep-sided erosional furrows (Figs. 5D, 5E, and Fig. 9). These features, much larger (20 m deep by 50–150 m wide and spaced 50–200 m apart) than the furrows seen superimposed on the mud waves, are clearly erosional, since deeper reflectors, seen on both surface and deep-tow records, continue unbroken under these features (Figs. 5B, 5C, 5E) or outcrop on furrow walls (Fig. 7C). Side-scan sonar records reveal branching, or "tuning-fork," patterns similar to those of the smaller mudwave furrows, with junctions opening into the currents (Fig. 9).

Photographs of the sides of these furrows show outcropping, more-resistant layers (Fig. 7C). Others show active erosion of material from beneath the more-resistant layers and downslope sliding of mud lumps. A few unidentifiable animals can be seen living on and in the precipitous side walls and may contribute to the process of erosion.

Current meters located 20 m above the bottom in the area of hyperbolic echoes recorded maximum current velocities of 8 cm sec^{-1} and average velocities of 5 cm sec^{-1} over a period of 153 h, directed steadily toward the North. Maximum velocities of 4 cm sec^{-1} (average velocities of 2 cm sec^{-1} over a period of 144 h) toward the north were recorded over the adjacent abyssal plain. These currents are also in rough agreement with the geostropic currents calculated by Amos *et al.* (1971).

Giant piston cores taken on the Bahama Abyssal Plain and in the area of hyperbolic echoes to the east show that the furrows are eroded into Pleistocene calcareous oozes and clays, similar to the sediment in area 1. The underlying sediment is a highly calcareous ooze with fragmental carbonate debris and graded bedding. The abyssal-plain sediments can be traced beneath the area of hyperbolic echoes to the east (Figs. 5B, 5C) and may determine the depth to which the furrows can be eroded.

On the abyssal plain, within several kilometers of the contact between the abyssal plain and outer ridge, smaller ridges with approximately 2 m spacings were photographed (Fig. 7D) and recorded on side-scan sonar records. These ridges are developed in hemipelagic muds overlying more typical abyssal-plain sediments. These bed forms are identical to longitudinal triangular ripples recorded in other areas of the ocean and are generally found to be parallel to the current directions. Here they are at an angle of 70° to the measured current directions. The orientation of these ripples changes as one moves away from the outer ridge, where they trend E–W to SE–NW. The significance of these features is not well understood.

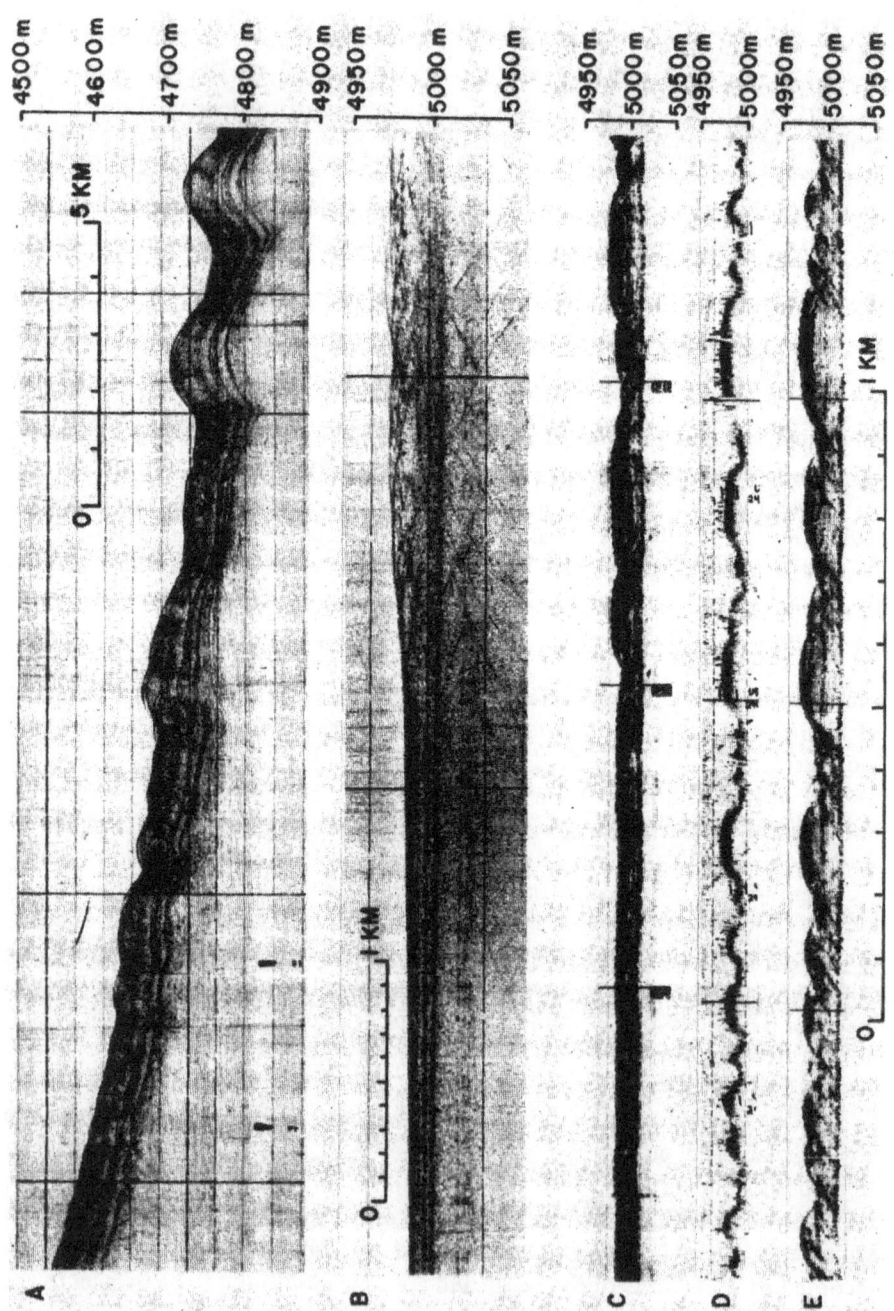

Fig. 5. A. Surface-ship 3.5-kHz record starting in area 1 and continuing upslope toward the crest of the Bahama Outer Ridge. The wave farthest to the right corresponds to the wave at 74°21.8'W, 28°16.9'N.

B. Surface-ship 3.5-kHz record of the Bahama Outer Ridge–Abyssal Plain contact.

C. Near-bottom 4-kHz record of the same contact shown in B demonstrating the abruptness of the contact. (Slight undulations in the record are caused by variations in fish elevation. The deeper reflectors are flat; see profile B.)

D. Near-bottom narrow-beam echo-sounding (40-kHz) record of furrows responsible for hyperbolae seen on profile B in area 2.

E. Near-bottom 4-kHz record taken simultaneously with profile D. Note edge effects caused by wide-beam angle.

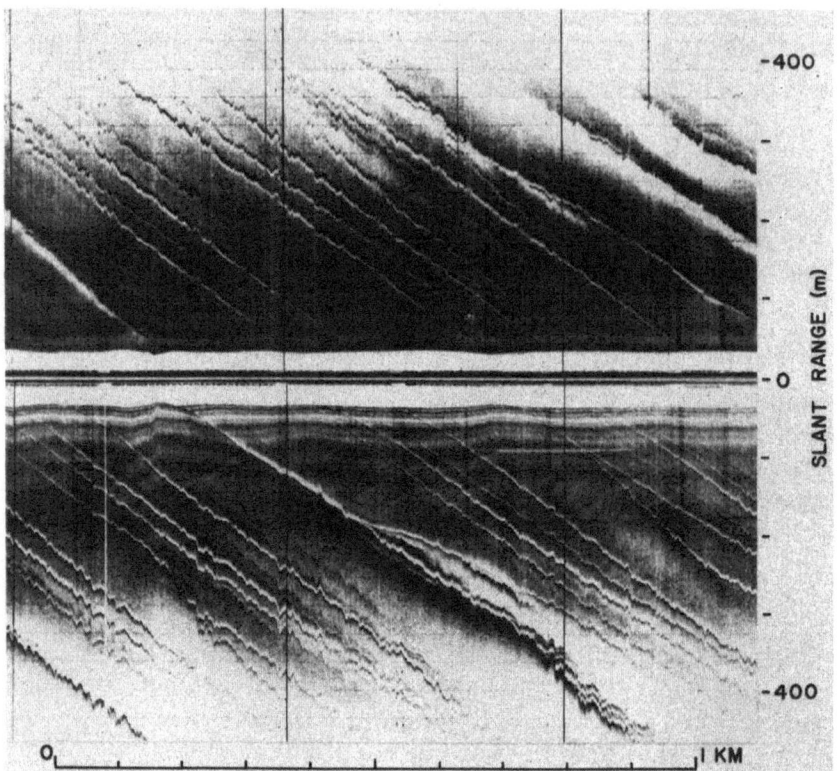

Fig. 6. Side-scan sonar record of the furrows in area 1. Line down the center of the record represents the fish track. Measured currents are along the furrows from lower right to upper left. Note "tuning fork" junctions opening into the current.

Origin of Abyssal Furrows

Bed forms, either erosional or depositional, oriented parallel to the flow, are widespread in many natural sedimentary environments, both subaerial and sub-aqueous, and have been produced in laboratory experiments involving a variety of flow conditions and sediment sizes. Such bed forms are in many cases partly or wholly masked by more robust transverse features or are not manifested because of variability in flow conditions, but they tend to be a ubiquitous fea-ture in sedimentary environments characterized by dominantly unidirectional flows over erodible substrates. Some of these bed forms are similar in only a very general way to the furrows reported here; others seem to be almost identical in essential features. These various kinds of similar bed forms are summarized briefly here, because, to date, most of the insight into the origin of abyssal furrows has come from study of longitudinal bed forms in other environments.

Extensive regions of longitudinal seif sand dunes exist in areas where steady winds blow over hot, dry, loose sand. These large dunes are thought to be the result of large helical vortex circulations in the lower mixed or unstable layer of heated air (Folk, 1971; Hanna, 1969). Such large vortices have been observed in the atmosphere (Angell et al., 1968) and banded cloud patterns are thought to be indicative of a similar type of circulation (Kuettner, 1959, 1971). These eolian longitudinal features are characterized by tuning-fork junctions opening into the prevailing wind direction (Folk, 1971).

Longitudinal bed forms are also developed in areas where water flows over a resistant substrate overlain by only a thin veneer of sandy material. These sand ribbons are known from flume experiments, shallow streams, tidal flats, and shallow seas. The spacing of the ribbons appears to increase with increasing flow depth (Allen, 1968).

Longitudinal bed forms closely similar to the abyssal furrows described here are found in shallow water in regions of strong currents. Furrows 5 m wide, 1 m deep, 4 km long, and 10–20 m apart with tuning-fork junctions that open into the stronger ebb tide have been reported eroded into the muds of a shallow estuary (Dyer, 1970). Stride et al. (1972) report possibly similar features in gravel sediments on the floor of the English Channel.

Patches of bifurcating furrows, also seemingly very similar to the abyssal furrows, have been recorded by side-scan sonar in a nannofossil ooze at a depth of 5.8 km in the Samoan Passage (Lonsdale, 1973), where the Antarctic bottom current is the principal agent controlling the sediment dynamics (Hollister et al., 1974b). However, these features do not appear to be presently developed enough to cause hyperbolic echoes on surface records.

Experimental and natural features formed in cohesive muds described by Dzulynski (1965) and Allen (1969) are parallel to water flow and have tuning-fork junctions opening upstream. The features produced by Allen (1969) show occasional well-defined median ridges where two furrows come together, and except for the very great difference in scale (they are spaced 6 mm apart and are about 1 mm deep) these "longitudinal rectilinear grooves" are quite similar to the features we have described in the abyssal Atlantic.

Another comparison which seems relevant is to compare the abyssal furrows with the wind slicks formed on the surface of water bodies which have winds blowing across them. These surface slicks appear to be caused by large helical circulations in the upper layer of the water body, called "Langmuir circulations."

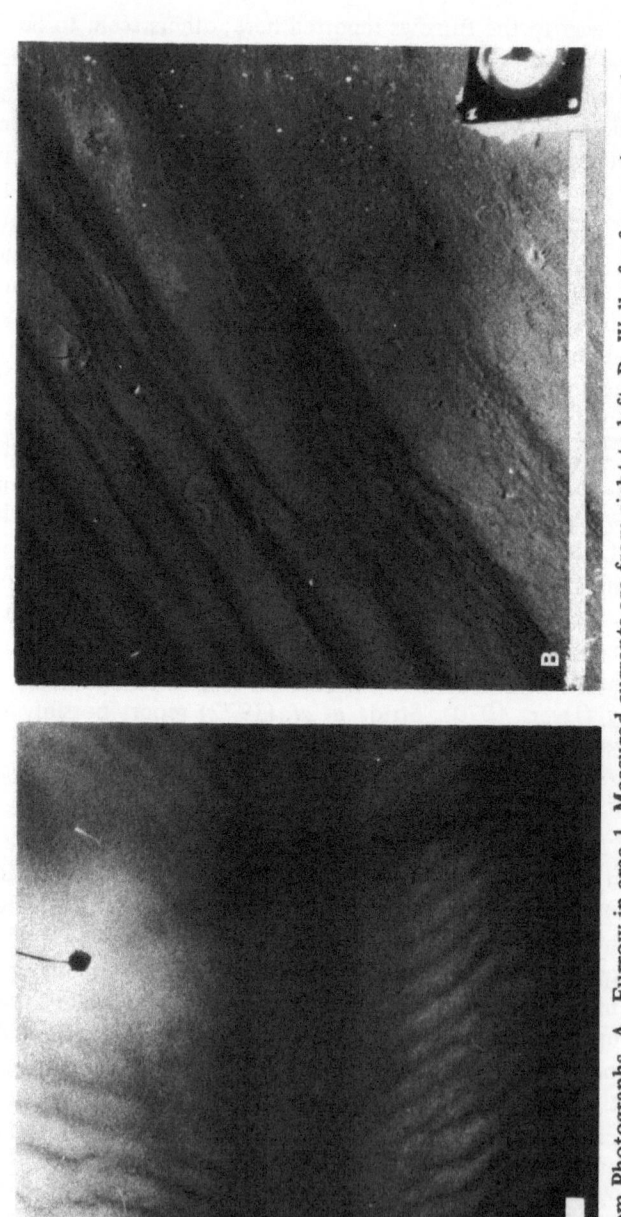

Fig. 7. Bottom Photographs. A—Furrow in area 1. Measured currents are from right to left. B—Wall of a furrow in area 1.

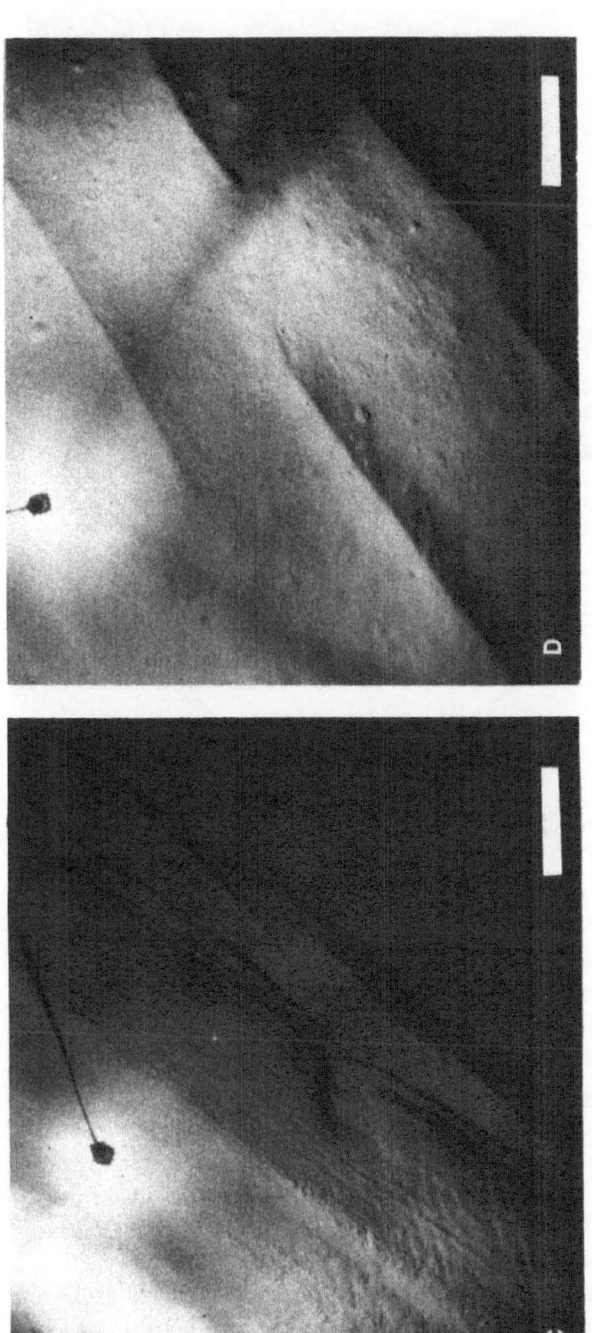

Fig. 7. C—Wall of a furrow in area 2. Note outcropping layers of more-resistant sediment. The wall shown is 9 m high. D—Bed forms observed near the edge of the abyssal plain (area of coherent echoes). The scale bars are 1 m long.

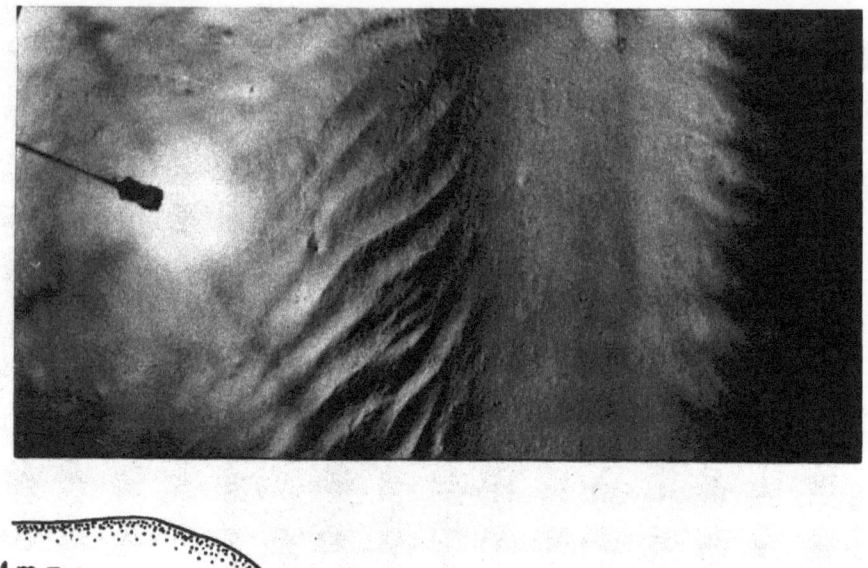

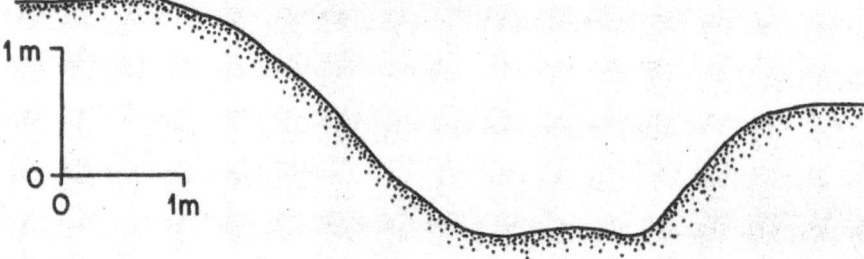

Fig. 8. Photograph and cross-section of a furrow (area 1). The cross-section is based on depths measured from stereo photographs of the furrow. No vertical exaggeration.

Photographs of the upwelling regions of the circulations taken by McLeish (1968) show some similarities to the abyssal furrows, especially in the tuning-fork junctions opening into the direction of the wind.

The similarity of the abyssal furrows to many bed forms described in other environments and the rough way in which the abyssal furrows appear to imitate the forms of cloud streets and wind slicks leads one to suspect that the mechanism responsible for the forms of these features may also be responsible for the form of the abyssal furrows. The current view regarding most of the bed forms described above is that they are formed by well-organized secondary circulations within the mean fluid flow—in the form of sets of streamwise-oriented helices with alternating sense of rotation—which modify the bed by localized erosion and deposition. The modified bed then tends to stabilize the flow pattern, which then further modifies the form of the bed. Helical circulations of this type have

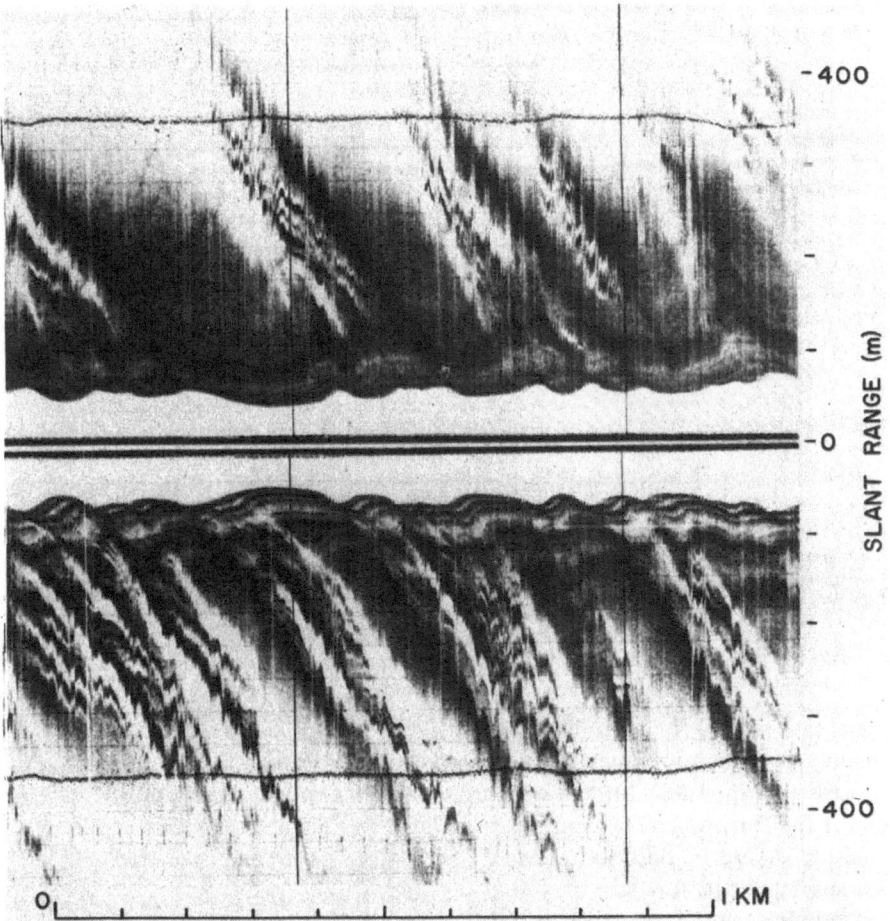

Fig. 9. Side-scan sonar record of the furrows in area 2. Line down the center of the record represents the fish track. Measured current directions are along the furrows from lower right to upper left. Note "tuning fork" junctions opening into the current.

long been postulated to exist in turbulent flows (Townsend, 1956), and the weight of present evidence strongly suggests that they are a characteristic feature of planetary boundary layer flows. They have been predicted and observed both in the atmosphere (Hanna, 1969) and in laboratory experiments (Faller, 1963).

One important result that has come out of investigations made into the nature of the planetary boundary layer is the relationship between the depth of flow and the wavelength of the instabilities present, which should be equivalent to half the spacing between the longitudinal bed forms. In all the cases reported,

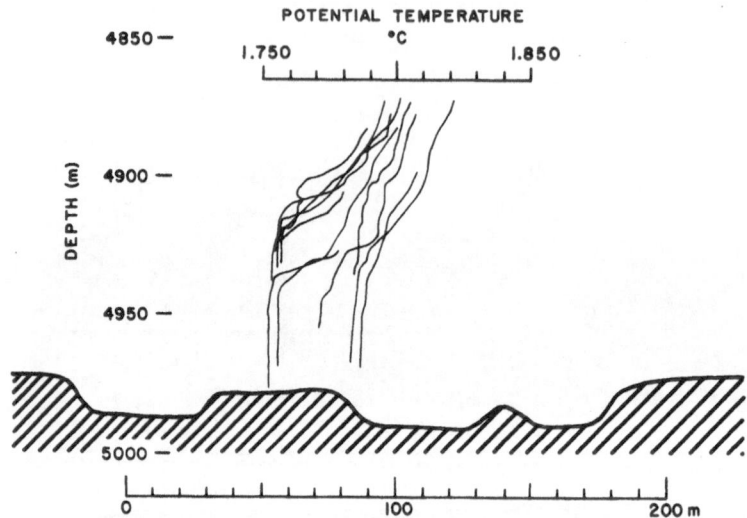

Fig. 10. Composite of vertical deep-tow temperature profiles showing a well-mixed bottom boundary layer in area 2. These profiles are plotted over a representative cross-section of the large furrows present in this area. Note the relationship between the well-mixed layer thickness and furrow spacing. (No vertical exaggeration.)

which are mostly for neutrally stratified flows, the spacings reported are approximately two to four times the boundary layer thickness. Thus, if this type of process is responsible for the bed forms observed on the Bahama Outer Ridge and if the furrows can be construed as being currently active, then a similar relationship should be observed between the bottom boundary layer thickness and the spacing of the furrows.

Salinity–temperature–density (STD) casts taken in regions of the Blake-Bahama Outer Ridge characterized by hyperbolic echoes and erosional furrows show that a layer of isothermal–isohaline water topped by relatively steep gradients of these parameters is usually developed near the bottom (Amos *et al.*, 1971). This well-mixed boundary layer may be the result of mixing produced by bottom friction and provides a layer in which secondary circulations could develop.

Deep-tow temperature measurements show that in area 2, at the transition between the outer ridge and the abyssal plain, this boundary layer averages approximately 60 m in thickness (Fig. 10). The furrows developed in this region are spaced from 50 to 200 m apart. In area 1, the large mud-wave area, a bottom isothermal layer of varying thickness was encountered. Deep-tow temperature measurements show that such a layer was usually no more than 20 m thick at the time of our survey; however, an STD cast taken by Amos *et al.*, (1971) shows a bottom isothermal layer approximately 100 m thick in a similar area.

The furrows in this area are spaced 20–125 m apart and the large mud waves are spaced approximately 2 km apart. Currents measured at 20 m above the bottom at the time of the deep-tow survey reached maximum velocities of 8–10 cm sec^{-1}. The tidal components of these currents are only 1–2 cm sec^{-1} and the currents are quite uniform during the extended periods of maximum velocity.

Thus the thickness of the boundary layer seems to be one-half the spacing observed between the furrows. This would be consistent with the view that secondary circulation could be responsible for the furrows and that the furrows could have been active at the time of our investigation. Much more information on the persistence and structure of the isothermal layer and the currents in these areas will have to be collected in order to test such a hypothesis. The combination of a steady current and a well-mixed bottom boundary layer would appear to provide almost ideal conditions for the development of secondary circulations.

Conclusions

1. Identification and description of bed forms which appear to result from large-scale secondary circulations in the bottom boundary layer provide important information on the nature of this boundary layer. This layer is of great importance in terms of sediment dynamics because most of the material deposited on or removed from the deep-sea floor must move through it.

2. If an estimate of the flow velocity, boundary-layer thickness, and length of time over which these currents must be flowing can be determined for a given furrow morphology or a given abyssal mud-wave profile, then a method will be available to determine what the geologically significant bottom currents have been in the past. This information can be used, along with the dating of suitable layers in cores, to get some idea of the history of deep circulation.

Our near-bottom investigations into the nature of the topography present on the Blake–Bahama Outer Ridge have shown the existence of a current-generated bed form—abyssal furrows—which has been previously unreported from the deep sea. The morphology of these furrows indicates that they have been formed by the interaction of large-scale helical vortices developed in the bottom boundary layer of the ocean and the sediments. Most of the furrows investigated in this study appear to have been recently active. As these furrows grow larger, sediments are eroded and made available for transport downstream.

Acknowledgments

Contribution No. 3700 of the Woods Hole Oceanographic Institution. Research supported by National Science Foundation Grant GA-39049. We thank

F. N. Spiess, J. D. Mudie, C. D. Lowenstein, and the technicians of the Marine Physical Laboratory, Scripps Institution of Oceanography, for providing and operating the deep-tow instrumentation; C. D. Lowenstein also served as co-chief scientist. Deep-tow development supported by NSF Grant GA-31377X and Office of Naval Research Contract N000014-69-A-0200-6002.

References

Allen, J. R. L., 1968, On the character and classification of bed forms, *Geologie en Mijnbouw 47:* 173-185.

Allen, J. R. L., 1969, Erosional current marks of weakly cohesive mud beds, *J. Sed. Pet. 39:* 607-623.

Amos, A. F., Gordon, A. L., and Schneider, E. D., 1971, Water masses and circulation patterns in the region of the Blake-Bahama Outer Ridge, *Deep-sea Research 18:* 145-165.

Angell, J. K., Pack, D. H., and Dickson, C. R., 1968, A Lagrangian study of helical circulations in the planetary boundary layer, *J. Atmos. Sci. 25:* 707-717.

Ballard, J. A., 1966, Structure of the lower continental rise hills of the western North Atlantic, *Geophysics 31:* 506-523.

Berggren, W. A., and Hollister, C. D., 1974, Paleogeography and the history of circulation in the Atlantic Ocean, in: (W. W. Hay, ed.), *Studies in Paleo-oceanography, Soc. Econ. Paleontologists and Mineralogists Special Publication 20:* Tulsa, Oklahoma, 126-186.

Bryan, G. M., and Markl, R. G., 1966, Microtopography of the Blake-Bahama Region, *Lamont-Doherty Geological Observatory, Technical Report No. 8, CO-8-66,* 44 pp.

Clay, C. S., and Rona, P. A., 1964, On the existence of bottom corrugations in the Blake-Bahama Basin, *J. Geophys. Res. 69:* 231-234.

Damuth, J. E., 1975, Echo character of the Western equatorial Atlantic floor and its relationships to the dispersal and distribution of terrigenous sediments, *Marine Geology 18:* 17-47.

Davies, T. A., and Laughton, A. S., 1972, Sedimentary processes in the North Atlantic, in: U.S. Government Printing Office, *Initial Reports of the Deep-Sea Drilling Project, Vol. 12:* 905-934.

Deardorff, J. W., 1972, Numerical investigations of neutral and unstable planetary boundary layers, *J. Atmos. Sci. 29:* 91-115.

Dyer, K. R., 1970, Linear erosional furrows in Southampton Water, *Nature 225:* 56-58.

Dzulynski, S., 1965, New data on experimental production of sedimentary structures, *J. Sed. Pet. 35:* 196-212.

Ellett, D. J., and Roberts, D. G., 1973, The overflow of Norwegian Sea deep water across the Wyville-Thompson ridge, *Deep-Sea Research 20:* 819-835.

Ewing, M., Eittreim, S. L., Ewing, J. I., and LePichon, X., 1971, Sediment transport and distribution in the Argentine Basin. 3. Nepheloid layer and processes of sedimentation, *Physics and Chemistry of the Earth 8:* 51-80.

Faller, A. J., 1963, An experimental study of the instability of the laminar Ekman boundary layer, *J. Fluid Mech. 15:* 560-576.

Flood, R. D., Hollister, C. D., Johnson, D. A., Southard, J. B., and Lonsdale, P. F., 1974, Hyperbolic echoes and erosional furrows on the Blake-Bahama Outer Ridge, *Trans. Am. Geoph. Union 55:* 284.

Folk, R. L., 1971, Longitudinal dunes of the northwestern edge of the Simpson Desert,

Northern Territory, Australia. 1. Geomorphology and grain size relationships, *Sedimentology 16:* 5-54.

Hanna, S. A., 1969, The formation of longitudinal sand dunes by large helical eddies in the atmosphere. *J. Appl. Meterology 8:* 874-883.

Heezen, B. C., and Hollister, C. D., 1964, Deep-sea current evidence from abyssal sediments, *Marine Geology 2:* 141-174.

Heezen, B. C., and Johnson, G. L., 1969, Mediterranean undercurrent and microphysiography west of Gibraltar, *Bull. Inst. Oceanogr.: 67.*

Heezen, B. C., Tharp, M., and Ewing, M., 1959, The floors of the oceans. 1. The North Atlantic, *Geological Soc. of Am., Special Paper 65:* 122 pp.

Heezen, B. C., Hollister, C. D., and Ruddiman, W. F., 1966, Shaping of the continental rise by geostrophic contour currents, *Science 151:* 502-508.

Heezen, B. C., and Schneider, E. D., 1968, The shaping and sediment stratification of the continental rise (abs.), *Marine Tech. Soc., Nat'l Symposium, Ocean Sciences and Engineering of the Atlantic Shelf Trans., March 19-20, 1968, Phila., Pa.,* Washington, D.C., pp. 279-280.

Hollister, C. D., Ewing, J. I., *et al.,* 1971, *Initial Reports of the Deep-Sea Drilling Project,* (A. L. Gordon, ed.), *Studies in Physical Oceanography,* Vol. 2, Gordon and Breach, New York, pp. 37-66.

Hollister, C. D., Ewing, J. I., *et al.,* 1971, *Initial Reports of the Deep-Sea Drilling Project,* *Vol. 11,* U.S. Gov't. Printing Office, Washington, D.C., 1077 pp.

Hollister, D. C., Silva, A. J., and Driscoll, A., 1973, A giant piston-corer, *Ocean Engng. 2:* 159-168.

Hollister, C. D., Flood, R. D., Johnson, D. A., Lonsdale, P. F., Southard, J. B., 1974a, Abyssal furrows and hyperbolic echo traces on the Bahama Outer Ridge, *Geology 2:* 395-400.

Hollister, C. D., Johnson, D. A., and Lonsdale, P. F., 1974b, Current-controlled abyssal sedimentation: Samoan Passage, Equatorial West Pacific. *J. of Geology 82:* 275-300.

Johnson, G. L., and Schneider, E. D., 1969, Depositional Ridges in the North Atlantic, *Earth Planet. Sci. Lett. 6:* 416-422.

Johnson, G. L., Vogt, P. R., and Schneider, E. D., 1971, Morphology of the northeastern Atlantic and Labrador Sea, *Deutsche Hydrographische Zeitschrift 24:* 49-73.

Jones, E. J. W., Ewing, M., Ewing, J. I., and Eittreim, S., 1970, Influences of Norwegian Sea overflow water on sedimentation in the northern North Atlantic and Labrador Sea, *J. Geophys. Res. 75:* 1655-1680.

Kuettner, J., 1959, The band structure of the atmosphere, *Tellus 11:* 267-294.

Kuettner, J., 1971, Cloud bands in the earth's atmosphere, *Tellus 23:* 404-426.

Laughton, A. S., 1972, The Southern Labrador Sea: A key to the Mesozoic and early Tertiary evolution of the North Atlantic, in: U.S. Gov't. Printing Office, *Initial Reports of the Deep-Sea Drilling Project, Vol. 12:* 1155-1179.

Lonsdale, P. F., 1973, Erosional furrows across the abyssal Pacific floor, *Transactions, American Geophysical Union 54:* 1110.

Markl, R. G., Bryan, G. M., and Ewing, J. I., 1970, Structure of the Blake–Bahama Outer Ridge, *J. Geophys. Res. 75:* 4539-4555.

McGeary, D. F. R., and Damuth, J. E., 1973, Postglacial iron-rich crusts in hemipelagic deep-sea sediment. *Bull. Geol. Soc. Am. 84:* 1201-1212.

McLeish, W., 1968, On the mechanism of wind slick generation, *Deep-Sea Research, 15:* 461-489.

Partheniades, E., 1965, Erosion and deposition of cohesive soils, *Proc. ASCE, Jour. Hydraulics Division 91* (HY 1): 105-139.

Rona, P. A., 1969, Linear "lower continental rise hills" off Cape Hatteras, *Jour. Sed. Petrol.* *39:* 1132–1141.

Ruddiman, W. F., 1972, Sediment distribution on the Reykjanes Ridge: seismic evidence, *Bull. Geol. Soc. Am. 83:* 2039–2062.

Schneider, E. D., Fox, P. J., Hollister, C. D., Needham, D. H., and Heezen, B. C., 1967, Further evidence of contour currents in the western North Atlantic, *Earth Plant. Sci. Letters 2:* 351–359.

Spiess, F. N., and Mudie, J. D., 1970, Small scale topographic and magnetic features, in: (A. E. Maxwell, ed.), *The Sea*, Vol. IV(1), John Wiley and Sons, New York, pp. 205– 250.

Spiess, F. N., and Tyce, R. C., 1973, Marine Physical Laboratory deep-tow instrumentation system, *Scripps Inst. Oceanogr. Ref. 73-4*, La Jolla, Calif., 37 pp.

Stride, A. H., Belderson, R. H., and Kenyon, N. H., 1972, Longitudinal furrows and depositional sand bodies of the English Channel, *Mémoire du B.R.G.M. 79:* 233–240.

Townsend, A. A., 1956, *The Structure of Turbulent Shear Flow*, Cambridge Univ. Press, Cambridge, 315 pp.

II. WORKING GROUP REPORTS

Introduction

Three major interfaces exist on the surface of the earth: the air–earth, air–sea, and sea–sea bottom interfaces. The air–earth interface is man's habitat and has been scientifically explored since antiquity. Processes at the air–sea interface have commanded man's attention since the beginning of navigation. Systematic scientific investigations of the ocean floor, however, began only in the middle of the last century.

This has been so for several reasons. Access to the sea bed is difficult, costly, and sometimes dangerous; its extent is more than twice that of the air–earth interface; and until recently there was little economic or social incentive beyond scientific curiosity. While continental shelf fisheries, coastal construction, and shallow-water oil and gravel extraction have been of importance for some time, it is only within the past decade that economically important activities have extended to the bed of the deep ocean.

Today, the importance to mankind of this vast area beneath the sea is changing rapidly and drastically. Man is farming some intertidal regions and may soon have to begin cultivation of animals on the sea bottom. Industrial construction—of atomic reactors, oil storage tanks, and off-shore harbors—is planned around the continents. Oil and gas fields are being exploited in progressively deeper waters on the shelf and under increasingly rough climatic conditions. Exploration for fuel resources even below the deep sea is underway. Therefore, means of drilling, production, storage, and transport will have to be developed for this environment. Mining of deep-sea manganese nodules, with high contents of several more valuable metals, will begin within the next few years. Waste dumping is becoming a problem in all water depths. We are therefore on the threshold of a new era, the expansion of man's impact upon the oceans to the deep-sea bed.

To assess the likely effects of these plans with their economic, safety, and environmental aspects, we must know the nature of the sea bed and we must understand the processes occurring in this region: the benthic boundary layer. During the last decade increasing numbers of scientists, pure and marine physicists, chemists, biologists, and geologists have become interested in filling the huge gaps in our knowledge about this interface. Some of them are attracted by the excitement of reaching this last frontier as divers or in submersibles; all are motivated by the attraction of the unknown.

It has become increasingly apparent that this zone of reaction and interaction between liquid and solid is most advantageously studied through integrated multidisciplinary approaches and by scientists willing to spend at least part of their efforts in fostering interdisciplinary research.

In order to examine our state of knowledge—or, rather, our state of ignorance—to clarify the most important problems, to formulate urgent questions, and to suggest methods to solve them, the Science Committee of NATO sponsored this conference on the benthic boundary layer, assembling scientists with different backgrounds and experience in these fields. The conference was planned to guarantee that "les extremes se touchent"—as on the sea bed.

As with all boundaries in nature (and in politics) the benthic boundary is a region of gradients, of exchanges, and of interactions, with more uniform and stable conditions existing away from the boundary. One of the aims of the conference was therefore to define the nature and range of these changes and the thickness of the benthic boundary layer above and beneath the sea floor.

This could only be done by considering the processes interacting at this interface. What kinds of processes interact? What, where, when, to what extent, and with what variability? What are the physical, chemical, biological, and geological effects of these processes on the sea floor? The answers to these questions will be of great value to sedimentologists and paleobiologists seeking to interpret the environmental conditions at ocean bottoms in the distant past, though we did not address ourselves to these interpretive problems.

We agreed that problem-orientated fields should be discussed, concentrating on the processes operating rather than on their more subtle regional differences, e.g., in polar or tropical locations. We also agreed to exclude the coastal zone from the discussions.

Lastly, methods and instrumentation were also to be considered. It is trivial to state that well-defined and well-planned projects are normally the most effective, but it deserves to be emphasized that progress in science often depends on new, better, and more accurate instruments. For the deep-sea boundary the required development of instrumentation necessitates teams of skilled electronic and mechanical engineers working alongside the scientists. This will be very expensive, but anything less will be wasted money.

Sixty-eight participants from eight countries met for a week at Les Arcs, France, in November, 1974. During plenary sessions brief lectures highlighted physical, chemical, geological, biological, and engineering aspects of the benthic boundary layer. The major part of the time was spent in small working groups, dealing with the following topics:

1. Gradients of velocity, physical, biological, and chemical properties

2. Velocity variations, turbulence, and stability
3. Sea floor deposition, erosion, and transportation
4. Solution–sediment chemical interactions
5. Organism–sediment relationships
6. Metabolism at the benthic boundary

These discussions were further deepened by close contact between working groups, and the resulting conclusions were finally refined by exposure in a plenary session of all the participants. Thus, the reports produced can be taken as the professional consensus of a considerable number of leaders in various aspects of work on the boundary layer. It is hoped that these results will be valuable to all working in this area.

The Organizing Committee
I. N. McCave (Chairman)
K. F. Bowden   E. L. Mills
S. E. Calvert   H. Postma
E. D. Goldberg   E. Seibold

A
Gradients of Velocity and Physical, Biological, and Chemical Properties

Group Leader: J. Gieskes

Members: M. Bewers, S. L. Eittreim, C. Hollister,
L. D. Kulm, D. H. Loring, P. G. Teleki,
S. Wellershaus, M. Wimbush

Rapporteur: K. Dyer

Introduction

The benthic boundary layer is a relatively thin layer extending just above and into the sea floor. It is a transfer zone, a transition region. It is a very important part of the sea and presents unique problems. Because of its thinness accurate measurements within it are difficult. Also, in the deep ocean, the water depth may exceed several thousand meters and in shallower water the currents can be strong and the gradients steeper. However, there is the advantage that instruments can be fixed in relation to the bottom, and long simultaneous time series of measurements of several parameters are possible when a close spatial framework is used.

Within the boundary layers there are measurable exchanges, particularly of momentum and heat. It is the zone in which all of the sediment entrainment, deposition, and most of the transport occurs. These processes must be governed by the dynamics of the boundary layer. There are also chemical reactions and biotic activity which are again basically controlled by the physical conditions.

The benthic boundary layer probably accounts for most of the dissipation of tidal energy in the oceans (though only 1% is dissipated in the deep-sea boundary layer). This produces mixing which can be detected by observable gradients of many properties and it is by these gradients that the boundary layer is generally defined. This does not mean that the boundary layer can be defined with a unique set of parameters, but since the dynamic viewpoint provides the physical insight enabling interpretation of most other studies, the following definition may be considered:

The benthic boundary layer in the water column is the zone near the sea bed where there are strong biotic, chemical, and physical gradients which are controlled by the presence of the sea bed. The dynamic thickness of this layer is the height of frictional influence. In deep water this is the Ekman scale $u_* f^{-1}$ (where u_* is the friction velocity, $u_* = (\tau_0/\rho)^{1/2}$, τ_0 being the bed shear stress, ρ is fluid denisty, and f is the Coriolis parameter), and in shallow water under wave conditions it is the minimum elevation at which the flow can be matched to potential flow. In tidal flows the boundary layer thickness is probably limited by stratification or occupies the whole water depth. Within the sea bed the boundary layer extends to a depth at which mixing processes are controlled by exchanges or activity occurring at the boundary.

Some of the problems can be separated conveniently into those of the deep sea and of the continental shelf, requiring sometimes very different approaches. In other aspects there is considerable overlap. On the shelf, the presence of waves and tides requires that unsteady, bidirectional flow properties be measured and analyzed.

There are many economic pressures for study of the benthic boundary layer. Beyond the continental shelf, waste disposal, the mining of manganese nodules and other mineral deposits, and the possibilities of deep-water fisheries are immediate problems. In shallow water, waste disposal, navigation, dredging, sand and gravel extraction, the siting of rigs, pipelines, and engineering structures, as well as fisheries, are important. These activities are carried out within the benthic boundary layer. Consequently studies need to be of a multidisciplinary nature; directed toward basic processes, for there is still inadequate information available to ensure a proper assessment of the complexities of the situation and to design the best experiments.

The State of the Art

The Deep Sea

There are very few observations of the physical, chemical, or biotic processes occurring in the deep-sea boundary layer. They suggest that the relatively low currents and shear stresses measured are not those that dominate the processes of transport and deposition of sediment; for example, current-formed topographic features occur in areas where the measured currents appear to be too low to have caused them. Obviously longer time series of measurements are needed. There is a tidal periodicity in the currents, but the mean currents are probably dominated in the long term by large-scale variations in the density fields above the boundary layer or by lateral variability. Some measurements indicate low mixing rates (mixing rates may even be dominated by geothermal fluxes rather than by bottom shear in extremely tranquil areas). High accuracies and sensitivities consequently are required of all instruments used in the deep-sea boundary layer. The fact that this layer can range from a few centimeters to hundreds of meters in thickness presents problems of measurement. Most past experiments have been on a spot-sampling basis and their interpretation is therefore uncertain.

Topography and Sediments

The regional topographies in the deep ocean basins and floors are generally low, but large roughness elements can be important, e.g., sea-mounts. It is only recently that adequate information about small-scale roughness is becoming available, even though there are extensive photographic data. These data show that the deep-sea floor relief can be more varied than had previously been suspected. The broad distributions of sediment size and mineralogy in many areas is reasonably well known. More detailed investigation may show significant variations in the distributions of, for instance, manganese nodules and current ripples which may affect the boundary roughness and the mixing scales. The geotechnical properties of these sediments are not well known.

Currents

Very few current measurements have been made within the benthic boundary layer. In tranquil areas, tidal components (both barotropic and baroclinic) appear dominant, but variability on time scales of months or longer are unknown. Even less is known about the vertical profiles of currents.

Turbulence

Turbulence measurements in the deep sea are rare and inadequate. They are necessary for calculation of Reynolds stresses, and for understanding suspension and diffusion processes.

Bed Shear Stresses

These have only been calculated from current velocity profiles.

Temperature

Very precise temperature measurements have shown that the temperature gradient in relatively tranquil areas is slightly superadiabatic. The positive deviation from an adiabatic gradient enables calculation of eddy diffusion coefficients, providing the geothermal heat flux is known. There are reported hyperadiabatic profiles whose interpretation is more difficult. Temperature measurements may provide the best way of examining the long-term variability of the deep-sea boundary layer.

Density

This is normally determined by temperature, with the salinity assumed homogeneous and invariant.

Sediment Transport

Suspended matter concentrations are determined by direct sampling and by the use of nephelometers, the calibration of which in concentration terms is difficult. Another major problem is that of residence times in suspension of the particles which may be deposited and reentrained at intervals. Compositional analysis and its variations require detailed consideration. Nephelometry provides a good method of analyzing the response of the suspended sediment to variations in current velocity.

Chemistry Related to the Sediment–Water Interchange

The most significant chemical gradient in the water appears to be that in excess radon. The thicknesses of layers with excess radon seem largely to correspond to zones of high concentrations of suspended sediment. Rates of vertical eddy diffusion can be calculated from these measurements.

Biology

There is little information on the dispersal—vertically, horizontally, and in time—of planktonic larvae or eggs of benthic and nektonic animals in the water and of organisms on and in the sediment. In shallow seas the dispersal during the larval life has been intensively studied; this has not been extended to the deep sea where some essential conditions, such as food supply, currents, and mixing density of adult benthic organisms, are very different from those on the shelf. The role of bacteria and fungi is almost unknown.

The Shelf

Tides are also important on the shelf, but there are the additional effects of superimposed wave-induced oscillations and wind-driven currents. They do not necessarily interact linearly and the techniques of measurement are consequently different than those used in the deep sea.

Long and intermediate period oscillations dominate the outer and inner continental shelf boundary layers, respectively. Consequently the shelf can be separated into tide- and wave-dominated areas. The boundary layer thickness under short-period waves is probably on the order of centimeters, whereas the boundary layer for tidal periodicity is several orders of magnitude thicker, or even the total water depth.

Certain characteristics of the boundary layer are reasonably well known, but many are relatively poorly known. Many of the physical parameters require both short-time observations and long-time series with area-distributed measurements.

Topography and Sediment Distribution

Of all the characteristics the best documented are the broad-scale topography, the surface sediment grain size and composition, and the distribution of certain biota. However, even in these terms there are many areas that require further exploration, such as the tropics and the southern hemisphere.

Currents

The water circulation patterns on the shelves have been studied using reasonably long-term synoptic measurements, but extrapolation of these results to the structure and variability of the boundary layer is not feasible. Consequently more information is required on the vertical profiles of currents, on instantaneous and mean velocities on three axes, especially near the boundary.

Turbulence

There are few studies of turbulence length and time scales, intensities, and Reynolds stresses, even under laboratory conditions. Theoretical studies have not been tested adequately in the field. The important properties of the time-dependent turbulent boundary layer that we must understand are the structure of the layer, the eddy diffusion coefficients, Reynolds stress distribution, and the production and dissipation of turbulent energy.

Bed Shear Stress

The optimum method of bed shear stress measurement is a direct one. The calculation of shear stress from velocity profiles or Reynolds stress, has inherent problems, but any measure is useful, at this stage, for the analysis of sediment transport.

Sediment Transport

Very little is known of the transport rates of either bedload or suspended load on the shelf. The practical implication of sediment transport to geology and ocean engineering are obvious. Perhaps less obvious is the effect that high suspended sediment concentrations and the movement of bed forms may have in determining boundary layer properties and stability.

Chemistry

Measurements have shown no significant gradients of most elements and compounds in the water column directly related to the benthic boundary layer.

Biological Studies

The organic constituents of suspended material are important food sources to many benthic organisms. The dispersal of benthic organisms by their planktonic larvae is not yet well known in all but a few areas. Little is known about epibenthic plankton, and the role of bacteria and fungi in the food chain has yet to be studied.

Continental Slope

These areas are morphologically complex and because of changes in bottom slopes the boundary layer is probably very variable. It is likely to be greatly influenced by lateral density fields, intermittent density flows, and other features such as breaking internal waves. So little work has been done in the boundary layer on the continental slopes that we find it difficult to make any definitive comments.

Required Measurements

The Deep Sea

Currents, Temperature, and Salinity

Typical physical conditions in the abyssal benthic boundary layer above a smooth bottom are taken to be $u_* = 1$ mm s^{-1}, the geothermal heat flux, $H = 1.5$ μcal cm^{-2} s^{-1}. If it is assumed that the logarithmic law for a turbulent boundary layer applies between elevations $z = 0.1$ m and 1.0 m, current velocity and temperature differences between these levels are expected to be of order 1 cm s^{-1} and 10^{-4} °C, respectively. Sensors capable of resolving these gradients should therefore have precisions of order 1 mm s^{-1} and 10^{-5} °C (or better). For current sensors whose precision decreases with increasing speed, 5% precision at speeds above 2 cm /s^{-1} would suffice. To measure the salinity component of the benthic density gradient with the same accuracy as the temperature component, a salinometer with 10^{-6} ‰ precision would be needed (this is three orders of magnitude beyond the capabilities of present technology).

If the horizontal current vector is sensed as speed plus direction rather than orthogonal components, vector averaging should be performed. Precision in recorded components of 5% should give 3° accurary in measurement of the change in direction with height. One would like to determine the kinematic Reynolds stress $-\rho \overline{u' w'}$ as a function of z. For this, vertical velocity w must also be measured to 1 mm/ s^{-1} within a few centimeters of the horizontal velocity sensors, and all velocity component sensors must have a uniform response time of 1 s or less. Either $-u'\,w'$ must be formed electronically from the sensor outputs, or frequent ($\Delta t \approx 1$ s) velocity samples recorded. The precision of the data recorded from a particular sensor is a function of the sampling integration time. Effective precision is increased by increasing this sampling time. However, one would not wish to rely on sampling times in excess of 1 h if the tidal fluctuations in the layer are to be resolved. Unless these precisions are also absolute *accuracies*, adequate mean gradient measurements may require an *in situ* intercalibration technique.

Integrating times should be as large a fraction of the sampling interval as possible, to give maximum precision and maximum protection against aliasing. All channels should be integrated contemporaneously for meaningful comparison of records.

Temperature sensors in the upper sediment ($z = 0$ to -5 cm) would be used to study downward propagation of temperature fluctuations from the overlying water. 10^{-5} °C precision is again required. Temperature sensors deeper in the sediment would be used principally to determine the upward geothermal heat flux, H, and 10^{-2} °C accuracy would be adequate to measure the gradient over Δz of 1 m.

Benthic Nepheloid Layers

High gradients in particle concentrations exist near the sea bottom, typically over a scale of a few hundred meters. Much needs to be learned about particle size distribution and composition. Coulter counter analyses of discrete water samples have shown in some areas a biomodal size distribution in which particles predominate at about 2 μm (presumably these materials are single mineral grains and biogenic detritus) and at 40-50 μm which may be the dominant "floc mode" or aggregate particle size.

Correlations exist between vertical particulate distributions, as determined by light scattering and gravimetric methods, and the scale of the frictional boundary layer defined or determined in physical terms. Similarly, excess radon profiles show high gradients over the same scale. Such correlations exist both in temporal and spatial terms. Deep oceanic benthic nepheloid layers are identified with strong bottom water flows (e.g., western boundary currents). Similar nepheloid layers are ubiquitous in inshore and continental shelf regions and may be important mechanisms of material transport across continental margins.

The predominant interest in benthic nepheloid layers is related to the delineation of geological and chemical transport mechanisms in coastal and deep-sea environments. These interests must be considered in relation to the overall particulate and chemical material cycles in the oceans, whether these are directed toward understanding present-day or past conditions in the sediments.

It is of primary importance that we be able to identify the nature of particulate material in the benthic nepheloid layer in relation to the composition or characteristics (chemical, mineralogical, and biological) of suspended material in overlying waters and in the surface sediments. This information is required in order to establish the sources and relative magnitudes of benthic nepheloid material. Subsequently, we need to establish the processes which control the production of material from these sources. Material entering the benthic nepheloid layer will include particles of aeolian, terrigenous, and biotic origin which settle through the overlying water column, and sedimentary particles either from bottom sediments or material brought in by horizontal advection from other sedimentary domains which has been resuspended by biotic activity or physical forces.

In order to carry out such budgetary and mass flux calculations, the following information is required. Some parameters can be calculated from others and therefore it is not essential that all measurements be carried out experimentally. The application of modeling concepts should be encouraged to gain maximum insight into processes from existing data.

Measurements should include:

1. Mass concentration of particles in units of mass/volume with precisions of $1 \ \mu g \ liter^{-1}$.

2. Size distribution of the particles, semiquantitative in the range 0.1–100 μm.

3. Average volume gradient of particles in units of volume/volume at precisions of 10^{-9}.

4. Mean particle density, which can be calculated from the above data.

5. Stokes and, ideally, *in situ* settling velocities for particles.

6. Mineralogical analysis of suspended and bed sediment particles. This will provide both compositional information and additional evidence for density estimates on particles of different size classes. Scanning electron microscope studies in conjunction with elemental analysis should be carried out on individual particles.

7. Chemical analysis of major and minor element composition of suspended and bed sediment particles. Particular emphasis should be placed here upon the identification of detrital and nondetrital (exchangeable) inorganic material. Such analyses can be fairly routinely carried out in sediments, but the major problem in their extension to suspended particle analysis is associated with the very small quantities of material which it is possible to collect. Sedimentation rates should be established by radiometric or other methods.

8. Temporal and spatial variability of the particle concentration gradients. This involves *in situ* measurements to which optical techniques are suited.

9. Residence times of particles. There are two scales of residence time involved: the times between initial entry and final removal from the layer, and the transport times of an average cycle between particle introduction and settling. It is probably overly optimistic to expect such data to be obtained from direct measurement due to the magnitudes of horizontal transport velocities of the material grains. It would be preferable to model such processes with basic physical and geochemical data.

10. Discrete sampling methods. Resolutions of the order of 1 m are required for sampling. While samplers such as Niskin bottles are adequate at distances in excess of 10 m from the bottom, it may be necessary to develop higher aspect ratio, well-flushed, large-volume (~30 l) samplers for sampling within 10 m of the bottom. The degrees of contamination and modification of the form of the material during the sampling, filtration, and analytical stages cannot be judged quantitatively at present. Repeatability or sampling precision estimates must be made by the employment of statistically designed experiments. On the other hand, the required analytical precisions for physical and mineralogical analyses of suspended samples collected by filtration of discrete water samples can be met by present techniques in a laboratory environment. It is doubtful, however, whether present analytical capabilities are adequate for the detailed chemical analysis of particulates.

11. *In situ* methods. *In situ* methods undoubtedly hold the greatest promise for assessing the degree of true near-bottom nepheloid variability. Optical techniques such as forward-angle scattering (Biscaye and Eittreim, 1974) should continue to be used and may be improved. Much work still needs to be done on the application of scattering theory to oceanic measurements. Since particulate scattering is predominantly in the forward direction, forward scattering gives the highest measurement sensitivity. Measurement of the volume-scattering function at a number of discrete forward angles in the range of 5–30° appears to be the best way to achieve the most useful information on particle concentration as well as reproducible optical results. A precision of better than ±5% should be attained. Gamma-ray and light-attenuation measurements might also be applicable for measuring particle concentrations.

Chemical Studies: Water Column

1. *Dissolved substances.* For an understanding of exchange processes between the sediments and the overlying water column we recommend the measurement of excess radon profiles (and perhaps those of other short-lived natural radio isotopes) from very near to the sediment–water interface to elevations where this quantity equals zero. Discrete samples can then be analyzed for other dissolved chemical substances that may be related to exchange processes with the sediments (nutrients, dissolved silica, oxygen, alkalinity, total dissolved carbon dioxide, dissolved, and particulate organic carbon).

2. *Sediments*. Investigation of possible disturbances of the surface layers of the sediments by biotic action, redeposition, and any other transport processes can be investigated by means of tracer techniques, particularly radioactive tracers or other appropriate methods. These processes are particularly relevant to exchange of particulate and dissolved material with the overlying water column.

Biology

Animal communities can be collected qualitatively for studies of physiology, taxonomy, diversity, and behavior, and quantitatively for estimation of biomass, etc.

1. *Plankton*. Plankton—including planktonic larvae—should not simply be collected by means of plankton net hauls. Alternative techniques are desirable, particularly within a few meters of the sea floor. For example, material can be collected by means of sucking the water through retrieving nets. Preservation of the samples *in situ* might be useful. For quantitative plankton samples 500 liter or more must be filtered, and for diversity calculations, 2000 liters or more. Measurements of vertical microdistribution require great care to avoid destroying stratification.

2. *Nekton*. Quantitative methods still must be developed. Qualitative collections can be made by traps, deep-sea trawls, and photography.

3. *Evaluation*. For all ecological work taxonomic (including larvae) and physiological research is essential. Counting of individuals and measurement of body size for consequent calculation of total biomass–volume will lead to ecological data. Chemical compounds of communities or individuals, both adsorbed and compositional, need to be analyzed. Behavior with respect to migration, reproduction, larval development, metabolism, respiration, etc. can be made by means of deep-sea photography, discrete collections, and aquarium experiments.

4. *Bacteria and fungi*. These cannot yet satisfactorily be collected. Methods are under development.

5. *Suspended matter*. Calorific analyses provide information on total energy content but not on the fraction available for metabolism. Studies of the decomposition and metabolic rates of temperature and pressure—measured *in situ* or under controlled laboratory conditions—are necessary.

The Shelf

Currents

Because of the tidal- and wave-induced velocities there are many measurement problems. These involve the response of current meters to changes of both in flow direction and in speed. The way in which the sensors are mounted can provide a filter on the measurements. Consequently systems with variable time constants must be used. For measurements of wave-induced currents a time constant no longer than 0.25 s is necessary. Vector averaging must be employed in tidal measurements and knowledge of surface wave heights, periods, and directions of propagation is necessary in wave oscillation measurement. This may be possible with improved instrumentation and through data analysis culminating in directional spectral outputs. Measurement using three-component sensors is preferable. A threshold of ~ 1 cm s^{-1} and an accuracy of 0.5 cm s^{-1} are necessary, with extreme values of at least 3 m s^{-1}. Direction needs to be measured to $\pm 2°$ unless derived from three-component analysis. Spectral analysis for the decomposition of the flow field must be used to separate the steady and unsteady components, the phase relationship within vertical profiles, and the instantaneous maxima of velocity and shear stress. Meteorological effects can produce surges and secondary currents that can only be accounted for by adequate wind and atmospheric pressure measurements.

Required techniques and measurements include: (1) Vector resolution at a point. (2) Construction of the mean velocity profile in the boundary layer from vector averaged values, (3) Determination of u, v, and w as functions of time and phase of waves. (4) Determination of the turbulent fluctuations u', v', w', and cross-correlated products. (5) Instantaneous maximum velocities as functions of wave parameters, height above the bed, bottom roughness, and sediment texture. (6) Measurement of the fluid mass transport in the boundary layer.

Shear Stress

Again there is the joint influence of mean values produced by the tidal currents and the superimposed wave effects. Measurements required involve: 1) Instantaneous maximum shear stress at the bed as a function of mean velocity, bed roughness, sediment size, and wave parameters. 2) Time- and space-averaged shear stress for application to bedload movement. 3) Vertical distribution of shear stress for phase relationships to wave profile and bed form development. 4) Determination of the friction and drag coefficients.

Turbulence

Information required includes: (1) Reynolds stresses at a point and as a function of elevation, bed roughness, and Reynolds number. (2) Generation and decay of turbulence and eddy length and time scales. (3) Diffusion and dispersion properties of the turbulent boundary layers as a function of time. (4) Determination of the thicknesses of viscous, overlap, and outer layers.

Density

In shallow-water regions density is generally governed more by salinity than by temperature. A temperature range of $-2\text{-}30°C$ and an accuracy of $0.1°C$ are necessary. Salinity ranges are $0\text{-}35\ \%_{oo}$ with a necessary accuracy of $0.01\ \%_{oo}$.

Sediment Motion

Very few quantitative measurements have been completed, and their explanation in terms of theory and laboratory studies is difficult. However, laboratory studies should be continued in conjunction with field measurements. Measurements necessary include: (1) Entrainment of homogeneous or heterogeneous sediments as a function of boundary layer characteristics. (2) Step length and residence times of individual particles. (3) Exchange of sediment particles between the bed and the boundary layer, the boundary layer flow, and various levels outside of it. (4) Rates of instantaneous mass movement, scour, suspension, dispersal, and settling with respect to boundary layer characteristics. (5) Areal variability of sediment movement. (6) Ratio of tractive to suspended load and its relationship to turbulence and shear stresses. (7) Determination of the depth of sediment movement in the bed. (8) Determination of total sediment transport rate. (9) Sorting of sediment as function of size and density. (10) The effect of high suspended sediment concentrations on the flow structure. (11) Assessment of nonlinear effects in the flow on the transport of sediments. (12) Mixing processes in sediment laden flows.

Chemistry

(1) Improved techniques for the observation and sampling of chemical gradients. (2) Studies of pore-water chemistry, diffusion, and diagenesis.

Biology

(1) Analysis of organic constituents of suspended matter–sediment as a food source for benthic and planktonic organisms. (2) Dispersal of benthic animals by their planktonic larvae in many areas and biotic mass transport related to it. (3) Epibenthic plankton: taxonomy and ecology. (4) Role of bacteria and fungi in the food chains: taxonomy and ecology.

Measurement Locations

The Deep Sea

Any experiments that are carried out in the deep sea should be representative either of large ($\sim 10^6$ km^2) regions that are not anomalous in any obvious way, or of areas where certain boundary layer processes may be especially well developed. Consequently they must be areas where significant amounts of data already exist or where existing oceanographic programs that could indicate significant features of the boundary layer are in progress. Individual workers with specialized experiments will continue to be the main source of information on the processes. Comparatively greater returns, however, could be achieved by teams working on allied investigations and by international programs.

With these criteria in mind we have chosen the following deep-sea areas as examples of suitable environments:

Deep Ocean: Low Energy (Two Sub-Sites)

Central Basin of the North Pacific.

1. *Approximate position*: 30°N, 160°W, 600 miles NNE of Hawaii.

2. *Regional topography*: Gently undulating abyssal mounds 100 m high and 1000 m wide, maximum local slopes: 2-5°; regional gradients: 1:500 to 1:1000.

3. *Benthic environment*: T (*in situ*) = 1.4°C; salinity = 34.7 ‰, [O_2] = 4 ml/liter. Current flow measured tens of meters above bottom = 2-8 cm s^{-1}, principally tidal (M_2). Biomass: 0.1-0.5 g m^{-2} (wet weight). Rate of sediment accumulation: mm/1000 years. Sediment type: brown clay, 90% < 10 µm; $CaCo_3$ < 1%; organic C < 0.1%. Water depth: 5500-6000 m.

4. *Comments*: One site could be within a field of manganese oxide concretions (nodules) which are about 2-10 cm in diameter and occur in fields of hundreds of km^2. Nodule coverage varies between 40-60% in these fields. The site nearby (within 100 km) should be devoid of nodules in order to compare structures of the BBL in similar areas with and without roughness elements of the scale of the viscous sublayer of the BBL. This region is presently the site of other on-going benthic studies and significant regional data from the area exists in the literature. BBL thickness is of the order of 10 m.

Deep Ocean: High Energy (Two Sites)

Continental Rise of the Western North Atlantic.

1. *Approximate position*: *Site 1*: 36°N 70°W, continental rise 300 nautical miles south of New York, 4500 m depth. *Site 2*: 57°N 57°W, continental rise off Labrador, 3000 m depth.

2. *Regional topography*: Smooth, gently sloping (gradient 1:500), maximum local slopes of 2–5°; bed forms consisting of current lineations and ripples with wavelengths of 1–10 cm.

3. *Benthic environment*: *Site 1*: continental rise off New York: T (*in situ*) = 2°C; Salinity = 34.9‰; $[O_2]$ = 5.5 ml/liter. Current flow measured meters above bottom $\sim$ 10–30 cm s^{-1}, principally unidirectional; tidal component 2–5 cm s^{-1}. Biomass: 1–2 g m^{-2} (wet weight). Rate of sediment accumulation $\sim$ 10–100 cm/1000 years. Sediment type: silty clay: 90% < 62 µm; $CaCO_3$: 20–40%; organic C: 0.5–1%. Water depth of rise: 3–5 km, width 200 nautical miles. *Site 2*. Continental rise off Labrador: T(*in situ*) = 1.8°C; salinity = >35‰; $[O_2]$ = 6 ml liter^{-1}. Current flow measured 100 m above bottom $\sim$10 cm s^{-1} principally unidirectional. Biomass: 2–3 g m^{-2} (wet weight). Rate of sediment accumulation $\sim$10–100 cm/1000 years. Sediment type: silty to sandy clay: 90% < 120 µm; $CaCO_3$: 30–50%; organic C: 0.5–1.5%. Water depth of rise: 2–3.5 km, width 150 nautical miles.

4. *Comments*: These two sites lie approximately beneath the axis of the deep Western boundary undercurrent that flows along the continental rise in depths between 2500–3500 m in the Labrador Sea, to 3500–5000 m (approximately 2000 km downstream) on the continental rise off the eastern United States. They represent two areas influenced by strong thermohaline circulation in areas of high rates of sedimentation. They are considered to be typical of high-energy deep-sea environments. Numerous data exist for the continental rise off New York, less for the Labrador Sea. Benthic boundary layer is of the order of hundreds of meters thick.

Other Nonshelf Areas of Benthic Boundary Layer Phenomena

1. Deep (4–5 km) passages that constrict the flow of thermohaline circulation and hence represent exceptionally high-energy benthic boundary layer phenomena, e.g., Samoan Passage, West Equatorial Pacific; Vema Channel in the Rio Grande Rise (Southwestern Atlantic); the Falkland Gap at 50°S (West-South Atlantic); and the Charlie–Gibbs Fracture Zone at 52–53°N in the North Atlantic.

2. Shallow (1 km) passages or sills that control the overflow of dense polar water that eventually becomes the zonal-flowing bottom water, e.g., the Iceland-Scotland Ridge.

3. Active, fast-spreading centers where high geothermal heat gradients may have an important influence on the structure of the benthic boundary layer, e.g., the East Pacific Rise.

4. Areas where endemic biotic communities can be expected, e.g., the Red Sea, deep-sea trenches.

Survey Requirements

Each site should have a precisely navigated (1–10 m), extensive near-bottor survey prior to instrument deployment. Long-time series experiments should t preceded by a site survey to determine optimum spacing of sensors and th position of the experiment.

The Shelf

Boundary layer studies on the shelf could be divided into two categories.

Fundamental Studies

These should consider selecting particular field sites both on the inner an outer portions of the continental shelf, where the processes operating respond t low-frequency driving forces (tides) and intermediate-frequency driving force (waves), respectively. Such studies should be planned for relatively simpl physical conditions in regard to the boundary properties, adjacent topography wave- and current-fields, stratification, etc. This is in order to maximize th generality of the data so that similarity considerations may be applied to th information obtained. Such studies should be supported by laboratory exper ments and as their first objective should determine the velocity profiles, th bottom shear stress, the shear stress distributions in the boundary layer, an the structure of the boundary layer. Fundamental studies should be conducte in shelf areas that successively reflect the increasing complexity of the processe operating at the sediment–water interface. Although it is difficult to select an one site, the first tests should be conducted in a nonstratified water colum with monochromatic waves (or at least a unique wave direction), level or graduall sloping topography, homogeneous sediment composition and texture at th bottom, minimal bioturbation, and no chemical gradients. Such studies shoul commence on the outer continental shelf—where wave refraction is minimal in an open ocean (i.e., without the influence of land bodies or submerged shoal which channelize currents). Increasing complexity can then be added to th experimentation.

It should be kept in mind that a synoptic test is preferred for cross-correlatio of boundary layer parameters and for flux computation, and that as man relevant physical, chemical, and biological parameters as possible should b recorded during the test. It would be important to have sediment transpor tests included.

"Climatological" Studies

These should be considered to serve the purpose of comparing existing boundary layer data with weather records, tidal measurements, etc., with a view to establishing long term boundary layer "climate."

Priorities

The most important physical measurements are the velocity profiles in three-dimensions: specifically, the closely spaced orthogonal sets of current measurements in the vertical (Eulerian), coupled with experiments designed for area reconnaissance of current distribution and structure (Lagrangian). Although velocity profiles can be used to determine the shear stress distribution in the vertical and its boundary value, the direct measurement of shear should be encouraged because of its importance as driving force to the motion of sediments.

Measurements of the second order are the transport of sediments; local temperature; salinity, and density values; the properties of bottom sediment composition, texture, fabric, and distribution; area variations in micro- and macroscale roughness; the eddy diffusion coefficients in the boundary layer; and the vertical mixing coefficient in sediments.

Recommendations

1. Long-time series of observations are required to determine the variability in the benthic boundary layer. The duration needs to be at least one day and preferably one year.

2. Synoptic measurements are also required, more especially on the shelf in areas of complex morphology.

3. Measurements of turbulence characteristics and of mean velocities are of the highest priority in providing estimates of bed shear stress and eddy diffusion coefficients.

4. Study of the zones outside the boundary layer whose long-term characteristics may be affected by long-term exchanges and by variability.

5. Measurement of fluxes in heat and momentum: various chemical compounds both organic and inorganic, and sediment to and through the boundary layer.

6. Investigations of the effects of biota and physical disturbances in stirring sediment and in altering properties of the sediment.

7. Studies on the biota; food chains, including exchanges with other biota;

horizontal and vertical distribution and spreading in epibenthic communities; transport of organic material and absorbed substances.

8. Studies on the effects of area variations in roughness and steadyness of flow on the characteristics of the boundary layer.

9. The continental slope requires intensive investigation.

10. Chemical, microbiological, and mineralogical studies of the suspended particles in the nepheloid layers must be carried out on a routine basis. Methods for the chemical microanalysis of suspended materials are urgently required for these purposes.

11. The improvement and extension of modeling techniques directed toward the understanding of particulate transport cycles should be encouraged.

Key References

Bagnold, R. A., 1946, Motion of waves in shallow water, interaction between waves and sand bottoms, *Proceedings of the Royal Society of London, Series A, 187:* 1–18.

Biscaye, P. E., and Eittreim, S., 1974, Variations in benthic boundary layer phenomena: Nepheloid layer in the North American basin, in: (R. Gibbs, ed.), *Suspended Solids in Water*, Plenum Press, New York, pp. 227–260.

Bowden, K. F., Fairbairn, L. A., and Hughes, P., 1959, The distribution of shearing stresses in a tidal current, *Geophysical Journal 2:* 288–305.

Broecker, W. S., Cromwell, J., and Li, Y. H., 1968, Rates of vertical eddy diffusion near the ocean floor based on measurements of the distribution of excess [222]Radon, *Earth and Planetary Science Letters 5:* 101–105.

Carstens, M. R., Neilson, F. M., and Altenbilek, H. D., 1969, Bed forms generated in the laboratory under an oscillatory flow, analytical and experimental study, *Coastal Engineering Research Center, Washington, D.C., Technical Memo 28*, 93 pp.

Collins, J. I., 1963, Inception of turbulence at the bed under periodic gravity waves, *Journal of Geophysical Research 70:* 4561–4572.

Eittreim, S., and Ewing, M., 1972, Suspended particulate matter in the deep waters of the North American Basin, in: (A. Gordon, ed.), *Studies in Physical Oceanography*, Gordon and Breach, London, pp. 123–167.

Gaertner, A., ed., 1968, Marine Mykologie, Symposium über Niedere Pilze im Küstenbereich in Bremerhaven, October 17–19, 1966, *Veröffentlichen Institut fur Meeresforschung Bremerhaven, Suppl. 3:* 1–159.

Gaertner, A., ed., 1974, Marine Mykologie. 2. Internationales Symposium in Bremerhaven, September 11–16, 1972, *Veröffentlichen Institut fur Meeresforschung Bremerhaven, Suppl. 5:* 1–159.

Grice, G. D., and Hülsemann, K., 1970, New species of bottom-living calanoid copepods collected in deep water by the DSRV Alvin, *Bulletin of the Museum of Comparative Zoology, Harvard University 139:* 185–230.

Heezen, B. C., and Hollister, C. D., 1971, *The Face of the Deep*, Oxford University Press, New York, 659 pp.

Hesthagen, I. H., 1970, On the near-bottom plankton and benthic invertebrate fauna of the Josephine Seamount and the Great Meteor Seamount, *"Meteor" Forschungs-Ergebnisse, Reihe D, No. 8:* 61–70.

Horikawa, K, and Watanabe, A., 1970, Turbulence and sediment concentration due to waves, *Proceedings of 12th Coastal Engineering Conference, American Society of Civil Engineers,* pp. 751–766.

Jannasch, H. W., Eimhjellen, K., Wirsen, C. O., and Farmanfarmian, A., 1971, Microbial degradation of organic matter in the deep sea, *Science 171:* 672–675.

Jannasch, H. W., and Pritchard, P. H., 1972, The role of inert particulate matter in the activity of aquatic microorganisms, *Proceedings of IBP-UNESCO-Symposium, Pallanza, 1972, Memoria Instituto Italiano Idrobiologico 29* (Suppl.): 289–308.

Johnston, T. W., and Sparrow, F. K., 1961, *Fungi in Oceans and Estuaries,* J. Cramer, Weinheim, 668 pp.

Jonsson, I. G., 1965, On the existence of universal velocity distribution in an oscillatory turbulent boundary layer, *Coastal Engineering Laboratory, Technical University of Denmark, Copenhagen, Report 12:* 2–10.

Kajiura, K., 1968, A model of the bottom boundary layer in waves, *Bulletin of the Earthquake Research Institute, Tokyo University, Japan 46:* 75–123.

Longuet-Higgins, M. S., 1953, Mass transport in water waves, *Philosophical Transactions of the Royal Society of London, Series A, 245:* 535–581.

McCave, I. N., 1973, Some boundary-layer characteristics of tidal currents bearing sand in suspension, *Mémoires Société Royale des Sciences de Liège, 6th Series, 6:* 187–206.

Sternberg, R. W., 1970, Field measurements of hydrodynamic roughness of the deep-sea boundary, *Deep-Sea Research 17:* 413–420.

Wellershaus, S., 1973, A new method for collecting near-bottom water in the deep sea, *"Meteor" Forschungs-Ergebnisse, Reihe A., No. 13:* 50–57.

Wimbush, M., and Munk, W., 1970, The benthic boundary layer, in: (A. Maxwell, ed.), *The Sea,* Vol. 4 (1), Wiley, New York, pp. 730–758.

Yalin, M. S., and Russell, R. C. H., 1966, Shear stresses due to long waves, *Journal of Hydraulics Research 4:* 55–98.

Zobell, C. E., 1970, Pressure effects on morphology and life processes of bacteria, in: (A. M. Zimmermann, ed.), *High Pressure Effects on Cellular Processes,* Academic Press, New York.

B
Velocity Variations, Turbulence, and Stability

Group Leader: K. F. Bowden

Members: Y. Desaubies, A. Führböter, C. H. Gibson,
C. M. Gordon, W. J. Gould, G. Gust,
A. D. Heathershaw, J. D. Smith, G. L. Weatherly

Rapporteur: R. C. Seitz

Introduction

The boundary layer above the sea bed, particularly in deep water, is one of the least understood of all geophysical scale flows, largely because few observational data within this region have been collected. The flow characteristics here are important in determining the vertical transport of suspended material and chemical substances to and from the sediments as well as the horizontal transport of sediments along the bottom boundary. A considerable body of knowledge dealing with boundary layer flows has been built up from theoretical models as well as from measurements in the laboratory, and in naturally occurring geophysical scale flows. Much more data in the layers at the sea floor, especially in the deeper part of the ocean, has to be obtained before one can establish a reliable basis for comparison of such benthic boundary layer flows with similar phenomena in the lower atmosphere and the laboratory. A further reason for developing an adequate description of the bottom boundary layer flow arises from the fact that it is necessary to understand the interaction between the flow in the interior of the ocean and the flow and stresses near the bottom boundary in order to model effectively the dynamics in the interior of the ocean.

Future efforts should be directed along two main lines, the first being an attempt to define the experimental approaches which will provide a basis for understanding the naturally occurring phenomenon and comparing it with other boundary layer flows. The second is to ascertain the sorts of experiments that will yield the information required by scientists working in other disciplines, in particular those who are concerned with chemical, biological, and geological processes within the boundary layer.

A working definition of the benthic boundary layer is the zone above the bottom boundary within which the effects of the boundary upon the flow can be detected. Such a definition recognizes that much work is still required with oscillatory, stratified, rotating boundary layer flows before a more precise definition can be attained.

Theoretical Background

Categories of Benthic Boundary Layers

There is reason to expect that the bottom boundary layer of the ocean usually will be turbulent and it is convenient to separate this region into four idealized categories. Comparison of various length scales can be used to characterize these categories one of which is the Ekman depth, which is defined as $L_E = u_*/f$ where $u_* = (\tau_0/\rho)^{1/2}$, τ_0 is the bottom shear stress, ρ is the density, f is the Coriolis parameter. The length scale for a stratified sheared boundary layer is the Monin–Obukhov length: $L_M = \rho u_*^3/g\kappa \overline{\rho'w'}$, where κ is Karman's constant and $\overline{\rho'w'}$ is the buoyancy flux at the boundary. This is the height at which the turbulent kinetic energy production is balanced by loss to the potential energy of mixing in a stably stratified medium. The Monin–Obukhov length is also applicable to unstably stratified flows, but changes sign under these conditions due to the change in sign of $\overline{\rho'w'}$. In the abyssal boundary layer, $L_E \approx 10$ m and $|L_M| > 100$ m. In an oscillatory flow the proper length scale is $L_\omega = \bar{u}_*/\omega$, where ω is the frequency. A final bounding length scale applicable to some shallow water flows is the fluid depth.

Neutral, Nonrotating, Steady Boundary Layers ($L \ll |L_M|, L_E, L_\omega$)

The most familiar boundary layer in hydrodynamics is found when a uniform flow of constant density fluid flows over a flat surface in a nonrotating system. The boundary layer thickness, L, increases continually with distance from the leading edge of the surface. Eventually the layer occupies the entire depth of the fluid. Such boundary layers have been carefully studied in laboratories and both the theory and experimental results are described by Schlichting (1968). In the lower part of this layer the shear stress is nearly constant and the velocity profile has the logarithmic form

$$\frac{U}{u_*} = \frac{1}{\kappa} \ln \frac{z}{z_0}$$

where z_0 is a length scale associated with the size of the boundary roughness and/or thickness of the viscous sublayer.

Stratified, Nonrotating, Steady Boundary Layers ($L \sim |L_M| \ll L_E, L_\omega$)

Where the stable stratification is strong or where $f \to 0$ near the equator the buoyancy may constrain the vertical extent of the turbulence to scales small compared to the Ekman depth u_*/f. The layer will be well mixed to scales of order L_M since the turbulence will prevail over buoyancy forces within a layer of this scale. Consequently vertical diffusion of scalar properties such as temperature, salinity, or chemical species may occur within the layer. However, diffusion across the density interface which forms at the top of the stable boundary layer may be strongly inhibited. Momentum transport across the density interface is less inhibited because entrainment is not mandatory as it is for salinity or temperature, and momentum may be transmitted by internal waves. Under unstable conditions the upward momentum flux and the turbulent mixing is enhanced by buoyancy forces resulting in a reduced boundary shear stress. In the stratified case the flow. near the boundary becomes $U/u_* = 1/\kappa$ [ln z/z_0 + $\beta z/L_M$], where β is approximately 4.0 in the atmospheric boundary layer over the sea.

Neutral, Rotating, Steady (Ekman) Boundary Layers ($L \sim L_E \ll |L_M|, L_\omega$)

Boundary layers formed by fluid in contact with a rotating surface are limited to a constant thickness depending on the stress, density, and component of the angular velocity normal to the x surface. For the earth this thickness is the Ekman depth u_*/f, which may be quite small for the small u_* values expected on the abyssal ocean bottom. The velocity profile for $Z < u_* (10f)$ is as given below.

Neutral, Nonrotating, Oscillatory Boundary Layers ($L \sim L_\omega \ll |L_M|, L_E$)

These boundary layers are formed under currents which oscillate with frequency $\omega \gg f$ (e.g., currents from surface gravity waves).

In the benthic boundary layer these idealized cases are rarely encountered in their pure forms. In the deep sea, tidal motions are often dominant. For these motions ω is of the same order as f ($L_E \sim L_\omega$); not much is known about this case. In some regions of the benthic boundary layer, L_M, L_E, L_ω may all be of comparable magnitude. Because of these complications and the fact that the layer is turbulent, the dynamics of the benthic boundary layer are poorly understood.

Mixing and Diffusion in Boundary Layers

Universal Similarity of Turbulent Velocity and Scalar Fields

Probability laws describing the differences of velocity between points in the fluid separated by distances that are small compared with the boundary layer thickness depend only on the Kolmogoroff length $L_K = (\nu^3/\epsilon)^{1/4}$, and time $T_K = (\nu/\epsilon)^{1/2}$, where ν is the kinematic viscosity, and ϵ is the local rate at which the turbulence loses energy per unit mass to viscous dissipation. In the logarithmic profile layer $\epsilon = u_*^3/\kappa z$. The Kolmogoroff length is the size of the smallest eddies and ranges from 0.2 to 0.02 cm at the base of the abyssal and shallow water benthic boundary layers, respectively.

An important consequence of the preceding "universal similarity hypothesis" is that parameters such as velocity spectra at sufficiently high wave numbers become universal when made dimensionless with L_K and T_K, so they can be compared with universal spectral forms established in the laboratory or in previous field experiments. By fitting various portions of measured spectra to such universal curves, it becomes possible to deduce ϵ for the particular flow. A similar approach can be applied to scalar fields such as temperature. This approach is quite useful in estimating characteristics of the turbulence in other portions of the spectrum, such as the diffusivity of heat or momentum or the separation or coagulation of various-sized solid particles separated by distances small compared with the elevation above the boundary in the universal range. Ventilation of living organisms smaller than L_K depends critically on the local rate of strain parameter $\gamma = (\epsilon/\nu)^{1/2}$. The thickness of thermal and chemical boundary layers on the surface of suspended particles or animals is given by $L_D = (D/\gamma)^{1/2} = L_K (D/\nu)^{1/2}$, where D is the molecular diffusivity. Hence the salinity boundary layer thickness on the sediment bed will be typically 30 times thinner than the viscous sublayer thickness. At the boundary L_K reduces to ν/u_* and experiments show the actual thickness of the viscous sublayer to be 10 L_K. In view of recent laboratory investigations (Eckelmann, 1974), a more exact estimate of the diffusion layer thickness may be possible.

Vertical Fluxes of Momentum, Heat, and Salinity

Direct measurements of vertical fluxes $\overline{u'w'}$, $\overline{T'w'}$, or $\overline{S'w'}$ are quite difficult, where u' and w' are horizontal downstream and vertical velocity fluctuations. It is often helpful to estimate these fluxes with eddy flux parameters such as the turbulent viscosity $\nu_T = \overline{u'w'}/\partial \overline{U}/\partial z)$ and the turbulent temperature diffusivity $D_T = \overline{w'T'}/(\partial \overline{T}/\partial z)$.

An alternative method of estimating ν_T is by using the relation $\nu_T = \ell b^{1/2}$ where b is the turbulent kinetic energy per unit mass and $\ell = -K(b^{1/2}/\ell)/\partial/\partial z$ $(b^{1/2}/\ell)$ (Weatherly, 1975), together with the full turbulent energy equation. In the log profile layer b becomes u_* and ℓ becomes κz so ν_T reduces to $\kappa u_* z$.

Results from Available Boundary Layer Models

Unsteady planetary boundary layer theories are more applicable to the benthic situation than steady ones because of the relative magnitude of tidal currents in most oceanic areas. Nevertheless, in some localities nontidal currents predominate and, in general, useful insights into the dynamics of the abyssal layer can be obtained from a brief examination of the models available for the atmospheric boundary layer. Recently McPhee (1974) has found these to be in good agreement with experimental data on the mean velocity and stress profile from the Arctic surface mixed layer, which is considered to be a potential model for a near neutrally stable, quasi-steady benthic "Ekman layer."

Most atmospheric boundary layer models are concerned with neutral and unstable rather than stable conditions, whereas the benthic layer is expected to range from slightly stable to very slightly unstable. For this reason, the greatest oceanographic concern is with neutral and stable models. Theories of the first type have been published by Deardorff (1972), Wyngaard *et al.* (1974), and Shir (1973).

Deardorff proposed a model based upon a numerical integration of the three-dimensional equations of motion for both neutral and unstable cases. Moreover, his neutral results are often used as a standard for other models due to the paucity of outer boundary layer Reynolds stress data. In contrast, Wyngaard *et al.* and Shir employ second-order closure schemes and evaluate the resulting "free constants" in terms of atmospheric surface layer measurements. Businger and Arya (1974) have presented an eddy viscosity model for a stable planetary boundary layer, and Lykosov and Gutman (1972) present a second-order closure model. The results indicate that the effect of increasing stability is to lower the geostrophic drag coefficient, increase the veering angle of the Ekman spiral, and decrease the penetration distance for frictional effects.

All of the above-described models, in conjunction with the McPhee data, indicate that u_*/f is the proper scaling for neutral and very slightly stable, quasi-steady boundary layers and that the angle between the geostrophic flow and surface stress is about $24°$. In addition, the predicted and measured mean velocity and stress profiles are generally in good agreement.

Recently Weatherly (1975) used the second-order closure model of Lykosov and Gutman (1972) to study effects of nonstationarity in a slightly stable stratified benthic boundary layer. This preliminary study is encouraging in that it reproduces some observed features unaccounted for in stationary turbulent Ekman layer theories.

Measurements in Turbulent Boundary Layers in Shallow and Deep Water

Table 1 gives the instrumental and environmental characteristics of recent measurements of high frequency velocity fluctuations in shallow water.

Turbulent Velocity Measurements in Shallow Water Boundary Layers

Spectral Measurements

Measurements of the energy spectra of one or more velocity components indicate that an (isotropic) inertial subrange of turbulent energy exists. This

Table 1.

Investigation	Date	Instrument	Measurement scale	Frequency response	Measurements and computations	Flow conditions
Bowden and Fairbairn	1956	electromagnetic	10 cm	1 Hz	Energy spectra	Continental shelf (12–20 m)
Heathershaw	1974	electromagnetic	5 cm & 10 cm	1 Hz	2-D spectra and cospectra 3-D spectra and cospectra	Shelf (10–60 m)
Gordon	1974	pivoted vane-ducted propeller meter			Two-dimensional stress measurements	Tidal-river (10 m)
Seitz	1973	acoustic-doppler	1 cm	(1–20) Hz	Three-dimensional spectra	Tidal-river (13 m)
Gust	1974	drag force probe	2.5 cm	15 Hz	Velocity profiles Velocity fluctuations	Tidal-shallow
Gust	1974[a]	hot wire probe	0.4 mm	25 Hz (3db)	Velocity profiles	Baltic sea (2.5 m)
Fuhrboter	1974	electromagnetic	12 cm	5–10 Hz	Horizontal velocity components (0.3)m	Surf zone
Gibson	1974[a]	hot wire	1 mm	1 Hz	Surface layer	Open-sea surf zone
Smith	1974	partially-ducted propeller meter	4 cm Rotor 2 cm sec^{-1} threshold $\Delta V = \pm .05$ cm sec^{-1}	5 Hz	Three-dimensional spectra and cospectra	1. Boundary layer below Arctic ice flow 2. Boundary layer above nonuniform bottom in tidal river

[a]Not yet published.

apparently extends into the dissipation scales of motion. The nonstationary characteristics of the tidal flows involved have made interpretation of the low wave number portions of all the spectral measurements difficult.

Cospectral Measurements

Reynolds stress values obtained from cospectral measurements are consistent with determinations from the mean profile and other methods. Very little contribution to the Reynolds stress occurs in the inertial subrange.

Reynolds Stress Measurements

Intermittency has been observed in the small-scale velocity structure in some of the boundary layer measurements. Data obtained from direct measurements of $\overline{u'w'}$ are well correlated with the levels of turbulent energy in all three directions, suggesting a possible indirect measure of Reynolds stress by means of the turbulent kinetic energy. Under nonuniform conditions such as those over uneven bottom topography special measurement and analysis problems arise (Smith *et al.*, unpublished).

Of importance to biologists, chemists, and geologists is the evidence that the "viscous sublayer" is periodically disrupted by a sequence of motions, collectively referred to as "bursting." The occurrence of this process has been clearly observed since the development of flow visualization techniques for studying laboratory boundary layers (Kline *et al.*, 1967). These experiments have shown that momentum transport (Reynolds stress) and turbulence generation are intermittent. These results suggest that the turbulent structure of boundary layers may best be described by statistical description of the actual motions involved. The implications of this quasiperiodic process for geophysical boundary flows have been pointed out by Grass (1971). Recent field measurements in a tidal estuary (Gordon, 1974; Seitz, 1973) and on the continental shelf (Heathershaw, 1974) have demonstrated that similar intermittent momentum-transporting events occur in the benthic boundary layer. Momentary values of $\rho u'w'$ exceeding 50 times the average $\overline{\rho u'w'}$ (Reynolds stress) have been observed. Future interpretations of chemical, geological, and biological processes will need to take intermittency into account.

Observations in Turbulent Ekman Layers

Measurements in Shallow Water

A well-developed Ekman layer has been observed beneath a wind-blown ice floe (Smith, 1974). Due to the absence of surface waves and the fact that the measurements were made with equipment attached to the ice, they are potentially applicable to the benthic boundary layer. The effects of stratification upon the Ekman layer were determined. The logarithmic variation of velocity extended 70 cm below the ice, and between 70 cm and 2 m merged into the Ekman profile which extended to a depth of 38 m. The variation of stress and eddy viscosity with depth were successfully compared with similar atmospheric boundary studies.

Measurements in the Deep Sea

Measurements in the deep-sea benthic boundary layer are very sparse. The few measurements that have been made show that the lower part of the layer is adequately described by existing theories of planetary boundary layers (Wimbush and Munk, 1970; Weatherly, 1972). In particular the velocity profile in the lower part of the layer is observed to be logarithmic. In the eastern Pacific the logarithmic layer overlies a hydrodynamically smooth bottom, and under the Gulf Stream, a marginally rough bottom. These logarithmic layers are characterized by friction velocities, u_*, of 0.1 cm s^{-1} in the eastern Pacific and 0.4 cm s^{-1} under the Gulf Stream. The observed thicknesses of these logarithmic layers are 1 m and 4 m, respectively.

Overlying the logarithmic layer is a turbulent Ekman layer. The mean total Ekman veering throughout the boundary layer is of order 10° or less, which is somewhat smaller than that expected for steady layers. The Ekman veering was observed to be highly variable with time. This variability is thought to be due to the oscillatory nature of the current outside the boundary layer. The depths of these observed turbulent Ekman layers are also variable in time. Representative thicknesses of these layers are 10 m for the eastern Pacific and 25 m under the Gulf Stream. Theoretical studies of time-dependent turbulent Ekman layers are beginning to appear (e.g., Weatherly, 1975). These should prove useful in interpreting the available data and data from future studies of nonstationary turbulent Ekman layers.

Measurements of longitudinal and transverse fluctuations of velocity, as well as of the mean current, were made at a depth of nearly 1000 m in the Mediterranean outflow in the Gulf of Cadiz (Thorpe et al., 1973). In this rapid flow the turbulent fluctuations were comparable in intensity to those measured in shallow water.

Vertical profiles of potential temperature and salinity reaching to the bottom of the abyssal ocean (~5400 m) in the northwest Atlantic are reported by Millard (1974). Homogeneous layers (to 0.001°C and 0.003°/oo) varying in thickness up to 140 m were found at the bottom at 49 out of 53 stations. The layer thickness was greatly variable in both space and time, with a mean value of around 40 m, and exhibited no systematic geographic trend. The potential temperature of the layers, however, appeared to correlate with the bottom topography.

Weatherly and Niiler (1974) report similar homogeneous layers on two sections beneath the Florida current. From 500 profiles taken over several years the frequency of occurrence of layers is 80–90% overall but is as low as 20% at certain selected positions. Both Weatherly and Niiler and Millard find that thickening of a layer is associated with a decrease in temperature, contrary to the result expected from a simple mixing theory. Weatherly and Niiler detected consistent variations of layer occurrence frequency with position (associated with bottom topography).

A model developed by Weatherly (1975) using cross-isobaric flow in the sections could be useful in explaining some features of the homogeneous layers.

Outstanding Problems and Recommendations

The benthic boundary layer of the deep ocean (abyssal boundary layer) poses a specific set of problems. Direct measurements of turbulent processes in this environment have been limited by difficulties associated with accessibility and instrumentation. Most of our knowledge has been inferred from mean gradient values.

The flow regimes in the deepest parts of the ocean (except in unusual areas such as the Gibraltar and Norwegian outflows) are characterized by low ($\leqslant 10$ cm s^{-1}) velocities of tidal or inertial period superimposed upon even smaller residual flows (1 or 2 cm s^{-1} when averaged over the dominant tidal-inertial periodicity).

Stratification in the deep layers of the ocean is a topic of special interest. The observations of Millard (1974) and Weatherly and Niiler (1974), as well as others in many parts of the abyssal ocean, continental slopes, and shelves, indicate the existence of homogeneous bottom layers often several tens of meters in thickness.

Long-Duration Deep-Sea Measurement

In order to test the validity of the extrapolation of shallow water and laboratory observations to the deep ocean, at least one definitive set of observations should be made in the abyssal boundary layer. Such a set of observations should, as a first priority, seek to measure the three components of turbulent velocity fluctuations as well as the turbulent fluctuations of temperature (and salinity) in the abyssal boundary layer. These measurements should be made at a sufficient number of heights above the bed to specify precisely the turbulent processes throughout the boundary layer. The measurements should be of sufficient duration and of adequately high frequency to provide good statistical data. The derivation of mean velocity, temperature, and salinity profiles from these or other measurements is also essential.

There are inherent difficulties in making measurements of flows in such low-velocity regimes. For the extremely low-velocity regimes which may be encountered further instrument development may be required.

Bottom Homogeneous Layers and Mixing Processes

Further theoretical and observational investigation of the known abyssal boundary layer phenomena such as the deep homogeneous layers should attempt to define both the frequency of occurrence and geographic distribution of such features (with particular emphasis on the role of bottom topography) and should ultimately seek to explain their origins. These layers are important since they represent, a possible region for significant mixing of bottom and overlying water masses and also form the lower boundary of the midwater ocean circulation. A final objective of these abyssal ocean experiments should be the parameterization of the interaction of the boundary layer with the external flow. The functional form of the geostrophic drag coefficient depends on such external parameters as the density stratification, the scale and distribution of the bottom roughness, the scale of the bottom topography, and the values of the exterior velocity.

It is also known that internal waves can be generated by a turbulent boundary layer. If such a process occurs at the benthic boundary layer, a loss of momentum from the layer is expected. A particularly important example of this is the interaction of the surface tides with the bottom boundary layer on the continental slopes, resulting in a conversion of their energy to internal waves of tidal frequency (Wunsch and Hendry, 1972) and in some local mixing. Experimental investigation of this effect would require near-bottom current and temperature records of sufficient length.

Spatial and Temporal Variation of Turbulence

In those areas of the sea bottom where currents are unsteady due to tidal influences, it is very important to study both spatial and temporal variation in the turbulent structure of the boundary layer over relatively long periods of time, in order to identify those processes occurring during the acceleration or deceleration of the flow. (Near-bottom current measurements at 800 m depth have shown that turbulence is intensified during decelerating tidal phases.) Such basic measurements, including effects of intermittency, are necessary for unambiguous interpretation of short-term measurements such as deep STD casts, and to develop appropriate sampling procedures. Since the temporal scale of variability in tidal and inertial motion is of the order of $1/f$ caution must be exercised in comparing experimental results with stationary theories of the turbulent Ekman layer.

Examination of the momentum balance equation of motion for boundary layers indicates that the vertical distribution of Reynolds stress is also influenced by accelerations of the fluid. Stress measurements in laboratory and shallow water confirm that this effect is considerable. Tidal phase dependence of stress distribution is of importance in our understanding of sediment entrainment and transport in unsteady flows. Velocity measurements in the benthic boundary layer should be of sufficient duration to match the predicted interval between "bursting" events and to establish their importance in vertical momentum transport in the deep sea.

Evaluation of Methods for Determining Bottom Stress

Future experiments should allow for a comparison of "profile," "eddy correlation," and other methods used in the derivation of bottom shear stresses and sediment transport related parameters. In particular the meaning of values of z_0 (roughness length) and C_D (drag coefficient) should be considered carefully in their relation to observable sedimentary features. Field observations of z_0 and C_D exhibit a high degree of variability which cannot be explained simply. Techniques should be developed for the *in situ* determination of suspended sediment concentrations. Such observations would enable direct correlation between sediment transport processes and the measured turbulent structure of the boundary layer. Laboratory experiments indicate that once sediment is in suspension it alters the dynamic characteristics of the fluid. The properties of the resultant two-phase fluid system may require reexamination of the methods used in calculating turbulence parameters and bottom shear stress.

Spectra

Spectral analyses of time series of turbulent motions have proved to be a valuable tool in interpreting natural boundary layer phenomena. The following extensions of this technique into the study of the benthic boundary layer regimes are proposed:

1. Attempts should be made to determine the influence of density stratification upon the three-dimensional turbulent kinetic energy distribution.

2. The terms in the equation for turbulent kinetic energy at various depths in the flow should be evaluated.
3. Cospectra measurements would be of value in understanding the structure of the Reynolds stress.
4. Spectral measurements should be made in the benthic boundary layer in order to obtain values for energy dissipation and its variation from very close to the bottom boundary up to heights of meters.
5. Further studies of the validity of the "frozen field" hypothesis in low-velocity regimes should be made.

Existence of Viscous and Conductive Sublayers

For chemical processes in the sediment and in the interstitial water it is important to know whether a conductive sublayer exists. Laboratory experiments have shown that a conductive sublayer can exist only if there is a viscous sublayer, which in hydraulically smooth flow has a purely viscous wall layer with a thickness on the order of a hundredth of a millimeter. It is only in this region that the molecular treatment of the flux is justified. In hydraulically rough flow no viscous sublayer exists. It is recommended that benthic boundary layer flow conditions should be examined to determine whether a purely viscous wall layer can exist, and, if so, how thick it is.

Relationships between Hydrodynamics and Biological Processes

There is a mutual relationship between the activity of benthic marine life and the hydrodynamic properties of the boundary layer. The structure of the flow has a strong influence on the chemical, sedimentary, and nutrient environment of plants and animals, while they in turn can produce a measurable effect on the bottom roughness and form–drag characteristics of the fluid–bottom interface. An extreme example of this type of interaction is provided by such biological phenomena as shallow-water kelp beds. At the present time this interdependence of hydrodynamics and marine ecology is poorly understood and requires detailed study.

Key References

Biermann, P., Gust, G., Hatze, G., and Ohm, K., 1974, A multi-channel measuring and recording device with high storage capacity and IBM-compatibility, *Report No. 5 of the Joint Research Programme*, SFB95, "Interaction sea-bottom" of Kiel University.

Bowden, K. F., and Fairbairn, L. A., 1956, Measurements of turbulent fluctuations and Reynolds stresses in a tidal current, *Proceedings of the Royal Society of London, Series A* 237: 422–438.

Businger, J. A., Wyngaard, J. C., Izumi, Y., and Bradley, E. R., 1971, Flux profile relationships in the atmospheric surface layer, *Journal of Atmospheric Sciences* 28: 181–189.

Businger, J. A., and Arya, S. P. S., 1974, The height of the mixed layer in the stably stratified planetary boundary layer, *Advances in Geophysics* 18A: 73–92.

Deardorff, J. W., 1972, Numerical investigation of neutral and unstable planetary boundary layers, *Journal of Atmospheric Science* 29: 91–115.

Eckelmann, H., 1974, The structure of the viscous sublayer and the adjacent wall region in a turbulent channel flow, *Journal of Fluid Mechanics* 65: 439–459.

Führböter, A., Büsching, F., and Dette, H. H., 1974, Field investigations in surf zones, *Proceedings of the 14th Coastal Engineering Conference, Copenhagen*, paper 1, pp. 2–4.

Gordon, C. M., 1974, Intermittent momentum transport in a geophysical boundary layer, *Nature* 248: 392–394.

Grass, A. J., 1971, Structural features of turbulent flow over smooth and rough boundaries, *Journal of Fluid Mechanics* 50: 233–255.

Gust, G., 1975, Das experiment "Schliwe 1" im Schlickswatt vor Süderhafen/Nordstrand, *Messprogram un Daten-Reports Sonderforschungsbereich* 95 *Universität Kiel*, No. 11.

Heathershaw, A. D., 1974, "Bursting" phenomena in the sea, *Nature* 242: 394–395.

Kline, S. J., Reynolds, W. C., Schaub, F. A., and Runstadler, P. W., 1967, The structure of turbulent boundary layers, *Journal of Fluid Mechanics* 30: 741–773.

Landau, L. D., and Liftshitz, E. M., 1959, *Fluid Mechanics*, Pergamon Press, London.

Lykosov, V. N., and Gutman, L. N., 1972, The turbulent boundary layer above a sloping underlying surface, *Izvestiya Atmospheric and Oceanic Physics* 8: 462–467 (799–809 in Russian original).

McPhee, M. G., 1974, The turbulent boundary layer under Arctic pack ice, Ph.D. Thesis, University of Washington, Seattle, Washington.

Millard, R., 1974, Bottom layer observations from MODE and IWEX, *MODE Hot Line News No. 60*, Woods Hole Oceanographic Institution, Woods Hole, Massachussetts, p. 5.

Schlicting, H., 1968, *Boundary Layer Theory*, 6th ed., McGraw-Hill, New York.

Seitz, R. C., 1973, Observations of intermediate and small-scale turbulent water motion in a stratified estuary, Parts I and II, *Chesapeake Bay Institute, John Hopkins University, Technical Report* 79.

Shir, C. C., 1973, A preliminary numerical study of atmospheric turbulent flow in the idealized planetary layer, *Journal of Atmospheric Sciences* 30: 1327–1339.

Smith, J. D., 1974, Turbulent structure of the surface boundary layer in an ice-covered ocean, in: Proceedings of a symposium on the physical processes responsible for the dispersal of pollutents in the sea, particularly in the near-shore zone, *Conseil International pour l'Exploration de la Mer, Rapports et Proces-Verbaux* 167: 53–65.

Smith, J. D., McLean, S. R., Chubb, J. E., and Begley, J. N., Flow over sand waves in the Columbia river (unpublished).

Thorpe, S. A., Collins, E. P., and Gaunt, D. I., 1973, An electromagnetic meter to measure turbulent fluctuations near the sea floor, *Deep-Sea Research* 20: 933–938.

Townsend, A. A., 1965, Excitation of internal waves by a turbulent boundary layer, *Journal of Fluid Mechanics* 22: 241–252.

Vager, B. G., and Kagan, B. A., 1969, The dynamics of the turbulent boundary layer in a tidal current, *Izvestiya Atmospheric and Oceanic Physics* 5: 88–93, (168–179 in Russian original).

Weatherly, G. L., 1972, A study of the bottom boundary layer of the Florida current, *Journal of Physical Oceanography* 2: 54–72.

Weatherly, G. L., 1975, A numerical study of time-dependent turbulent Ekman layers over horizontal and sloping bottoms, *Journal of Physical Oceanography*, 5: 288–299.

Weatherly, G. L., and Niiler, P. P., 1974, Bottom homogenous layers in the Florida current, *Geophysical Research Letters* 1: 316–319.

Wimbush, M., and Munk, W., 1970, The benthic boundary layer, in: (A. Maxwell, ed.), *The Sea*, Vol. 4 (1), Wiley, New York, pp. 730–758.

Wunsch, C. I., and Hendry, R. E., 1972, Array measurements of the bottom boundary layer and the internal wave field on the continental slope, *Geophysical Fluid Dynamics* 4: 101–145.

Wyngaard, J. C., Coté, O. R., and Rao, K. S., 1974, Modelling the atmospheric boundary layer, *Advances in Geophysics* 18A: 193–211.

C
Sea-Floor Deposition, Erosion, and Transportation

Group Leader: A. F. Richards

Members: M. de Vries, C. Einsele, P. D. Komar, R. B. Krone,
M. W. Owen, J. B. Southard, F. N. Spiess,
D. Taylor-Smith, J. Terwindt

Rapporteur: G. H. Keller

Introduction

A knowledge of the materials, processes, and environments of the sea floor is fundamental to many scientific and engineering disciplines. The continuity of natural processes within the benthic boundary layer with respect to deposition, erosion, and transportation is shown in Fig. 1. In this group report, problems concerned with the properties and processes relating to sea-floor deposits are first considered, followed by a discussion of erosion and transportation of cohesive and cohesionless sediments.

Geotechnology of Sea-bed Deposits

The sea-bed deposits of the benthic boundary layer and their subsequent history can be discussed by referring to their geotechnical properties. Marine geotechnology is the study of sea-floor sediments and rocks with the goal of determining their static and dynamic mechanical properties. To accomplish this, one must consider the acoustical, biological, chemical, mechanical, and physical properties affecting the electrolyte–gas–solid sedimentary system of the sea floor and the response of this system to applied static and dynamic loads.

Stress and Failure

A knowledge of the state of stress in sediments is necessary to understand their behavior under natural or structural loads, which may be static and/or dynamic. Total stress analysis, which generally is adequate for static loading, is a standard practice. Since the magnitude of *in situ* pore-water stresses is almost unknown in sediments, effective stress analyses, which are particularly important for dynamic loading, can only be approximated. This is especially true in the consideration of slope stability. While some submarine slope failures close to land have been studied in detail, little attempt has been made to consider the problem elsewhere in the marine environment. In an indirect sense, slope failures affecting the deep-sea floor can be inferred from the wide distribution of turbidite sequences throughout the ocean basins.

Two types of slope failure are common: one where there has been little or no pore pressure build-up and the failure is a result of depositional oversteepening, perhaps initiated by erosion at the toe of the slope; and the other where rapid sediment loading or dynamic loading by earthquakes, severe storm waves,

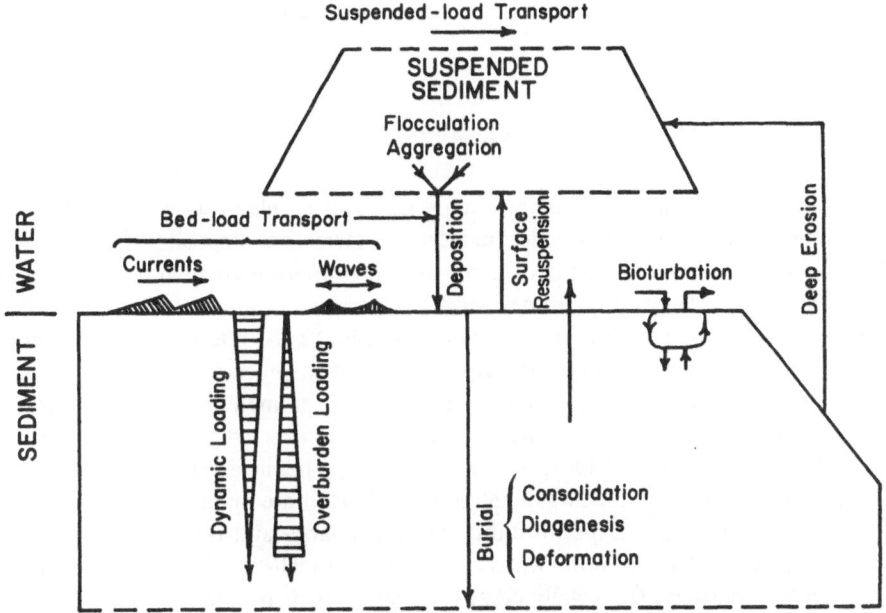

Fig. 1. Benthic boundary sedimentary processes.

etc., causes large excess pore-pressures, with consequent loss of strength and slope instability.

Studies are required of areas susceptible to slope failure both in the deep sea and on continental margins. Initial efforts should be directed toward such areas as delta fronts, where, because of rapid deposition, the probability of failure is high. Data to be collected *in situ* include the angle of friction in sands and the shear strength in clays, as well as bulk density and pore-water stress in both materials. The effects of microbially produced gas, bioturbation, cementation, and other modifications of the texture and fabric should also be examined. The changes in geotechnical properties under static or cyclic loading should be monitored *in situ*. Complementary to *in situ* observations, carefully collected samples should be subjected to cyclic loading tests in the laboratory to enable a clearer assessment of the critical condition beyond which they become unstable.

Slope failure can also be regarded as one source of material for density flows along the sea floor. Such flows range widely in sediment concentrations from low-density nepheloid layers, to turbidity currents, to high-concentration flow slides or slumps. There is only a limited understanding of the dynamics of density flows and what governs their capability to be self-sustaining. Hypotheses have been developed, but there is a need for laboratory study and appropriate field measurements.

Factors Controlling Their Geotechnical Properties

Sediments deposited in and maintained in an electrolyte appear to display some different characteristics relative to terrestrial deposits. The controlling factors in the development of geotechnical properties of such deposits are complex. They are partially understood for silica and carbonate sands, where the relationship of grain-to-grain contacts (packing) is of importance in the control of shear strength, compressibility, and acoustic properties. These factors are, however, poorly understood for cohesive, fine-grained sediments. In these materials, an influence of unknown magnitude is exerted by such factors as the mechanical effects of bioturbation; the biogeochemical effects of colloidal and gel-like organic matter in general and of invertebrate mucus in particular (which tends to decrease the permeability of both coarse- and fine-grained sediments); the geochemistry of pore water; and the physical–chemical properties of sedimentary particles (especially clay) controlling cementation in different ways and to varying degrees. The influence of dynamic loading on the sediment strength—particularly in relation to the permeability of sands—and the long-term effects of creep are poorly known. The relationships between water content, burial depth, and pore-water stress in sediments is not well known. A higher or a lower water content than might be expected, based on the overlying thickness of overburden, may occur in sediments that do not represent environments of rapid deposition or erosion. *In situ* measurements of permeability and pore-water stress would help to explain these relationships. Acoustic properties are seemingly related to the same sediment characteristics (grain size, porosity, structure, etc.) as are the other geotechnical properties. To understand their basic relationships and the various factors that influence them, it is essential to study a few simplified but representative situations. Such studies might make it possible to establish direct relationships among the geotechnical properties. Beyond the resulting basic understanding, this would have the added advantage of allowing the use of acoustic systems for remote sensing of the other properties, both laterally and in depth.

Gas

Microbially produced gas in a dissolved or free state is a very common constituent in most sediments. The generally small quantity appears to have an insignificant effect on geotechnical properties. However, off deltas and other areas of rapid deposition, particularly where the quantity of organic matter is large, the presence of free gas may drastically alter geotechnical properties. A decrease in the actual or potential stability of the sedimentary deposit results from the decreased shear strength and increased compressibility. The very large attenuation (and/or scattering) of acoustic energy may cause gas-rich deposits to appear acoustically opaque on echograms and prevent penetration below them for subbottom profiling purposes. Gas-rich areas require studies of the effects of gas on sedimentary texture and structure, on contributing chemical factors, and on microbial alteration affecting geotechnical properties. *In situ* measurements of shear strength, bulk density, pore-water stress, and acoustic propagation phenomena are essential. Cores collected should be maintained and tested in the laboratory under ambient sea-floor pressures and temperatures.

Variability of Properties

Grain size is a fundamental property of sediments. For fine-grained materials, the *in situ* grain sizes of natural floccules or aggregates are very poorly known. Laboratory grain-size measurements of the $\lesssim 5$ μm-size fraction are a function of how sediments are chemically dispersed and mechanically disaggregated. The actual arrangement of particles under *in situ* conditions is unknown.

The variability of geotechnical properties (taken to include grain size, Atterberg limits, water content, bulk density, shear strength, and compressibility, as well as such acoustic properties as velocity, attenuation, and impedance) in the continental shelves and ocean basins is not clearly known with respect to burial depth and area extent in specific environments. Because the effects of different test methods and procedures must still be evaluated, comparisons of data emanating from different organizations can only be approximated. All sediment samples taken from the sea floor are disturbed to some degree and the magnitude of the disturbance cannot yet be quantified. A definitive comparison between the geotechnical properties measured in cores and those measured *in situ* needs to be made for the principal kinds of sediments and depositional environments. From such comparisons, as well as from comparisons between the properties themselves, it may be possible to predict the variability expected in any environment once the relevant parameters of that environment are known.

Erodibility and Deposition of Cohesive Sediments

One of the most important points in understanding the physical properties of deposited sediment, and erosional as well as depositional processes, lies in the nature of the aggregation and the dynamics of the deposition of the sediment particles. Fundamental progress in this area would thus be of profound value in dealing with all of the aspects of cohesive sediments discussed below. Continued basic laboratory studies of the dynamics of flocculation and deposition are essential. Specifically, development of new techniques for observing the fine structure of flocs and aggregates, if possible, would pave the way for substantial progress.

The state-of-the-art in investigations on the erodibility and deposition of cohesive materials is presented by Krone in this volume and by Migniot (1968), and Task Committee on Erosion of Cohesive Materials (1968), Owen (1971), Einsele *et al*. (1974). The following statements and suggestions deal with important problems peculiar to cohesive sediments.

Erodibility of Cohesive Bed Surfaces

At present there is no satisfactory way to predict the critical shear stress of a cohesive bed surface or erosion rates as a function of excess stress. A correlation between erodibility and other more easily measured properties of the sediment, such as cation exchange capacity, density, and oxidation–reduction potential (Kandiah, 1974), would be useful if their reliability and limitations were known. A new direct method should also be developed to measure erodibility parallel to the sea bed on small undisturbed samples. Such a method would facilitate studies of bed structure. Furthermore, inverted *in situ* sea-floor flume measurements should be encouraged, as well as detailed measurement of naturally occuring erosion and eroding conditions.

The characteristics of the near-bed flow, especially the instantaneous and local shear stresses on the sea bed, need to be described in order to understand and predict the erosion and deposition of cohesive material. The effects of suspended particles and mechanical properties of the bed may affect near-bed flows. Oscillatory and unidirectional unsteady flows may be especially important.

A knowledge of the processes responsible for the formation of both small-scale bed features (such as scour marks or ripple-like forms) and large-scale bed forms (such as giant abyssal sediment waves or erosional furrows) would permit the interpretation of existing and past flow conditions, and possibly of inhomogeneities of the bed. The mechanisms of the formation of small-scale bed forms on cohesive materials should be explored in the laboratory as well as in the field. The processes responsible for large-scale bed forms can perhaps be best studied by direct measurements in nature, but laboratory studies or models might also prove fruitful.

Modes of erosion appear to depend greatly on the structure of the sediment bed. Mass erosion may occur when the applied tractive stress exceeds the shear strength of the material. The shape of such failure surfaces may well depend on inhomogeneities within the material. At lower shear stresses above the critical value for erosion, material is entrained aggregate-by-aggregate or grain-by-grain, depending on the structure of the exposed surface. Stronger and overconsolidated material may be weakened at a newly exposed surface so that erosion can proceed in the course of time. Weakening may occur as the result of unloading, swelling, biological activities, chemical processes, etc. The change of strength with time at a newly exposed sediment surface needs study.

Effects of the transportation of coarse material over clay beds have not been studied. Such effects may include aggravated erosion rates or armoring.

Benthic organisms can change apparent grain size, roughness, strength, and other physical and chemical properties of the sea bed. Therefore their effects on erodibility need to be investigated. The collaboration of marine biologists is essential in these studies.

The erodibility of layered sediments, which may contain a change from cohesive to noncohesive materials, needs additional investigation. Furthermore, many marine deposits contain shells and shell fragments of different sizes that influence bottom roughness and may even cause local variations in shear stress. Studies are needed that will yield information on the erosion of the wide variety of naturally occurring beds.

Transportation and Deposition Processes

The settling velocity of suspended aggregates, including biogenic material of all kinds in low-density suspensions, depends on the history of their aggregation, including consolidation of previously deposited and resuspended aggregates. Observed settling velocities of aggregates are highly variable. Therefore, measurements of the settling velocities of aggregates under natural conditions are essential for quantitative descriptions of transport and deposition processes. High-density suspensions differ from low-density suspensions in their settling behavior, deposition, and consolidation; similar *in situ* studies are needed.

Some high-density "fluid mud" suspensions appear to be locally stable, even under high shearing stresses. The processes involved are not known.

The large sizes and high concentration of suspended particles provided by erosion can enhance the aggregation and settling of finely dispersed suspended material. In particular, the settling behavior of mixtures of cohesive and noncohesive materials needs more study.

The flow conditions during deposition affect the structure and physical properties of the bed surface. Investigations of the depositional processes of a variety of materials under a range of flow conditions are needed. In addition, the effect of the bed and the suspended matter on the near-bed flow urgently needs investigation because of the central importance of near-bed flows in both deposition and erosion processes.

Transportation of Cohesionless Sediments

Threshold of Grain Movement

As velocity of flow over a sediment bed is increased, there comes a stage when the fluid exerts a stress on the particles sufficient to cause movement. Considerable attention has been given to the threshold under unidirectional currents (as opposed to wave motions). Most available data are from laboratory studies, although Sternberg (1971) has measured initial movement in tidal channels. Most of the data are for an initially flat sediment bed, and the threshold under those conditions is fairly well established; additional data are needed for beds roughened by preexisting ripples or irregularities produced by benthic organisms. Also, nearly all threshold studies refer to a long-term average stress exerted by the flow. Insufficient consideration has been given to threshold where stresses may be temporarily much greater than the average stress.

Present understanding of the threshold of cohesionless sediment under waves refers to an evaluation of the maximum value attained by the bottom orbital velocity (Komar and Miller, 1973). We can thus predict whether threshold will be achieved under simple oscillatory waves, but we cannot determine at what instantaneous velocity and stress within the wave orbit the threshold will be achieved. A related problem is that of a unidirectional current superimposed on wave motions, both contributing bottom stress. A knowledge of the instantaneous stress at which sediment is placed in motion is important in evaluating sediment transport under such conditions.

Continued study of threshold both in the laboratory and in the oceans is important. The effect of sediment-binding organisms on threshold is perhaps more widespread than is commonly believed, and special studies are badly needed. Field investigations of threshold for both unidirectional and oscillatory motions are especially needed. Such studies would better provide data on the instantaneous stresses necessary for sediment movement.

Transport Processes and Transport Rates

Modes of Grain Transport

An understanding of the modes of grain movement is essential in a consideration of sediment transport. Moreover, there is a strong interrelationship between textural properties (packing, fabric, and porosity) of cohesionless sediments and the small-scale aspects of grain transport. Probably because of the difficulty of making unambiguous observations of such a ubiquitous and seemingly straight-forward phenomenon, there has been a great proliferation of hypotheses largely unsupported by definitive observations. Bagnold's (1966) concept of the impor-tance of intergranular collisions represents the most ambitious attempt to pro-vide a conceptual basis for mode of grain transport, and has gained currency among a number of sediment dynamicists. But most would not disagree that definitive experiments are lacking and that indirect evidence is inconclusive.

Experimental studies of basic modes of grain transport are more promising than theoretical developments because of the complexity of two-phase turbulent flow. The most fruitful direction will probably involve laboratory studies of specialized and judiciously simplified flow systems involving artificial materials designed to afford maximum observability of basic effects without sacrificing the important aspects of grain motions. There is also a need for basic work on other closely related problems of grain transport, e.g., the lift and drag on sedi-ment particles on or immediately above the bed; the nature and statistics of grain trajectories; the thickness of the moving grain layer; the effects of grain shape and density on modes of transport (as in foraminiferal sands or shell beds); and the modification of turbulence in the layer of flow immediately adjacent to the bed due to the presence of transported grains. These studies are more likely to be successful by using flume experiments rather than by observations in the oceans, because the essential aspects can be duplicated in the laboratory under readily observable and controllable conditions.

Transport under Unidirectional Currents

Prediction of transport rates from the hydrodynamic conditions is still inac-curate. Formulas for both bed-load and total-load transport are applicable only to circumstances for which they have been tested. There seem to be three rea-sons for this: the physical processes are not well understood; the measuring techniques used to check various models (transport formulas) are insufficient; and the theoretical models in use are inadequate. There is a logical interaction between these reasons. For instance, there is little need to improve models as long as the measuring techniques are incomplete. On the other hand, a good measuring technique can be developed only when the physical processes are well understood. It therefore seems that the understanding of physical processes and the development of adequate measuring techniques are both key problems.

Conventional instruments and tracer techniques must be improved. While radioactive tracers have been fairly well developed (IAEA, 1973), improvement is needed here also. Most interpretation techniques are based on the assumption that the flow field is homogeneous; this leads to errors when (as is usually the case) the flow field is not homogeneous (Price, 1968).

Transport under Surface Water Waves and Currents

On continental shelves waves are an important agent in sediment transport, so that in general one must consider waves with superimposed unidirectional currents. Very few studies have been made of these complex conditions. Transport will be strongly dependent on ripple geometry, and especially on ripple asymmetry because this governs the amount of sand thrown into suspension under the wave crest versus under the wave trough (Inman and Bowen, 1963). Inman and Bowen even found instances where the net sand transport was opposite in direction to the wave travel and superimposed current. Kennedy and Locher (1972) summarize work on sediment suspension under waves. Research is in such a primitive stage, and so much needs to be done, that it is not yet possible to propose well-defined narrow problems.

Bed Configurations in Cohesionless Sediments

Beds of relatively coarse sediment are molded by unidirectional turbulent currents into a great variety of longitudinal and transverse configurations at widely different physical scales. Such configurations are observed in many shallow-marine areas and in some deep-water areas. It has been difficult to characterize the dynamics of these features, both because of the difficulty of relating local transport rate to flow conditions and because of the strong coupling and mutual interaction between flow structure and bed geometry. Because of the greatly different scales and the dominant unsteadiness of natural flows, correspondence between the results of flume experiments and natural observations has been incomplete with respect to large-scale features such as dunes, megaripples, and sand waves.

Knowledge of marine bed configurations is desirable for several reasons: their essential but largely unknown role in the relation between flow conditions and transport rates; the possibility of specifying flow conditions by examining bed-form geometry; their role in the generation of sedimentary structures in a depositional regime; and their effect on structures placed on the sea bed.

Some of the most important research goals involve clarification of relationships among large-scale features. Questions in need of consideration are: What flow velocities are characteristic of the various kinds of configurations? Under what flow conditions are smaller sand waves superimposed on larger sand waves? How many orders of superimposed sand waves are possible? How does effective flow depth limit sand-wave size? Large-scale longitudinal bed forms, which are less common and less well known than transverse features, need to be investigated from the standpoint of what kinds of secondary circulations are responsible. Further advances are most likely in two directions: continued observations in marine areas where currents are fairly steady; and experiments in flumes or other controlled artificial channels at greater flow depths than previously attained. Dynamic scale modeling at modest scale ratios, to extend the range of effective flume depth into a range more nearly representative of the oceans, seems practicable and valuable.

Since most oceanic bed configurations are generated by unsteady flows, observations on steady-flow configurations must be viewed largely as providing a baseline for understanding or sorting out disequilibrium effects. Two logical lines of work, both in the laboratory and in the field, on the effect of unsteady flow are: first, study of bed configurations produced by one-dimensional reversing flow that reproduces a tidal cycle (including typical velocity asymmetries); and second, study of bed configurations under the influence of a velocity vector periodically changing in both magnitude and direction.

Modeling of Erosion and Sedimentation

Net erosion or deposition can involve resuspension or settling from suspension under conditions of spatially constant sediment transport rate. By simple mass–balance considerations, erosion or deposition also occurs whenever the transport rate varies along the streamwise direction. Quantitative treatment of this latter case is difficult because of the complicated spatial variations in flow structure and mode of sediment transport involved. Most attention has been given to short-term problems in tidal environments.

For relatively coarse sediment the transport rate can be considered to depend on the local hydrodynamic characteristics. This is not the case for relatively fine noncohesive and cohesive sediments. Here the concentration distribution over the vertical requires time or space to adjust to the prevailing flow conditions.

For the modeling of coarse material it can be expected that the transport rate depends on the velocity or shear stress near the bed. Due to the distinct difference between the time scale of water movement and of morphological processes, a two-step computation has to be carried out; 1) tidal computation for a known bed level, and 2) computation of bed-level changes due to a number of tides; only after relatively large changes occur does step 1 again become necessary.

A one-dimensional model for these coarse sediments has been fairly well established (de Vries, 1973). There is a need for further research as far as large ranges of grain size are concerned. Also the influence of time-dependent roughness must be considered.

For fine cohesive and noncohesive materials the distribution of the sediment concentrations in the vertical must be considered, and must be integrated to obtain the instantaneous transport rate. For simple one-dimensional or two-dimensional (in the horizontal plane) models, integration of a theoretical vertical distribution of sediment is adopted (Ariathurai, 1974). The most accurate results would be obtained from models which include depth as one of the dimensions. This very difficult step can be approximated by treating the water column as a series of horizontal layers (Odd and Owen, 1972).

In summary, only for simple situations are mathematical models with some degree of accuracy available. For the more complex situations a wide field of research is still open. There is a need for difficult laboratory research and for the collection of extensive field data on settling velocities, currents, and suspended concentrations, in order to test the reliability of these models.

Application of these morphological computations involve short-term changes of the bed due to natural and human effects, but the same approach may be valuable in problems of erosion and deposition on much longer (geological) time scales.

Recommendations

Many recommendations have been made in this report. The more important of these are that studies be made of:

1. Areas particularly susceptible to sediment instability.

2. Factors (e.g., fabric and primary structures) controlling the basic geotechnical parameters (e.g., shear strength, compressibility, elastic wave speed, and attenuation) and their interrelationships.

3. Sediment density flows, including laboratory and field measurements.

4. *In situ* bulk properties of gas-bearing sediments.

5. Comparative geotechnical properties measured in cores and *in situ*.

6. Aggregation of sediment particles of different sizes and kinds, their settling velocities under natural conditions (including flow), and their influence on the structure and physical properties of the sea bed.

7. The influence of suspended particles and bed forms on the near-bed flow, and vice versa.

8. Effects of benthic organisms, embedded shells, shell fragments, and the transportation of coarse-grained material over the bed surface, on the erodibility of cohesive material and layered deposits.

9. The importance of the weakening of newly exposed bed surfaces on the critical erosion shear stress and rate of erosion.

10. Sediment threshold under a current where the bottom is roughened by preexisting ripples or irregularities produced by benthic organisms.

11. Instantaneous bottom stress within the wave orbital motion where sediment threshold will be achieved.

12. Modes of grain transport essential to consideration of sediment transport mechanisms.

13. Sediment transport due to combined wave action and unidirectional flow.

14. The development of bed forms, their origin, and their relationship to flow.

15. Improved instrumentation for the collection of field data essential for testing the reliability of sediment transport models.

16. The importance of mucous and other excretory products of biological activity on interparticle cohesion and resulting aggregate and floccule strengths.

Key References

Ariathurai, C. A., 1974, A finite element model of cohesive sediment transportation, Ph.D. Thesis, University of California, Davis, 210 pp.

Bagnold, R. A., 1966, An approach to the sediment transport problem from general physics, *U.S. Geological Survey, Professional Paper 422I*, 37 pp.

Einsele, G., Overbeck, R., Schwarz, H. U., and Unsold, G., 1974, Mass physical properties, sliding, and erodibility of experimentally deposited and differently consolidated clayey muds, *Sedimentology 21:* 339–372.

Gibbs, R. J., ed., 1974, *Suspended Solids in Water*, Plenum Press, New York.

Hampton, L., ed., 1974, *Physics of Sound in Marine Sediments*, Plenum Press, New York, 569 pp.

IAEA, 1973, Tracer techniques in sediment transport, *International Atomic Energy Agency, Vienna, Technical Report No. 145*, 234 pp.

Inderbitzen, A. L., ed., 1974, *Deep-Sea Sediments: Physical and Mechanical Properties*, Plenum Press, New York, 497 pp.

Inman, D. L., and Bowen, A. J., 1963, Flume experiments on sand transport by waves and currents, *Proceedings of the 8th Conference on Coastal Engineering*, pp. 137–150.

Kaplan, I. E., ed., 1974, *Natural Gases in Marine Sediments*, Plenum Press, New York.

Kandiah, A., 1974, Fundamental aspects of surface erosion of cohesive soils, Ph.D. Thesis, University of California, Davis.

Kennedy, J. F., and Locher, F. A., 1972, Sediment suspension by water waves, *in:* (R. E. Meyer, ed.), *Waves on Beaches*, New York, Academic Press, pp. 249–295.

Komar, P. D., and Miller, M. C., 1973, The threshold of sediment movement under oscillatory water waves, *Journal of Sedimentary Petrology 43:* 1101–1110.

Migniot, C., 1968, Etude des propriétés physiques des différents sédiments tres fins et de leur comportement sous des actions hydrodynamiques, *La Houille Blanche 23* (7): 591–620.

Noorany, I., 1972, Underwater soil sampling and testing: a state-of-the-art review, in: *Underwater Soil Sampling, Testing and Construction Control, American Society for Testing and Materials, Special Technical Publication 501:* 3–41.

Noorany, I., and Gizienski, S. F., 1970, Engineering properties of submarine soils: state-of-the art review, *Proceedings of the American Society of Civil Engineers, Journal of the Soil Mechanics and Foundations Division 96* (SM5): 1735–1762.

Odd, N. V. M., and Owen, M. W., 1972, A two-layer model of mud transport in the Thames Estuary, *Proceedings of the Institute of Civil Engineers, Paper No. 7517S*, 30 pp.

Owen, M. W., 1971, The effect of turbulence on the settling velocities of silt flocs, *International Association for Hydraulic Research, Proceedings of the 14th Congress, Paris, Paper D4*, 6 pp.

Price, W. A., 1968, Variable dispersion and its effects on the movements of tracers on beaches, *Proceedings of the 11th Coastal Engineering Conference, Paper 125*, 4 pp.

Scott, R. F., and Zukerman, K. A., 1970, Study of slope instability in the ocean floor, *U.S. Naval Civil Engineering Laboratory, Report CR-70.007*, 72 pp.

Sternberg, R. W., 1971, Measurements of incipient motion of sediment particles in the marine environment, *Marine Geology 10:* 113–119.

Swift, D., Duane, D. B., and Pilkey, O. H., eds., 1972, *Shelf Sediment Transport: Process and Pattern*, Dowden, Hutchinson and Ross, Stroudsburg, Pa., 656 pp.

Task Committee on Erosion of Cohesive Materials, 1968, Erosion of cohesive sediments, *Proceedings of the American Society of Civil Engineers, Journal of the Hydraulics Division 94:* 1017–1049.

de Vries, M., 1973, Riverbed variations: aggregation of degradation, Lecture, International Seminar on Hydraulics of Alluvial Streams, New Delhi, *Delft Hydraulics Laboratory Publication No. 107*, 20 pp.

D
Solution-Sediment Chemical Interactions

Group Leader: R. Chesselet

Members: R. A. Berner, S. E. Calvert, R. C. Cooke, A. J. de Groot, J. C. Duinker, A. Lerman, J. M. Martin, N. B. Price, E. Suess, R. Wollast

Rapporteur: F. L. Sayles

Introduction

Historically, many studies of the oceans and underlying sediments have centered on processes on one side or the other of the sediment–water interface, largely ignoring their inherent interdependence. Reactivity between sea water and the sediments is most pronounced in the benthic boundary layer and identification of chemical reactions and their consequences is most readily undertaken in this zone. Because it is an interface, processes occurring here influence processes on either side of it. A knowledge of processes in this region is essential for an understanding many of those processes occurring in the oceans and in sediments at depth.

There are also a number of practical goals to be achieved in promoting an understanding of the benthic boundary layer:

1. Regeneration of nutrients is a principal factor in primary productivity and is therefore of major consequence to marine fisheries. Regeneration in the sediments on a local scale, at least, has been found to be important in nutrient cycling. The extent to which this is true on a global scale is unknown. It is possible that the boundary layer plays a significant role in primary biological production.

2. Man-made wastes are continually added to the oceans, many eventually being incorporated in the benthic boundary layer. The ultimate fate of these products is governed by processes in this zone and, in turn, these same processes may be strongly modified by the amount and nature of material added.

3. The search for raw materials has entered the sea and, to a large extent, currently centers on the sea bed. In particular, recent efforts have involved consideration of manganese nodules as a source for a number of metals. The distribution and formation of this potential resource is a direct consequence of authigenesis in the benthic boundary layer. Examples such as these serve to emphasize the fact that processes in this zone can and do have a direct and practical influence on man and his activities as well as being affected by him.

Studies of chemical processes in the benthic boundary layer unavoidably are interrelated to those of other disciplines. Chemical reactions fundamentally influence, and are influenced by, biological and physical processes in and above the sediments. The nature of deposited material and the character and rate of chemical reactions may modify the ecology of the benthic boundary layer, strongly influencing the type and activity of organisms. This includes the release of toxic compounds, utilization of oxygen, modification of pH, and release of nutrients. The activity of the organisms in turn affects the chemistry of the environment through bioturbation and mediation of the breakdown of organic detritus. Reaction in the sediments releases a number of components to yield concentrations that far exceed those found in seawater. Migration of these components into overlying waters provides tracers that may be utilized in the determination of stability and rates of mixing at the interface and in bottom waters. The existence of concentration gradients in the sediments can, under some circumstances, provide information on the rate and extent of bioturbation. The stability of the interface and its susceptibility to mixing through turbulence also directly influence reaction and chemical migration through accelerated mixing in the uppermost sediments. Such mixing enhances transport to far above that occurring through molecular diffusion. Reaction in the benthic boundary layer also affects the physical properties of the sediments. Breakdown of the organic matter that binds aggregates of detrital material leads to disaggregation and a reduction in grain size. Precipitation of inorganic salts such as carbonates, phosphates, and Fe–Mn compounds can drastically alter physical properties through cementation. The following discussion is, of necessity, centered largely on the chemical aspects of the benthic boundary layer, but it should be borne in mind, as emphasized in the preceeding, that these processes do not occur independently. Rather, they are intimately interrelated to the biological and physical processes of this zone.

Organic Matter in Sediments

Chemical reactions taking place in the benthic boundary layer are fundamentally influenced by the presence and composition of organic matter. This component in many sediments is the single most important source of chemical energy. Because this material is far from equilibrium, it is highly reactive and undergoes alteration mainly at the benthic boundary layer. It is an energy source and substrate for microorganisms which mediate many physical–chemical reactions.

The Nature of Organic Material

The bulk composition of organic matter in marine sediments is poorly understood. Characterization of its constituents is rudimentary, although many attempts are being made to isolate and identify specific groups, e.g., pigments, amino acids, carbohydrates, etc., in different marine environments and sediment profiles. In recent years there has been a tendency to characterize specific molecules, which are themselves quantitatively unimportant, resulting in an obvious neglect of the nature of the bulk material. We require reliable methods for the isolation and characterization of this material.

Reactivity of Organic Material

It is important to estimate the relative proportions of living, readily metabolizable, and refractory organic materials. These to a large extent control microbial biomass, which in turn reflects the potential reactivity of the bulk material. Estimation of the $C:N:P$ ratio in sedimentary organic matter is currently used to describe the source and subsequent reaction pathways during alteration. Unfortunately, such an approach fails to provide a complete picture of organic processes.

Laboratory experiments involving whole-sediment and labeled (C, H, N) compounds could provide important insights into the degree of reactivity with respect to the total organic matter present. Allied to this is the requirement to monitor the changes in functional groups during burial.

Organic materials of various sources and having different compositions are susceptible to different rates and pathways of degradation. We require methods for estimating the influence that the relative proportions of planktonic, benthonic, and terrestrial organic constituents have at the benthic boundary layer.

Organo–Metallic Complexes

The presence of organo–metallic complexes may, by promoting or inhibiting reaction, have an important influence on the utilization of organic matter. Apart from a few instances, adequate methods for their identification are not available. It is important to identify the major group or groups carrying metals, their total complexing capacity, and their specificity for particular metals. At the present time it is not known whether many natural or man-made organo–metal chelates are stable in the environment of the benthic boundary layer.

Inorganic Consequences of Organic Matter Reactions on Pore Waters

The microbiological decomposition of organic matter affects the concentrations of a large number of inorganic species dissolved in sediment pore waters of the benthic boundary layer. These species may attain concentrations far different than those found in the overlying water, and thus give rise to appreciable fluxes between sediment and sea water. Notable examples are O_2, the nutrient elements nitrogen, phosphorus, and silicon, SO_4, H_2S, CH_4, CO_2, Fe^{2+}, and Mn^{2+}. These species may not only be involved as products or reactants of reactions with organic matter, but also may be involved in mineral precipitation, dissolution, or transformation.

Nutrient Regeneration

The nutrient elements phosphorus and nitrogen are directly formed by the bacterial degradation of organic compounds containing these elements or are transformed from one oxidation state to another by other bacterial reactions. Of special interest is the possible production or utilization of dissolved N_2 in sediment pore water, for which data are generally lacking.

Rates of decomposition are functions of bacterial metabolic rates, which in turn are affected primarily by the presence or absence of dissolved oxygen and the nature of the organic material itself. Knowledge of rates is important in that concentration gradients and consequent fluxes between sediment and overlying sea water are dependent upon the rate of build-up in the sediment. These fluxes may provide a significant contribution to the nutrient budget and consequently the primary productivity of marine waters. These rates are poorly known. Techniques for measuring rates include utilization of labeled HCO_3^- and specific organic phosphorus and nitrogen compounds in laboratory or, ideally, in *in situ* experiments, using whole sediments. Rates can also be evaluated by modeling of depth-versus-concentration profiles in the sediment, provided critical parameters, such as rates of deposition and diffusion coefficients, are known.

Silica may be released to solution by reactions involving organic matter. The dissolution of biogenic opal, a common source of dissolved silica, may be inhibited by organic envelopes around each particle. Destruction of this coating, in some cases by bacteria, is an important step in the occurrence of dissolution.

Sulfate Reduction

Considerable effort has been placed on studying the microbiological reduction of sulfate to hydrogen sulfide in sediments. Hydrogen sulfide, which is a poison to aerobic organisms, may be released to the overlying water. Rates of sulfate reduction need further study in terms of direct measurements and modeling, as in the case of nutrients.

Dissolved Gases

Dissolved gases may be either consumed or produced in sediments. Oxygen is consumed and carbon dioxide produced by respiration, whereas methane and hydrogen are formed by fermentation. Methane may also be produced by CO_2 reduction. Little is known about the distribution of dissolved gases in pore waters, let alone their rates of production or consumption. This is primarily because of the difficulty of sampling without loss or contamination by atmospheric gases. Dissolved gases may reach saturation with respect to bubble formation in sediments of the benthic boundary layer. Virtually nothing is known about the nucleation, entrapment, and migration of these bubbles. Bubbles of one gas moving upward may entrain other gases, leading to the release of both from the sediment. Bubbling may also affect the physical and mechanical properties of the sediments.

Redox Potential and pH

In the absence of specific chemical measurements, the chemical state of pore waters affected by organic decomposition reactions can be characterized, in a crude fashion, by the measurement of redox potential (Eh) and pH. Eh measurements are fraught with many inherent difficulties, but have proven useful in characterizing, in a general manner, certain redox couples. These include $NO_3^- - N_2$, $Fe^{2+}-Fe(OH)_3$, $Mn^{2+}-MnO_2$, and HS^--S^0. Measurement of Eh with the Pt electrode is also useful in the determination of the depth of the boundary between oxidizing and reducing conditions in sediments. Measurement of pH is considerably more accurate than that of Eh. Improved techniques for the measurement of Eh and pH are needed.

Chemical Reaction between the Solids and Interstitial Solutions of Sediments

Chemical reactions between the solids and interstitial solutions of sediments are important because of their effect on the marine geochemical cycles of many elements. Major problems are the identification of authigenic minerals, the thermodynamic properties of minerals and aqueous species, the reactivity of sedimented material, and the kinetics of reactions.

Identification of Authigenic Minerals

The identification of newly found (authigenic) minerals in many situations is difficult because of their poor crystallinity, their lack of distinction from detrital components, and their presence in low concentrations.

Possible approaches to identification are: solubility of poorly crystalline minerals; radioactive dating, stable isotope measurements, and textural analysis; and physical enrichment and separation techniques.

Thermodynamic Properties

The use of chemical thermodynamics enables the prediction of the direction of chemical reactions and the state of equilibrium. There are many sedimentary substances for which thermodynamic data—namely, free energies of formation—are lacking, especially for solid solutions and disordered phases. An outstanding example is the clay minerals.

The surfaces of minerals may differ in thermodynamic properties from their interior portions, and it is the surface which, via equilibration, may affect the composition of pore waters. Techniques for measuring the surface composition and properties of solids have been developed recently and should be applied to sedimentary minerals.

Reactivity of Sedimented Materials

The detrital components of any given sediment, including terrigenous and volcanic mineral, and rock particles plus biogenic skeletal debris, have highly variable reactivities in the benthic boundary layer. Weathering products from high latitudes appear to be much more reactive than similar products from low latitudes; the reactivity of volcanic glass appears to be highly variable, and biogenous carbonate and silica tests exhibit variable solution rates among different taxa.

In order to use the information on the rates of formation of the reaction products of these detrital particles, it is necessary to acquire some understanding of the source, or sources, of these variations. Are they due to supplies of fresh, as opposed to weathered, materials? Is the degree of order of the individual minerals or particles itself highly variable? Or does the interstitial environment control the reaction rates and pathways?

Kinetics

Many phases can be demonstrated to be out of thermodynamic equilibrium in sediments within the benthic boundary layer. Little is known about the rates and mechanisms of chemical reactions involving sedimentary constituents. Reactions include nucleation, crystal growth, dissolution, adsorption, ion-exchange, and diffusion to and from mineral surfaces. Laboratory studies are needed to elucidate basic controlling mechanisms as well as to obtain a general idea of rates. Rates obtained in the laboratory should not be blindly extrapolated to sedimentary situations because of complicating factors such as differing surface reactivities and the possible presence of rate inhibitors (poisons) in natural water. The nature and role of inhibition, however, can be studied in the laboratory and used to predict the qualitative behavior of minerals in pore waters. In the absence of *a priori* knowledge of reaction rates, it is still possible to estimate them through the use of models of concentration–depth data.

Migration of Chemical Components in Sediments

Pore-Water Chemistry

Pore waters provide a continuous connection throughout the benthic boundary layer and are the medium and usually the participant in reactions. Knowledge of their composition is central to the understanding of chemical reactions in this zone. Their composition also reflects the direction of transport and flux of dissolved constituents.

Studies of pore water composition have been carried out since at least 1895 with few modifications of sampling techniques. Interstitial solutions have generally been obtained by extraction from bulk sediment after recovery of sediment samples, usually by coring methods. Evidence of rapid reaction between the solids and interstitial solutions indicates that some modification of these techniques is necessary. Pressure and temperature changes incurred during the recovery of samples have been found to alter significantly the *in situ* concentrations of several major components. Further, sample distribution within cores should more closely reflect rates of change of properties. In the relatively reactive sediment near the interface, concentration gradients often are steepest, and it is essential that these be adequately delineated.

In certain instances sediment coring and subsequent extraction of pore water can provide adequate samples. These include those circumstances where pressure and temperature changes during recovery are minimal. Also, some components have been found to be unaffected by the above conditions, including Cl^- and SO_4^{2-}. On the other hand, the major cations, silica and bicarbonate, are seriously affected. Analyses of dissolved gases and other labile constituents are also probably subject to error. Effects on components not studied to date should be identified and considered prior to utilization of coring and extraction.

Several approaches can be taken to improve the determination of the reactive constituents of pore waters. *In situ* sampling techniques can, at least in part, prevent reactions during recovery. The effects of interface disturbances have not been completely eliminated by the instruments presently available. Perhaps the most desirable approach would be the development of *in situ* measuring techniques, which would eliminate all possibility of alteration during recovery. It should further be possible to minimize interface disturbance.

Diffusive fluxes of dissolved material across the sea water–sediment interface depend upon the gradients at the interface. As noted above, the uppermost sediments of the benthic boundary layer are the most highly reactive, imparting to the pore waters in many situations steep concentration gradients of a number of components. Evaluation of fluxes across the interface requires accurate knowledge of these gradients. In order to avoid the uncertainties inherent in estimating rapidly changing properties by extrapolation, sampling in the uppermost layer (0 to 15 cm) should be closely spaced, ideally on a scale of one centimeter or less.

Pore-water studies have included analyses of components ranging from the most abundant to the exceedingly rare. Two basic constituents are conspicuously absent from the list: dissolved organic matter and aluminum. Both of these can be expected to be important influencing the course of reactions in the benthic boundary layer. Development of analytical capabilities for both, although difficult, ought to be pursued.

Causes and Mechanisms of Migration

The migration of reactants and products in the benthic boundary layer is caused by (1) physical processes in the water column immediately above the benthic boundary layer, and (2) biological, physical, and chemical processes within the sediment. These processes center on:

1. Turbulence and resuspension of sediments. Mechanical energy transferred from the ocean to sediments may result in resuspension of sediments and dispersion in the pore water within the upper part of the layer. Turbulent stresses affect the physical structure of the sediment–water interface, and may enhance the transfer of dissolved species from the upper parts of the sediment column. Knowledge of these interrelationships is lacking.

2. Mixing of the sediment by deposit-feeders and other moving organisms may result in homogenization of the chemical composition of the upper layers of sediment. The rates of mixing are functions of organism type and community structure. These may vary with the type of sediment and the bathymetry of the ocean floor. There is a great need for methods of quantifying biological mixing, such as the use of chemical or radiochemical tracers.

3. Diffusion of dissolved species, both reactants and products, is a migrational mechanism of primary importance. Considerable attention should be devoted to the estimation of the diffusion coefficients of the reacting species, both by direct measurement and by calculation from basic theory. Both theory and experimentation should incorporate the effects of temperature, thermal gradients, tortuosity, porosity, and permeability of sediments.

4. Advection, or bulk flow, is a migrational mechanism causing net displacement of mass relative to the sediment–water interface. Gravitational compaction without deposition results in net outflow of pore water. If the place of reference used is the sediment–water interface, then, under conditions of continuous deposition, the sediment particles and pore water move downward away from this datum plane (the interface). At some depth within, the sediment compaction will attain a steady value and porosity will remain constant (and greater than zero) with depth. If the compaction is either exceeded or balanced by the sedimentation rate, the net fluxes of water and solids are downward relative to the sediment–water interface. Another type of advection is the underflow of water through the continental margins. The magnitude and extent of this flow are not well known.

Fluxes across the Sediment–Water Interface

The flux of various species across the sediment–water interface may be one of the more important processes controlling the chemical composition of sea water. The fluxes of various species are also important to the geochemical cycles of several elements.

Fluxes within the sediment of the benthic boundary layer may be grouped into two categories: (1) the flux downward due to deposition and burial of sediment, and (2) diffusional fluxes which may go in any direction depending on the chemical and physical conditions of the reacting system. In general, the magnitudes and nature of the fluxes are poorly known.

The rationale behind models of migration is to allow description and prediction of the fluxes and chemical behavior in the system in general. Further development of models should consider the effects of bioturbation and turbulence, and should include more accurate reaction-rate parameters. Also, greater attention should be devoted to two- and three-dimensional flux models.

Fluxes in the sediments of the benthic boundary layer can be estimated by computation or by direct measurement. Computation requires accurate knowledge of the transport and chemical reaction-rate parameters. These are often obtained from models of migration. Direct measurement, when and if possible, provides independent estimates of the fluxes and checks the accuracy of the models. The experimental difficulties in measuring fluxes make direct measurement the less common but nevertheless the more important method.

Depositional Controls on Chemical Processes in the Benthic Boundary Layer

The chemical reactions taking place in the benthic boundary layer have been discussed in this report without regard to variations in the input of sedimentary material, both organic and inorganic. The most important overall variable considered here is the total sedimentation rate, which varies widely throughout the ocean on both regional and local scales. Thus, authigenic iron sulfides are being precipitated in many rapidly accumulating near-shore environments, while hydrous manganese and iron oxides are forming on the surfaces of slowly accumulating oceanic sediments. Coupled with variations in the total accumulation rate is variation in the input of sedimentary constituents (e.g., fluctuations in the ratio of organic to various inorganic constituents). Discussion of chemical processes in the benthic boundary layer and their effect on other processes in the environment should therefore take into account these additional considerations.

In addition to sedimentary variability, as discussed above, the influence of the input of man-made wastes can have consequences similar to those produced naturally. We have in mind, for example, the additional input of dissolved metals into coastal or estuarine environments which may, through organic and inorganic reaction, profoundly affect natural sediment diagenesis.

Recommendations

Specific recommendations are given throughout the text of this report. The most important areas to which research should be directed are the following:

1. Intensified effort toward the identification of organic and metal–organic compounds in the sediments, both solid and dissolved.

2. Study of the physical–chemical properties and the mode of formation of authigenic phases.

3. Determination of fluxes across the seawater–sediment interface as well as within the sediments, and of chemical gradients in the sediments, both on the basis of model computations and by direct measurement.

4. Determination of rates and mechanisms of chemical and biochemical reactions in the sediments.

Key References

Aizenshtat, Z., Baedecker, M. J., and Kaplan, I. R., 1973, Distribution and diagenesis of organic compounds in JOIDES sediment from Gulf of Mexico and Western Atlantic, *Geochimica et Cosmochimica Acta 37:* 1881–1898.

Berner, R. A., 1974, Kinetic models for early diagenesis of nitrogen, sulfur, phosphorus, and silicon in anoxic marine sediments, in: (E. D. Goldberg, ed.), *The Sea*, Vol. 5, Wiley, New York, pp. 427–450.

Goldhaber, M. B., and Kaplan, I. R., 1974, The sulfur cycle, in: (E. D. Goldberg, ed.), *The Sea*, Vol. 5, pp. 569–656.

Lerman, A., 1976, Chemical reactions and migration processes in sediments, in: (E. D. Goldberg, I. N. McCave, J. J. O'Brian, and J. H. Steele, eds.), *The Sea*, Vol. 6, Wiley, New York (in press).

Manheim, F. T., 1976, Interstitial waters of marine sediments, in: (J. P. Riley and R. Chester, eds.), *Chemical Oceanography*, Vol. 5, Wiley, New York (in press).

Menzel, D. W., 1974, Primary productivity, dissolved and particulate organic matter, and the sites of oxidation of organic matter, in: (E. D. Goldberg, ed.), *The Sea*, Vol. 5, Wiley, New York, pp. 659–678.

Nissenbaum, A., Baedecker, M. J., and Kaplan, I. R., (1972), Organic geochemistry of Dead Sea sediments, *Geochemica et Cosmochimica Acta 36:* 709–728.

Price, N. B., 1976, Chemical diagenesis in marine sediments, in: (J. P. Riley and R. Chester, eds.), *Chemical Oceanography*, Academic Press, New York.

Wollast, R., 1974, The silica problem, in: (E. D. Goldberg, ed.), *The Sea*, Vol. 5, Wiley, New York, pp. 359–392.

E
Organism-Sediment
Relationships

Group Leader: J. E. Webb

Members: D. J. Dörjes, J. S. Gray, R. R. Hessler,
Tj. H. van Andel, F. Werner, T. Wolff,
J. J. Zijlstra

Rapporteur: D. C. Rhoads

Introduction

Productivity in the surface waters of the sea is closely linked with water-sediment-organism processes at the benthic boundary.

Organic material, particulate and in solution, from surface and midwater food webs reaches the bottom and provides an energy input for the organisms associated with the sediments. Breakdown products returned from the benthic layer to the sea are eventually used by autotrophs in the euphotic zone.

Sediments range from rocks through gravels and sands, mainly in the shallower waters of the continental shelf, to clays on the ocean floor. But in all of these there is an organic component that contributes to the structure of the

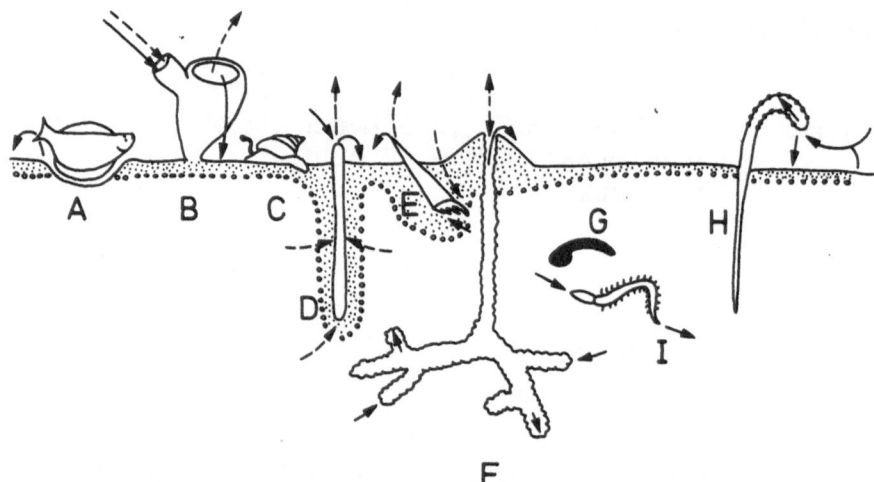

Fig. 1. A diagram of some of the ways organisms affect the benthic boundary layer. Key: ⟶ solid transport; ----→ fluid transport; ooooo redox potential discontinuity; ⋮⋮⋮⋮⋮ oxidizing sediments above RPD. A: Surface dweller (fish) disturbing surface; B: Epifaunal suspension feeder (tunicate) converting suspended solids into deposit feces; C: Epifaunal deposit feeder (gastropod) disturbing the surface, laying down a mucus trail, and increasing particle size by fecal deposition; D: Infaunal suspension feeder (polychaete) circulating interstitial water; E: Infaunal deposit feeder (polychaete) transporting particulates upward and water downward; F: Burrower (crustacean) transporting sediment upward and horizontally; G: Animal with mineral hard parts (bivalve) converting dissolved ions into sedimentary particles; H: Tubiculous animal (polychaete) concentrating specific components of the sediment; and I: Burrower (polychaete) disturbing and sorting sediment.

benthic layer and modifies and sometimes initiates the physical and chemical processes that take place at the boundary. The ways in which these interrelate one with another are complex, and the organisms that partake are highly diverse. Microorganisms are ubiquitous and, apart from providing a food source for animal life in the sediment, have enzymes capable of catalyzing many chemical reactions in the benthic boundary layer. They also modify the physical structure of the sediment by increasing grain-to-grain adhesion and at the same time by reducing friction. Larger organisms, living mainly in the interstitial spaces in burrows or tubes or moving freely through the sediment, continually modify the structure of the sediment (bioturbation) by mixing, sorting, and aggregating small particles into pellets and by pumping water into and out of the sea bed. These organisms also change the chemistry through their circulatory, respiratory, and excretory behavior and notably alter the redox potential discontinuity (RPD) of the sediment.

On the other hand, the dynamic mosaic of sediments which forms the sea bed exerts a powerful influence on the distribution of animals according to their preferences both in the larval and adult states.

The total interaction between organisms and of organisms with the inorganic fraction of the sediment is poorly understood by reason of its complexity, although many of the broad processes are known to biologists. The function of this report is twofold. First, to provide an outline of our present knowledge for scientists in other disciplines who may not be familiar with biological phenomena, and second, to offer some opinion on biological areas in this field where further investigation is badly needed and should have interdisciplinary impact.

To this end, a scheme of organism–sediment interactions has been prepared to serve as a framework, or context, in which the various processes can be considered and our recommendations for new work evaluated.

An Event–Response Scheme of Organism–Sediment Interactions

The event–response scheme as presented here is obviously oversimplified. The following limitations should be kept in mind:

1. Effects are listed in the place where the phenomenon is most pronounced, even though similar effects may well occur to a more modest extent in adjacent members of the sediment gradient.
2. Sediments are regarded as homogeneous and well sorted even though this is the exception rather than the rule.
3. In the real world, there is seldom a one-to-one correlation between sediment type and fauna.

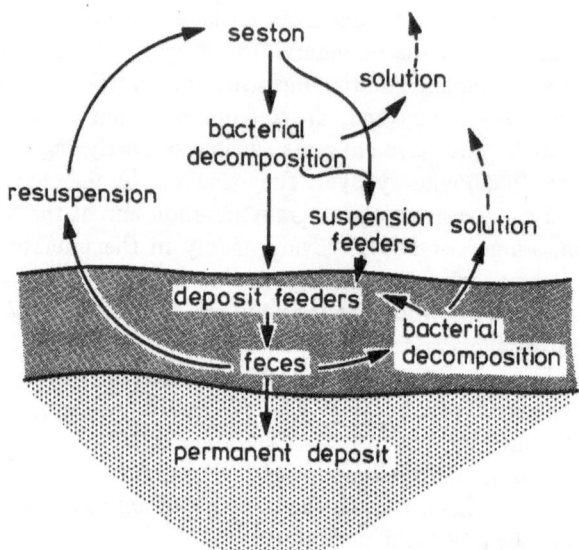

Fig. 2. Diagram to show the cycling or organic material at the benthic boundary
layer. (From Young, D. K., 1971. *Vie et Milieu, Supp. 22,* pp. 557–571.)

4. The functional types of organisms (e.g., suspension- and deposit-feeders)
   are not always sharply defined.
5. Sequence of colonization may not always be in the listed order.
6. The progression of events need not be irrevocable, but may be retarded or
   reversed by a variety of processes.

However, these defects do not hinder the utility of the scheme. Even if the
sequence of invasion is out of order, it is unlikely that reordering will result in
substantial changes in the kinds of organism–sediment interactions that occur.
In spite of lack of clear boundaries, the relationships do reflect central tendencies
that have validity even if the range is left unspecified. Thus, the conditions listed
in columns II and III of Table 1 typify the ways in which certain categories of
substrata react to certain categories of organisms.

Modification Due to Light

When the photic zone is considered, a further order of complexity results:
the colonization by uni- and multicellular algae and vascular plants.
   The first-order effects of the addition of algae are:

1. Increased spatial complexity
2. Increased food source, which includes the production of dissolved organic
   material

Table 1. An Event–Response Scheme of Organism–Sediment Interactions

Starting conditions:
1. An azoic bottom
2. A spatial gradient in grain size ranging from extensive solid rock surfaces to gravel, intermediate grain size (1000–63 μm) and fine sediment (<63 μm)
3. No light
4. Constant temperature and salinity (35‰)
5. Sufficient water motion to supply pO_2 (saturated), erosion and deposition not dominant effects, and sedimentation rate constant
6. All organic and inorganic compounds present in water and sediment to sustain life
7. Organic content of sediment inversely proportional to mean grain size

Hypothetical Sequence of Colonization	Initial (Early) Sediment–Organism Responses	Higher-Order (Later) Sediment–Organism Responses
A. I. Uniform settlement on all surfaces by all species of bacteria and fungi	A. II. 1. *Solid rock:* surface coatings of bacteria and exudates 2. *Gravel:* surface coatings of bacteria and exudates 3. *Intermediate sed.:* surface coatings of bacteria and exudates 4. *Fine sed.:* surface coatings of bacteria and exudates	A. III. 1. *Solid rock:* bacterial coatings may prepare surface for subsequent larval settlement 2. *Gravel:* bacterial coatings may prepare surface for subsequent larval settlement 3. *Intermediate sed.:* grain–grain adhesion increased; friction decreased. Coatings may prepare surface for subsequent larval settlement

(Cont.)

Table 1. (*Continued*)

Hypothetical Sequence of Colonization	Initial (Early) Sediment–Organism Responses	Higher-Order (Later) Sediment–Organism Responses
A. I.	A. II.	A. III. 4. *Fine sed.:* grain–grain adhesion increased; friction decreased. Coatings may prepare surface for subsequent larval settlement
B. I. Increase in bacterial biomass proportional to grain surface area available for growth	B. II. Highest population densities on fine sediment	B. III. a. Bacterial respiration lowers pO_2, esp. in fine sediments b. Bacterial exudates and dead bacteria accumulate fastest in fine sediment
C. I. Movement of bacteria into bottom	C. II. A RPD[a] developed in intermediate and fine sediment	C. III. Vertical stratification of bacterial metabolic types along the RPD gradient
D. I. Colonization by heterotrophic microfauna and meiofauna	D. II. a. Vertical stratification of species according to metabolic types along the RPD gradient b. Some sediment burrowing	D. III. a. Increased community respiration further elevates the RPD, esp. over fine deposits b. *In situ* deposition of dead organics increases BOD;[b] RPD moves toward sediment surface c. A consumer food web formed

E. I.
Macrofaunal colonization by detritus feeders and predators

E. II.
1. *Solid Rock:* dominated by epifaunal suspension feeders, surface grazers, and boring suspension feeders

   a. Substratum destruction by borers and grazers (excavation, abrasion, dissolution)

   b. Construction of carbonate overgrowths by $CaCO_3$ secretion of epifauna

2. *Gravel:* same as solid rock except pore spaces occupied by macrofauna

E. III.
1. a. Increased complexity of food web

   b. Deposition of mineral detritus by biological weathering

   c. Deposition of organic detritus by fecal accumulations

   d. b. and c. above produce a granular substratum

   e. Increase in habitat complexity produces increased biologic complexity

   f. Surface evolves toward sediment-covered rock or carbonate biostrome, depending on rates of deposition of detritus vs. carbonate overgrowth

2. a. Same as above with addition of increased trophic complexity as infaunal deposit feeders move into mud-filled pore spaces

   b. Bioturbation of pore muds

   c. Sediment sorting decreases; gravel filled with mud and shell

(Cont.)

Table 1. (*Continued*)

Hypothetical Sequence of Colonization	Initial (Early) Sediment–Organism Responses	Higher-Order (Later Sediment–Organism Responses
	3. *Intermediate sed.*	3. a. Fecal deposits increase organic content
	a. Colonizers tend to increase the stability of the bottom with tubes and mucus (increase boundary layer thickness downward)	b. RPD moves upward
		c. Deposit feeders enter. Mobile forms mix the surface, decreasing physical stability of the deposit
	b. System dominated by suspension feeders	d. Accumulation of mineralized tissue may affect pore-water chemistry ($CaCO_3$, $SiO_2 \ldots$)
	c. Tube irrigation locally depresses RPD	e. Bioturbation may decrease mean grain size
	d. Some bioturbation by surface grazers and infaunal predators	
	e. Fecal pellets prod. at surface	
	4. *Fine sed.*	4. a. Intensive bioturbation lowers critical erosion velocity. Suspension feeders may disappear (trophic group amensalism). Deposit feeders dominate; trophic diversity lowered
	a. Colonizers tend to bind bottom by presence of tubes and mucus	
	b. Local bioturbation; tube irrigation lowers RPD	b. Resuspension of sediments by active injection
	c. Both suspension and deposit-feeders present	

c. Extensive horizontal and vertical particle transport by bioturbation. Decrease in bulk density, increase in porosity, permeability, and water content of sediment, esp. in upper 5–10 cm

d. Biogenic activity may produce surface topography resulting in spatial patchiness in biology, sediment composition, and stability

d. Pelletization of surface increases mean grain size

[a] RPD = Redox Potential Discontinuity
[b] BOD = Biological Oxygen Demand

3. Increased trophic diversity, by allowing for greater specialization
4. Increased organic content of sediment via primary productivity
5. Elevation of the RPD layer

Higher-order effects resulting from the critical effects listed above are:

1. Increased sedimentation rates
2. Stabilization of the sediment by algal mats, root structures, etc.
3. Encrusting algal and coral overgrowths may produce hard substrate and lead to reef development
4. Calcareous algae give rise to aragonite deposition
5. Algal secretions may induce flotation of sand by rendering it hydrophobic
6. Large algal holdfasts may assist in transporting sediments via wave and current action

Modification Due to Depth

The scheme presented above assumes constant depth. If we allow depth to vary from outside the surf zone to the deep sea, we find that increasing depth (and distance from land) is accompanied by the following effects. In most cases the first- or higher-order effects involve issues about which our knowledge is still inadequate. These issues are given in parentheses:

1. Decreasing hydrodynamic energy, and therefore decreasing resuspension of bottom sediment, as well as a decreasing average grain size of the non-biogenic material that is supplied
2. Decreasing environmental variability, including regular cycles (annual, tidal, daily)
3. Decreasing, but more uniform supply of nutrients
4. Decreasing temperature and increasing pressure

The first-order result of these differences will be:

1. Decreasing standing crop and potentially decreasing turnover rates (Do deep water organisms, including bacteria, have lower metabolic rates than shallow water organisms?)
2. Decreasing average size of individuals (Does this miniaturization apply to all taxa? What are the causal factors?)
3. Alteration of trophic dimensions (What affects the relative success of predators, deposit feeders, and suspension feeders?)
4. Decrease in fluctuation of biological activity (Are deep water organisms rhythmic in any sense, global or otherwise?)
5. More complete utilization of organic materials (Are deep water animals metabolically and ecologically more efficient?)

6. Less-frequent downgrading of communities due to physical factors (Are deep water animals correspondingly more sensitive to physical change?)

Higher-order effects will be:

1. Reduction reactions and products of diminished importance; RPD greatly depressed
2. Shallower depth of biological activity (Is this true? How deep in the sediment do deep water organisms live?)
3. Biogenic effects in sediment occurring at slower rate and longer time scale (What are these rates?)
4. Altered relative importance of life styles resulting in differential effects on sediment, both quantitatively and qualitatively (What are these differences?)
5. Although biological processes are slower, the correspondingly lower sedimentation rate results in bioturbation as complete or more so than in shallow water (This inference is based on a variety of data, including the very low organic carbon content and the observed smearing with time of initially sharp datum planes. However, the truth of the inference remains to be established.)

The Response of Organisms to Sediments

Most classical marine ecology (Petersen, Thorson, etc.) implies that similar groups or species consistently occur on similar substrata and that the type of sediment in part controls the species distribution. Current ecological thought suggests that populations of species occur log-normally distributed along gradients in an overlapping manner and form continua. For sediments, preferenda are usually held to a grain-size range; other factors relating to grain size (e.g., porosity, permeability, oxygen content) may be the important factors, but grain size is the most obvious correlative factor.

Segregation of species to narrow grain-size ranges may result from substratum selection by larvae and/or mobile adults. With larvae, behavioral responses to light, current, pressure, etc. will segregate species in different water masses so that a restricted number of species is available to colonize any sediment type. The responses of larvae and mobile adults may be merely to a broad grain-size distribution or to a restricted response, such as bacteria of a favorable species within a given grain-size range. Clearly, settlement is critical for those sedentary species making a once-and-for-all selection of a site for tube building or feeding or to ensure cross-fertilization.

In mobile species, substratal selection is less critical and sediment responses

Table 2.

Sediment type	Sediment origin	
	Recent (in hydrodynamic balance)	Relict (in hydrodynamic imbalance)
Mud	(i) Water content high. For shelf sediments mainly near shore. (ii) Organisms predominantly deposit feeders. No interstitial species; micro- and meiofauna in superficial layer. Burrows unstable.	(i) Hard (includes peat). For shelf sediments usually not near shore. (ii) Organisms predominantly suspension feeders. No interstitial fauna; micro- and meiofauna in superficial layers. Often burrowing species with many permanent burrows.
Sand-clay	(i) Under tidal and wave influence. Sand–clay layered, not mixed. Usually high primary production and large organic matter content. Organic matter decreases with depth.	(i) Not under tidal influence. Sand mixed and covers 60% of shelf sediments. More constant environment than Recent sand-clay. Primary production lower than Recent sand-clay.

can be more general. Thus one can assume that the restriction of certain species to certain types is produced by elaborate substratum selection mechanisms.

It is not felt relevant to consider species differences in sediments (Thorson, 1957; Hedgpeth, 1957). Instead, some generalizations will be made on how sediments may control organisms. In order to consider sediment–animal relationships, a primary division has been made between recent sediments (in hydrodynamic balance) and relict sediments (in hydrodynamic imbalance) because the antiquity and greater stability of the latter have important effects on the organisms therein. The data are presented in tabular form in Table 2.

A number of points arise from the above:

1. How do the properties of different sediment types control the kinds of organisms that can live there? How have the functional morphology and behavior of a species solved the problems imposed by different sediment types?

2. The behavior of sediment-living species and their role in bioturbation processes needs to be much more thoroughly investigated. Problems arise in defining the role of: (a) feeding types, (b) mobility of organisms,

Table 2. (*Continued*)

Sediment type	Sediment origin	
	Recent (in hydrodynamic balance)	Relict (in hydrodynamic imbalance)
	(ii) All feeding types present. Micro-, meio- and macro-fauna abundant. Populations may be very large but subject to great cyclical changes. Species diversity increases with depth. Secondary production highest among sediment types.	(ii) Probably all feeding types present. Reduced organic matter reduces population sizes. Species diversity high due to lack of dominance and greater predictability of environment. Highly bioturbated giving increased structural complexity of environment.
Fine sand	(i) Hard packed due to wave and tidal influence. Characterized by small ripple marks. Most easily moved about of all sediments. Small interstitial voids. (ii) Poor insterstitial fauna. Difficult to burrow; organisms that can may be very common. Suspension feeders dominate.	(i) Soft, not well packed. Most easily moved about of all sediments. (ii) Interstitial fauna poorly understood. Suspension feeders dominate.
Medium–coarse	(i) Unstable due to tidal currents and waves. Characterized by mega-ripples. Much water exchange. (ii) Rich interstitial and meiofauna. Poorly developed macrofauna. No epifauna.	(i) Stable environment. No sedimentation. (ii) Less diverse meiofauna than Recent sediment. Well-developed macrofauna. Suspension and deposit feeders common.
Gravel	(i) Highly unstable. (ii) Almost no fauna.	(i) Stable, pore spaces filled with smaller grains. (ii) Well-developed epifauna and interstitial fauna.

(c) secretions and excretions, and (d) tube formation and burrowing activities

3. The general biology of the epifauna of superficial sediment layers is almost unknown and yet may play a significant role in bioturbation processes.

4. How stable are benthic populations in the different sediment types and at different water depths?

Problems of Importance and Experiments

Substratum Stabilization

Many studies of bioturbation show that this process tends to lower the critical erosion velocity of the sea floor. Less well studied are those biogenic activities that bind or otherwise stabilize the sea floor. It is reasonable to assume that the different stages of community succession and sediment types will be characterized by different substratum stabilities.

Substratum binding probably takes place in the following ways: (1) organic films (exudates) may serve as agglutinates and/or decrease skin friction; (2) structures such as vertically oriented tubes or filaments may serve to elevate the boundary layer; and (3) mineralized tissue may cover and cement the surface.

Most marine plants and animals secrete organic compounds as high molecular weight polysaccharide or mucoprotein (mucus), which coat virtually all surfaces with mucus. Dense populations of tubes or marine algae, kelps, or vascular plants also bind the sea floor and elevate the boundary layer. The formation of calcareous crusts by algae and invertebrates may also serve to cement a granular bottom.

Questions (numbers in parentheses refer to the Scheme in Table 1):

1. What role does mucus have in:
   (a) aggregating particles and binding surfaces? (A.II)
   (b) chelating metals or serving as a substrate for chemical reactions with dissolved compounds?
   (c) providing a food source for consumers (recycling)? (B.III.b)
2. What is the significance of tubes and crusts in binding the sea floor (E.II.2.b; E.II.3.a)
3. What are the stages of community succession for which these processes of stabilization are characteristic? (EIII).

Experiments

The role of binding is best studied in a flume, complemented with appropriate field observations. The activity of mucus as a substrate for chemical reactions may be studied *in vitro* with innocula of pure mucus.

Spatial Heterogeneity in Organisms

Rationale

Observations show that most marine species occur in aggregated rather than random or regular patterns, although the factors producing such spatial heterogeneity in faunal populations have been studied in only a few cases. This patchiness may range in scale from bacterial aggregations measured in microns to regional differences in faunal associations.

In selecting substrates, organisms respond to a wide range of stimuli ranging from selection of grain size to contact with a particular structure such as a tube. A restricted stimulus will produce an aggregated pattern, but for most species we have no knowledge of settlement behavior.

Heterogeneity may also result from differential mortality in randomly or regularly distributed populations. Again, data are lacking in factors producing mortality. Organisms are also known to create mosaic distributions of sediments. Such differences stimulate faunal heterogeneity both spatially and temporally.

Questions

To what extent do spatial and temporal sediment heterogeneity influence the structure of benthic population communities, and vice versa?

Applicability

Spatial heterogeneity of animal populations can clearly lead to intense local variation in flow rates, the flow of chemicals, etc., and consequent changes in the boundary layer.

Increased spatial heterogeneity is known to lead to higher species diversity. While spatial heterogeneity of sediments is difficult to define, mixed (heterogeneous) sediments seem to encourage higher organism diversity than do uniform sediments. It may well be that in highly reworked deep-sea sediments spatial heterogeneity is a major contributory factor leading to high deep-sea species diversity.

Succession in Benthic Communities

Rationale

The event–response scheme developed for sediment–organism interactions tends to suggest a succession of stages leading to a climax both in organisms and sediments. In general terms this climax would have intermediate to fine sediments, with a relatively high porosity and permeability, since it would be dominated by deposit-feeders. Organic matter would be almost completely consumed, so that no redox gradient would be present. The deep-sea bottom could well fit into this picture, but in shelf areas events tend to suggest cyclic phenomena, due possibly to:

1. High physical stress, disturbing sediment–organism interactions periodically or sporadically
2. Catastrophic events in sediment–organism interactions, leading to a complete or partial return to the starting situation

However, even in shelf areas it is possible that the succession is predictable, depending on the spatial and time scales.

Questions

Does a predictable succession exist in the sediment–organism interaction, starting from a given abiotic circumstance and leading to a climax stage? Do spatial distribution patterns reflect temporal patterns?

Applicability

If present, a predictable succession would allow a classification of the maturity of the system and of the system's functions (e.g., production, bioturbation, etc.). Also sediment parameters could then be predicted from the observed successional stages.

Experiment and Observations

Spatial heterogeneity of organisms and successional changes can be studied in the same type of experiment.

Field Studies. The settlement and succession following natual and artificial destruction of communities can be studied to show whether events follow a fixed pattern. Observations should be made repeatedly in the same area. The scale should be sufficiently large to cover the known range of spatial heterogeneity of the community.

Photographic surveys and quantitative samples in the deep sea and deep shelf areas can elucidate spatial patterns of benthic organisms.

Laboratory Work. Laboratory experiments should be large scale and can be divided into two classes:

1. Natural organisms should be distributed at random in a homogeneous sediment. Spatial patterns and successional changes can then be observed.
2. Introduction of naturally occurring larvae into homogeneous sediments allow settlement patterns and successional changes to be followed.

Since these experiments will be highly dependent on seasonal factors they should be repeated to encompass natural cyclic phenomena.

Behavior and Functional Morphology

Detailed knowledge of structure and behavior in particular species with regard to kind and rate of locomotion, feeding, and excretion, and type of housing and protection facilities, etc., is needed to give an estimate of the intensity of bioturbation, biogenic sedimentation, and erosion. Furthermore, such knowledge would answer questions about rate of stabilization or destabilization of the sediment, rate of recycling of sediments and organic matter, and the changing of other physical and chemical processes. All these effects can be caused by a single species, by a whole population, or by a community.

In a small number of animal species only, and especially those from shallow-water environments, behavior and functions are well known. These data show already the existence of local correlative adaptations, e.g.:

1. Burrow type and water depth
2. Shell thickness and surf influence
3. Feeding behavior and sediment type
4. Feeding behavior and surf
5. Mean size of organisms as related to environmental parameters

Further investigations is this field are necessary to find other general correlations between animal adaptations and their environmental conditions.

The knowledge of any of these correlations seems to be important for the recognition and interpretation of different marine environments. The knowledge of rate and kind of activity in animal species in different marine environments may be useful for ecologists (animal communities), paleontologists (trace- and body-fossils), geologists (biogenic structures), and chemists and physicists (water–bottom interaction).

It still remains to be determined what kind of correlations exist between the behavior and the functional morphology of animal species on one hand and between specific water depths and sediment types on the other.

Processes and Products of Bioturbation

In nearly all environments bioturbation is sufficiently active to modify significantly the nature of sediments in the benthic boundary layer and their chemical and physical processes. These changes in turn affect biological activity. Bioturbation results from a complex of biological activities, resulting in inter-acting chemical and physical processes. Some such activities are:

1. Formation of holes or burrows with one or more entrances
2. Formation of tubes, frequently with coated walls
3. Movement of siphons of bivalves, etc.
4. Plowing through sediments, notably by predators such as gastropods
5. Tracks of invertebrates moving on the sediment
6. Disturbance by demersal organisms (especially fish) in search of food
7. Production of feces: (a) within the sediment by stationary or migrating organisms, (b) at the surface by stationary organisms feedng in the sediment, (c) on the sediment by stationary or migratory organisms,
8. Circulation of water resulting from biological pumping and physical pumping by tides and waves

The biological processes of bioturbation are partly understood in some shallow environments, but are virtually unknown in most others, especially in the deep sea. The physical and chemical consequences are nearly unknown, as are the complex interactions.

For convenience, the problem can be considered in two parts, each containing several questions. The list is not exhaustive.

Bioturbation Affecting the Water-Sediment Interface

Two distinct results of surface bioturbation can be recognized: (1) construc-tion of a microrelief on an initially smooth depositional surface (common on fine deposits), and (2) replacement of an existing microrelief, e.g., ripple marks, by another one. In either case the product is a microrelief of several centimeters that erases the record of depositional events and constitutes significant boundary roughness in hydrodynamic terms.

Questions (numbers in parentheses refer to the scheme in Table 1):

1. What is the rate of obliteration of preexisting surface reliefs in a wide range of environments and what are the typical dimensions of the new reliefs? (E.II, 3.d, 4; E.II, 3.c, 4)

2. What are the textural and compositional changes resulting from the

process of bioturbation, as either an increase or decrease in homogeneity, and what effect do they have on erosion or deposition? (E.II, 3.e, 4.d; E.II, 3.c, 4.a-d)

Bioturbation Affecting Vertical Gradients within and just above the Sediments

The various activities listed above can modify in a major way and to appreciable depth the physical and chemical gradients within the sediment. The nature of the deposit and the processes operating in it are thereby commonly and profoundly altered.

Questions:

1. Under what condition are the deposits homogenized or, conversely, differentiated by bioturbation, how deep does this process penetrate, and at what rate does it proceed? *(E.II, 4; E.II, 2.c, 3.c, 4.a-d)

2. What are the rates of the various chemical and physical consequences of bioturbation, and how are they related to distinct and separate biological activities?

3. What are the separate and distinct organism–sediment interactions which together result in bioturbation; what are the organisms, processes, and rates involved, and what is the effect of bioturbation on the benthic community?

4. What are the rates of resuspension of sediment from the bottom, and what is the quantitative role of this process in erosion and sediment transport? (E.II, 3.c, 4.b)

Applicability

It is clear that these problems must be investigated in a broad spectrum of environments and using a broad and interdisciplinary approach. The complex of bioturbative processes plays an essential role in geochemical, physical, and geological studies of the benthic boundary layer. Inadequate knowledge in this area constitutes a major hindrance to the development of a full understanding of the benthic boundary layer.

Experiment on Integrative Measurements of Biological Activity and Bioturbation

In an analysis of the effects of organisms on sediments, it is critical to learn the rates of processes as well as their mechanics. Rates resulting from the activity

---

*A special case of particular importance is the degree to which bioturbation in the deep sea broadens initially sharp datum planes of a physical or chemical nature. This process reduces the temporal resolution of the pelagic sediment record. It is of major consequence for numerous studies (e.g., in geochemistry, biostratigraphy, paleoceanography).

of individual organisms or selected species are insufficient because it is unacceptable to extrapolate these rates to the whole community, and it is impracticable to measure all species individually. Measurements are needed that integrate the activities of the entire community through time. These measurements would preferably employ techniques that can be applied uniformly to all environments. Furthermore, they should be made *in situ* to avoid the uncertainties inherent in laboratory studies.

We suggest three potential approaches. The first is *in situ* measurement of community respiration. This is probably the best available measurement that can be related to total calorific utilization and therefore in some sense to total community activity. Within limits, it may also be the best index of the input rate of nutrients. We may expect that the reworking rate may be related to the mean body size of the organisms present, smaller organisms being proportionately more active than larger ones. This measure is not by itself sufficient as a bioturbation index. Communities of widely differing life-style compositions could have the same respiration rate. Moreover, although it would be appropriate to compare shallow- and deep-water deposit-feeding communities, it would be invalid to compare one of these with a suspension-feeding community. Therefore, this measure would have to be accompanied by biotic analysis to determine whether the communities are amenable to this kind of comparison.

The second is measurement of chemical gradients with depth in the sediment. Each chemical system will have characteristic profiles, in many cases related to the redox gradient. As a result, each system will have a unique combination of gradients within the sediment. Bioturbation will alter these gradients in a way which is proportional to its rate. Thus a study of chemical gradients in presence and absence of burrowers can yield a measure of the integrated effect of whole communities. The way in which different chemical gradients are affected may yield information on depth of bioturbation, transport vectors, and the relative impact of pore water exchange compared with particle advection. Gradient studies could involve introduction of exotic tracers such as dyes, heavy minerals, and fluorescent particles, or a study of substances already present in the environment. Natural radioisotopes may be useful in this respect. The rate of change of the magnetic grain fabric may also be used to estimate reworking rates in some sediments.

Third, time-lapse photography may be used to study surface reworking. This may be especially useful for the deep sea.

Paleoecology: Preservation of Bioturbation Structures

Some structures within sediments that are useful paleoenvironmental indicators result from biological activity. These indicators not only aid in the interpretation of ancient depositional conditions, but, having been subjected to postdepositional processes as well, they also have uses in the study of diagenesis. If the initial processes of formation are thoroughly understood, postdepositional modifications (diagenesis) can be derived from the nature of the structures exhibited in ancient deposits. Utilization of biogenic structures as paleoenvironmental indicators requires a thorough understanding of the origin of biogenic structures in recent environments. In this context the following problems require attention:

1. What does the structural inventory of various environments look like when the sea bottom is accreting under a balanced sedimentation rate? In other words, when biogenic structures move vertically through time in response to an elevation of the sediment, what assemblage of structures results.

2. What kind of preservational selection takes place with respect to type of structure (e.g., dissappearance of pelletal structure, and filling and/or compaction of biogenous cavities)? (These processes also refer to mass properties and chemical gradients.)

3. In many sediments the maximum vertical transfer rate of particles seems to be due to only one or a few burrowing species. Particularly regarding preferential preservation of larger burrows, measurements of burrowing activities of these animals should therefore be made.

4. In may sediments bioturbation activity is expressed mainly by burrow structures which cannot be assigned to defined types of *lebensspuren*. Nevertheless, they have morphological characteristics which can be measured. Therefore, what is the environmental significance of these "cloudy-turbulent" structures?

5. Do preserved deformational structures adjacent to burrow walls indicate initial rheological properties of the bottom?

Recommendations

Our recommendations are expressed in the questions we have asked in this report. A summary of these is given below. The questions must be answered by research so that our understanding of the important animal–sediment interactions can increase:

1. What role does mucus have in
   a. aggregating particles and binding surfaces?
   b. chelating metals or serving as a substrate for chemical reactions with dissolved compounds?
   c. providing a food source for consumers (recycling)?
2. What is the significance of tubes and crusts in binding the sea floor?
3. What are the stages of community succession for which particular processes of stabilization are characteristic?

4. To what extent do spatial and temporal sediment heterogeneity influence the structure of benthic population communities and vice versa?

5. Does a predictable succession exist in the sediment-organism interaction, starting from a given abiotic circumstance and leading to a climax stage? Do spatial distribution patterns reflect temporal patterns?

6. What kind of correlations exist between the behavior and the functional morphology of animal species on one hand and specific water depths and sediment types on the other?

7. What is the rate of obliteration of preexisting surface reliefs in a wide range of environments and what are the typical dimensions of the new reliefs?

8. What are the textural and compositional changes resulting from the process of bioturbation, as either an increase or decrease in homogeneity, and what effect do they have on erosion or deposition?

9. Under what condition are the deposits homogenized or, conversely, differentiated by bioturbation, how deep does this process penetrate, and at what rate does it proceed?

10. What are the rates of the various chemical and physical consequences of bioturbation, and how are they related to distinct and separate biological activities?

11. What are the separate and distinct organism-sediment interactions which together result in bioturbation, what are the organisms, processes, and rates involved, and what is the effect of bioturbation on the benthic community?

12. What are the rates of resuspension of sediment from the bottom, and what is the quantitative role of this process in erosion and sediment transport?

13. What does the structural inventory of various environments look like when the sea bottom is accreting under a balanced sedimentation rate? In other words, when biogenic structures move vertically through time in response to an elevation of the sediment surface, what assemblage of structures results?

14. What kind of preservational selection takes place with respect to type of structure (e.g., disappearance of pelletal structure, and filling and/or compaction of biogenous cavities)? (These processes also refer to mass properties and chemical gradients.)

15. In many sediments the maximum vertical transfer rate of particles seems to be due to only one or a few burrowing species. Particularly regarding preferential preservation of larger burrows, measurements of burrowing activities of these animals should therefore be made.

16. In many sediments, bioturbation activity is expressed mainly by burrow structures which cannot be assigned to defined types of *lebensspuren*. Nevertheless, they have morphological characteristics which can be measured. Therefore, what is the environmental significance of these "cloudy-turbulent" structures?

17. Do preserved deformational structures adjacent to burrow walls indicate initial rheological properties of the bottom?

Key References

Ager, D. V., 1963, *Principles of Paleoecology*, New York, McGraw-Hill, 371 pp.

Berger, W. H., and Heath, G. R., 1968, Vertical mixing in pelagic sediments, *Journal of Marine Research 26:* 134–143.

Dörjes, J., 1974, Recent biocoenoses and ichnocoenoses in shallow-water environments. A. Biological aspects, in: (R. Frey, ed.), *Studies of Trace Fossils*, Springer, New York.

Frey, R. (ed.), 1974, *Studies of Trace Fossils*, Springer, New York.

Gray, J. M., 1974, Animal–sediment relations, in: (H. Barnes, ed.), *Oceanography and Marine Biology Annual Review*, Vol. 12, George Allen & Unwin, London, pp. 223–261.

Heezen, B. C., and Hollister, C. D., 1971, *The Face of the Deep*, Oxford University Press, New York, 650 pp.

Hedgpeth, J. W., (ed.), 1957, Treatise on marine ecology and paleoecology, *Geological Society of America, Memoir 67:* 1296 pp.

Hesse, R., Allee, W. C., and Schmidt, K. P., 1951, *Ecological Animal Geography*, 2nd ed., John Wiley & Sons, New York, 715 pp.

Hessler, R. R., and Jumars, P. A., 1974, Abyssal community analysis from replicate box cores in the central North Pacific, *Deep-Sea Research 21:* 185–209.

Moore, D. G., and Scruton, P. C., 1957, Minor internal structures of some recent unconsolidated sediments, *Bulletin of the American Association of Petroleum Geologists 41:* 2723–2751.

Reineck, H. E., and Singh, I. B., 1973, *Depositional Sedimentary Environments*, Springer, Heidelberg, 439 pp.

Rhoads, D. C., 1974, Organism–sediment relations on the muddy seafloor, in: (H. Barnes, ed.), *Oceanography and Marine Biology, Annual Review*, Vol. 12, George Allen & Unwin, London.

Schafer, W., 1972, *Ecology and Paleoecology of Marine Environments*, Oliver & Boyd, Edinburgh, 568 pp.

Scoffin, T. P., 1970, The trapping and binding of subtidal carbonate sediments by marine vegetation in Bimini Lagoon, Bahamas, *Journal of Sedimentary Petrology 40:* 249–273.

Thorson, G., 1957, Bottom communities (sublittoral or shallow shelf), in: *Treatise on Marine Ecology and Paleoecology, Geological Society of America, Memoir 67* (1): 461–534.

Webb, J. E., 1974, The distribution of *amphioxus, Symposium of the Zoological Society of London 36:* 179–211.

Zenkevich, L. A., (ed.), 1969, *Biology of the Pacific Ocean, Vol. 2, The Deep Sea Bottom Fauna, Pleuston* (in Russian), Nauka, Moscow, 353 pp. (Translation published by U. S. Naval Hydrographic Office, 1970).

F
Metabolism at the Benthic Boundary

Group Leader: A. D. McIntyre

Members: J. M. Davies, P. J. de Wilde, P. Lasserre,
E. L. Mills, M. M. Pamatmat, J. M. Teal,
H. Thiel, B. Zeitzschel

Rapporteur: B. T. Hargrave

Introduction

From the biologist's point of view the benthic boundary marks the lower limit of pelagic organisms, receives material dropping out of the pelagic phase, provides a substratum for the activities of epibenthos, and constitutes a habitat for burrowing forms.

On muddy ground where stiff clay is present most of the organisms are usually restricted to the top 10 cm or so, in soft mud to 30 cm or deeper, but in sandy areas active metabolism has been detected as deep as several meters. Some animals are more or less restricted to the epibenthic mode of life, but others use the sediment surface as a base from which to make excursions sometimes hundreds of meters upwards, while certain species of the upper waters utilize the bottom as a feeding or resting site. It is therefore difficult for the biologist to accept a narrow definition of the benthic boundary layer, but he would probably agree that much of the activity is concentrated on the top few centimeters of sediment and in near-bottom water.

In shelf areas, the benthic boundary is the site of valuable commercial fisheries and of mineral extraction. It is the region mainly affected by man's dumping of a wide range of water waste substances. In the deep water also, where our limited knowledge suggests a highly diverse benthic community of great ecological interest, there is an increasing possibility of extensive mineral exploitation, and the use of this region for disposal of highly toxic wastes is widely accepted. For these reasons a good understanding of the biological processes which take place on the sea floor is desirable.

Metabolism at the benthic boundary depends on the supply of material produced in the euphotic zone utilizable by living organisms. This is processed and transferred within the community and will be gradually lost from the system. An understanding of the metabolism will thus depend, first, on an accurate quantitative estimate of the input and of its nature and relative usability; second, on an appreciation of the way in which the community makes use of the material— implying a knowledge of the interactions of the various components of the community; and third, on how material is lost from this system.

In the following discussion, each of these three aspects is considered, and an attempt is made to indicate the current state of knowledge, to list the major relevant topics, and to suggest the most fruitful lines of research. Difficulties caused by lack of suitable techniques are noted and certain areas where input from other disciplines would be appropriate are identified.

Input of Metabolizable Organic Matter

Particulate Matter

Mechanisms of Transport

Transport mechanisms to the bottom can be listed as follows:

1. *Passive settlement of detritus.* This includes particles ranging in size from those retained by filters to rather large floccules. Their settling rate is about 1 m day^{-1}, and their specific gravity is only slightly more than sea water.

2. *Sinking of fecal pellets.* Fecal pellets of pelagic animals, especially copepods, are released as units of encapsulated material. They are on the order of several hundred micrometers long and their specific gravity is considerably higher than the particles in the first category; their settling rate is about 100–200 m day^{-1}.

3. *Macroparticle transport.* This category includes animal carcasses, crustecean exuviae, and large plant remnants (sargassum, eelgrass, and wood), as well as human artifacts. The amount of these macroparticles in the deep sea and their quantitative importance is unknown.

4. *Downwelling.* Particulate matter may be taken down by physical processes, e.g., downwelling of water and convergence.

5. *Vertical transport via animals.* Migrating macrozooplankton and fish may actively transport particles from the surface to greater depth. This mechanism might be especially effective when animals in successively greater depths transport organic material in stepwise fashion during vertical migrations.

6. *Larval settlement.* Pelagic larvae are an important input to shelf sediments. There is only limited knowledge of their importance in the deep sea.

7. *Horizontal transport.* There is a continuous horizontal transport of particles which is more a process of redistribution than a net import of organic material. Underwater landslides redistribute, expose, and stir sediments, but their overall significance is unknown.

Quantitative Sampling

Historically, cylinders fixed at various depths above the bottom have been used to collect detritus thought to be sedimenting through the water column. It is now apparent that in all areas except those with low horizontal velocity a trap of this shape considerably overestimates the settling rate. An open cylinder left in the corner of a circulating tank of turbid water will accumulate most of the particulate matter because the velocity profile of the water is changed during passage over the open mouth. Eddies created in the open mouth of such cylinders result in particle entrapment, leading to an overestimate of settling rate. This problem decreases with increasing particle settling velocity and decreasing horizontal flow velocity.

A cylindrical collector with a convex sloping upper surface is preferable since it will interfere less with the velocity profile over the opening than will cylinders with the normal mouth opening. We suggest that for future studies a standard design of settlement trap should be developed and tested so that data from different areas can be compared more meaningfully.

Chemical Composition

The composition of the detrital input to the sediment varies greatly with location. Analyses show that detritus samples from deep waters have C/N ratios greater than $10:1$, while in shallow waters, where direct settlement of detritus of plant origin may occur, the ratio may be as low as $5:1$.

Direct microscopic examination of the material provides information as to the type and source of the detritus, e.g., whether it consists predominantly of fecal pellets or of senescent plant cells. Fluorescence microscopy may be used to enumerate bacteria associated with the particles and to estimate the percentage of surface coverage. Measurement of particulate organic carbon and ATP allows calculation of the ratio of living to dead organic carbon.

Detrital material is usually analyzed for pigments, carbon and nitrogen, and ash weight, but instances of detailed biochemical or caloric analyses are rare. Histological techniques have been used to detect protein, carbohydrate, and fats in detrital particles, but are not quantitative. Measurement of the percentage composition of protein, carbohydrate, and fat on an ash-free dry weight basis of settled material from many deep-sea and coastal stations would be valuable.

Various methods have been used to determine how much of the detrital material is metabolically useful. Biological oxidation and enzyme treatment have been used to estimate the total biodegradable material. The real question is how much of the total is biodegradable and what proportion of that is readily utilizable. One possible way to determine these two factors is to follow the rate of respiration of a sample of the material to determine at what point the initially high respiration rate decreases, corresponsing to exhaustion of the readily metabolizable portion of the detritus. The possible decomposition of material accumulated in sediment traps during exposure must always be considered and quantified.

Dissolved Organic Matter

Separation of particulate and dissolved organic matter in water is carried out by filtration. In practice, material passing a 0.2 μm or 0.45 μm membrane, silver, or Nuclepore filter is generally taken as the dissolved fraction. Although this procedure is operationally necessary, it combines colloidal and subcolloidal material of high adsorptive surface area with what may be considered as material truly in solution. Degradation of these various size categories of organic material at the sediment surface is probably markedly different since the availability to organisms depends on both chemical composition and state of dispersion.

Mechanisms of Transport

Observations of concentration gradients of dissolved organic matter in over-lying and interstitial water alone cannot be used to infer the rate of entry of these compounds into benthic systems. Specific experiments to quantify and separate adsorption (physical adhesion due to cation exchange and surface charges) and active biological accumulation are needed. Dissolved organic compounds incorporated in the crystal lattice structure of clay, for example, may have a completely different "availability" for biological degradation than similar compounds adsorbed on the outer surface of silicate particles. Properties of adsorbing surfaces are thus critical for an understanding of these adsorption phenomena. Stirred and intact sediment systems must be combined in experiments to quantify these effects, since rates of movement through intact sediment surfaces proceed at a fraction of the rate when particles are in suspension. Small amounts of stirred surface sediment that are allowed to resettle reduce the amount of dissolved organic matter in solution, indicating adsorption on particle surfaces.

The addition of labeled organic compounds to intact aquatic sediment systems and the study of their metabolic fate, followed most easily as $^{14}CO_2$ production, may provide a generally useful method for consideration of the degradation of specific organic compounds. It must be stressed that the measured fluxes apply only to the specific compound added (in pure soluble form) and that extrapolation to large pools of more structurally complex material is not possible. The addition of "homogeneously" labeled mixtures of soluble organic material (e.g., derived as algal exudates produced in the presence of HCO_3^- during photosynthesis) might provide a more ecologically meaningful substrate for use in such experiments.

Collection of Samples

Samples of water from various depths above deep ocean sediments can be taken by routine water-bottle methods; but gradients of dissolved organic matter within the benthic boundary layer have yet to be identified. A serious problem of contamination for certain dissolved organic compounds may also occur due to adsorption on open bottle walls during vertical passage through the sea-surface film and water column. This has been observed with soluble hydrocarbons, which are in high concentrations at the air–sea interface. The samplers used for such studies of vertical distribution should be made of inert material and extreme care should be taken to avoid contamination from the surface film.

For any meaningful measurements of the flux of various types of dissolved organic material, accurate assessments of concentrations in supernatant and interstitial water are a prerequisite. At present there is no sampling device that can retrieve undisturbed sediment cores from the deep-sea bed with overlying water held at *in situ* pressure and temperature. The maintenance of these conditions after sample retrieval is essential if measurements of concentrations of organic matter are to have any ecological reality. An alternative to removal of cores is an *in situ* submersible system working entirely at the sampling depth. *In situ* experiments discussed below could be accomplished in this manner.

Analysis of Soluble Organic Matter

While several methods for the identification and quantification of various types of dissolved organic matter exist, few measurements have been made in the deep ocean and usually only total dissolved organic carbon has been measured in water immediately above deep-sea sediments. Free amino acids, sugars (single and polymerized), and fatty acids occur in shallow coastal ocean waters and in interstitial waters. Concentrations are usually low and it is not always clear that such compounds are truly dissolved. Dialysis techniques or other separation methods (molecular sieves) may provide a more accurate assessment of the type and nature of such organic compounds. Separation on some basis other than filtration is necessary before detailed studies of the flux of particular dissolved compounds can be made.

Measurements of dissolved organic carbon by several techniques (e.g., wet oxidation, freeze-drying, plasma furnace injection) have been used to establish the overall amount of dissolved organic carbon in sea water. Unfortunately, this measure gives no indication of the availability of such carbon compounds for metabolic oxidation. The large size of the pool suggests a relatively slow turnover rate. Measurements of dissolved organic nitrogen may provide a better index of usable organic matter.

Metabolism

Respiration

A large portion of organic input to the sea bed is lost through respiration, but the fraction may change with conditions affecting the oxidizability of organic matter, e.g., water depth, oxygen tension, and the nature and source of organic matter.

Community Metabolism (Community Respiration)

Oxygen Uptake. At present, community metabolism, the sum total of respiration of all organisms present in the sediment column, is commonly estimated as the rate of oxygen uptake by the undisturbed sediment. Two techniques in use are (1) bell jars or other *in situ* enclosures, and (2) sediment cores brought to the ship or the laboratory. Both methods introduce an undertermined bias because of disturbance of the sediment and interference with natural turbulence. In both methods the change in oxygen concentration of enclosed water overlying the sediment is either monitored continuously with oxygen electrodes or analyzed before and after incubation. Oxygen uptake after poisoning the water represents a rate of inorganic chemical oxidation. The difference between rates of total uptake and inorganic chemical oxidation is aerobic respiration. Inorganic chemical oxidation, often taken as a measure of anaerobic metabolism, actually underestimates it. Anaerobic metabolism must be measured independently. Annual oxygen uptake by the sediment surface is probably a good estimate of oxidative loss of organic input during the year, but there is always a residue which is buried and oxidized over a longer period by anaerobes, or lost through other processes. The sum of these losses is an indirect measure of organic input to sediment communities. For the sake of accuracy, the term "benthic community metabolism" should be applied only to total respiration of aerobes and anaerobes (if present) in the sediment column. If anaerobic metabolism is evident from the presence of reduced substances in the sediment or measureable rates of chemical oxidation, measurements of oxygen uptake by the sediment surface should not be called community metabolism. The term "community oxygen uptake" is suggested, which implies that subsurface anaerobic metabolism is excluded.

In situ and shipboard or laboratory techniques give the same estimates of oxygen uptake to depths of at least 180 m. In deeper waters, the accuracy of shipboard measurement remains to be seen. This is best determined by comparison with simultaneous *in situ* measurements because, although tests in a pressure bomb may be revealing, there will remain the question of possible irreversible effects by rapid initial decompression as the sample is brought aboard ship. If the effect of pressure can be shown to be negligible or correctible, and *in situ* and shipboard measurements in the deep sea give the same estimate of community oxygen uptake, the latter method is a logical choice for routine ecological studies. The *in situ* method would be the only choice for areas which cannot be sampled with a coring device, such as hard bottoms, rocky areas, and coarse gravel. Fine sand can be successfully cored for metabolism studies. There is no bottom that is too soft for coring with the right kind of equipment.

Carbon Dioxide Production. For the purpose of relating community oxygen uptake to the cycle of organic carbon in the ecosystem, it is desirable to measure CO_2 production as well. Estimating CO_2 production from assumed respiratory quotients is risky because community RQ varies greatly. Simultaneous determination of CO_2 production and O_2 uptake over an annual cycle should be made. Laboratory experiments could be conducted by replacing all but a few millimeters of water above the core with a gas. Measurements of CO_2 production also indicate a rate of sediment surface metabolism.

Anaerobic Metabolism. This may be estimated by direct calorimetry, but the technique is extremely difficult. The method measures net change in entropy of sediment plus organisms. To estimate heat production due to anaerobic metabolism, a control blank from which metabolic activity has been eliminated (e.g., by heating to kill organisms, assuming this treatment does not alter other thermochemical reactions) should be subtracted from the total heat production. If metabolic heat production is low and there is a concomitant large increase in entropy, the sample may actually absorb heat.

It is also possible to estimate the anaerobic metabolism of specific metabolic types of microorganisms (e.g., SO_4^{2-} reducers, NO_3^- denitrifiers) by mathematical modeling.

Respiration of Groups of Organisms

Community metabolism ideally would be partitioned into all the component species populations in order to understand the energy flow through the community, but this is probably not feasible. Partitioning has been done into broad groups of organisms (e.g., macrofauna, meiofauna, microfauna, bacteria) and a few dominant macrofauna species.

Macrofauna and Meiofauna. Total respiration of macrofauna and meiofauna has been computed from standing crop and known relationships between respiration, biomass, and sometimes temperature. This computation may lead to overestimation of the importance of macrofauna in the community because the method is based on experimental values that tend to overestimate respiration. More realistic rates should be determined by conducting experiments with animals in their natural living conditions, e.g., in sediment if they are infaunal. Some infauna may be facultative anaerobes whose true respiration must be estimated by means other than oxygen consumption, such as by direct calorimetry while buried in sediment.

Microfauna. Total metabolism of all microfauna has never been estimated anywhere because of lack of data on both biomass and respiration. Since microfauna (e.g., ciliates) are partly aerobic and partly anaerobic, the foregoing statements for macrofauna and meiofauna also apply.

Bacteria. Total bacterial respiration has been estimated from the drop in rate of total oxygen uptake following the addition of antibiotic to the water inside bell jars or sediment tubes. The method is unreliable because the antibiotic is not 100% effective and it can affect higher organisms. The technique misses anaerobic bacterial metabolism which must be estimated independently. This has been done with the use of 2,3,5-triphenyl tetrazolium chloride (TTC), an artificial electron acceptor, but this assay should be calibrated by direct calorimetry.

Physiological Ecology

The application of physiological and biochemical tools to give some indication of different metabolic pathways is a necessary adjunct to studies of metabolism in groups or communities of organisms.

The Substances Metabolized

For most benthic organisms we do not know the specific kinds of material incorporated as food, although our knowledge of the range of potential food items is probably adequate. In particular, we do not know what food items, of those taken into the guts of metazoan animals, are actually assimilated and support the animals' metabolism. The first gap in our knowledge (concerning the kinds of food items and their relative importance to benthic animals) is very significant in evaluating the input of food to the sediments from the water column. The second point (concerning the food items actually assimilated) bears very strongly on the partition of energy among coexisting species. Food items may be considered as follows.

Particulate

The fraction of detritus utilized by organisms, either as nonliving organic substrate or attached microorganisms, is seldom known. A way must be found of labeling detritus so that its nutritional value may be investigated, possibly by incorporation of radioisotopes. Modeling of high-precision, stable-isotope ratio measurements may provide a complementary understanding of metabolic processes. Assessment of the role of bacteria and other microbiota which may be selectively consumed by some or all metazoan benthic animals and by protozoa would then become possible.

Fecal pellets may be abundant in the sediments. They often serve as a substrate and source of organics for bacteria which may enter food chains. We need to know more about the dynamics of this situation and need to separate the activity of bacteria in the guts of organisms from their activity on fecal pellets in the sediments. Plant cells may be present in surface sediments and settle into the deep ocean. Their role as food items should be evaluated by labeling techniques.

The interactions of other metazoan animals are inadequately known. Some macrofauna species may eat meiofauna. The meiofauna itself in some habitats may constitute a system relatively independent from the larger organisms. The source of meiofauna nutrition is poorly known. Gametes, eggs, and larvae may be a significant food source, but few quantitative measures are available for benthic animals in shallow-water marine areas. The importance of this food source decreases in deep water.

Dissolved Organic Matter

The evidence is reasonably good that metazoan and protozoan organisms, and certainly bacteria, can take up dissolved organics. For the nonbacterial organisms there is scanty evidence that a net energy gain results from the uptake of dissolved organics; more experimental work aimed at trophic dynamics is needed.

Growth and Production

A portion of the organic input to the sea bed is temporarily stored in the different elements of the community through growth of the organisms, which consist of such groups as microorganisms, meiofauna, and macrofauna. Over periods of a year or longer, most benthic communities do not show a net increase in standing stock, i.e., the communities are at steady state. This means that total growth or production by the community is just balanced by losses due to outside predation and emigration. Estimates of growth and production have been made for dominant macrofauna species, mostly in intertidal and near-shore waters, while some knowledge of growth and generation times in meiofauna has been gained from laboratory cultures. Virtually nothing is known about growth and production of organisms in the deep sea. Some studies have given estimates of the relative magnitude of the aerobic respiration of microorganisms and meio- and macrofauna. Meiofauna and microfauna may be very active in energy flow, but there is little evidence that meiofauna transfer significant amounts of energy to larger organisms. Estimates of microorganism production carried out by different techniques give vastly different results.

Little is known about protozoa other than foraminifera and some ciliates. Data on meiofauna (including foraminifera) and on macrofauna are not usually considered in the same investigation and lack direct comparability. Equal amounts of wet weights of macrofauna and meiofauna have been found in abyssal depths, and the assumption of higher turnover rates in meiofauna as compared with macrofauna suggests that biological activity is higher in the smaller organism size group. This trend to smaller size has been shown within the macrofauna, but it remains an open question whether it can be extended to the activity of protozoa, fungi, and bacteria in deep sea communities.

Energy flow studies of whole populations show a significant relationship between production and respiration. Deep-sea organisms may have low metabolic rates and correspondingly low production. If the relation between respiration and metabolism is the same as in shallow water, a common regression equation could be used to estimate production by benthic populations.

Loss of Metabolizable Organic Matter

A variety of processes carry biologically useful energy away from the boundary layer, affecting potential future metabolic activity. Chemicals which serve as plant nutrients have a special importance but they are not considered in this report since they typically do not carry biologically useful energy from the bottom.

Loss of Particulate Organic Matter

Food Catching

Organisms from outside the boundary layer may enter the layer only to feed. These are probably mostly fish, though *Cephalopoda*, *Crustacea*, and perhaps even some *Holothuria* may be involved. Animals which swim up into the water at night to feed or to reproduce (e.g., cumaceans and annelids) are especially vulnerable to such removals. Besides catching and eating whole animals, predators crop portions of benthic animals (e.g., clam siphons) and annelid heads or tails. Some mobile animals ingest sediment as well as animal tissue, and we do not know to what extent they are selective.

Reproduction Losses

Whole animals (e.g., annelids) may swim from the boundary layer for spawning, after which they may be lost by predation or carried away by currents. Gametes released into the water may be similarly lost. Larvae may be shed into the water and are in some species adapted for very long larval life, practically assuring loss at least from the local community in which they were produced. Many forms are involved, including decapods, galatheids, annelids, molluscs, etc. Such loss of reproductive forms is significant in some shallow-water systems, but we do not understand its importance in deeper waters.

Migrations

Some animals leave the bottom on either regular or sporadic excursions. They are then vulnerable to predation or may be carried away. There are also more permanent migrations, such as seasonal movements of shrimps from one area to another if not out of the boundary layer. Some cephalopods and fish lay eggs on the bottom which upon hatching permanently leave the boundary layer.

Transport by Currents

Though not primarily a biological mechanism, transport of particles by currents may be initiated by animals disturbing sediments. Burrowing, rooting in the bottom while feeding, pumping water for respiration and/or feeding, ejecting pseudofeces, and moving on or above sediments can all initiate erosion by currents otherwise too weak to remove materials from the boundary layer.

Burial

Finally, permanent loss of particulate organic matter may occur by biologically mediated burial. Animals feed from the water and deposit feces and pseudofeces on the surface. Others feed below the surface and deposit their castings on the surface and so bury particles. The collapse of burrows and feeding holes may bury objects. The death of deep burrowers may effectively eliminate them from the boundary layer. We know little about the animals that burrow deeply in shallow water, and virtually nothing about those in the deep sea. The burial of the organic matrix of shells and calcified chitinous exoskeletons might be a significant loss.

The existence of organic matter in deep-sea subsurface sediments suggests that some organic compounds are so refractory to degradation that their presence represents not burial but the absence of efficient processes capable of utilization. Detection of sulfate reduction at considerable depths in such sediments, however, provides evidence for at least some susceptibility of these compounds to decomposition.

Loss of Dissolved Organic Matter

Losses of dissolved organics is principally by current transport and mixing into overlying waters, but biological actions can both solubilize substances and inject them into the water. Organisms continuously excrete, secrete, and leak substances into the water (e.g., mucus) and losses are greatly increased at death. They may lose these organics into pore water or actively pump them into the overlying water. Pumping activity will also increase the loss-rate of substances dissolved in sediment pore water.

Burial is also a potential mechanism for removal of dissolved organics, although the compounds may be largely adsorbed onto particles when lost. Loss of organics in true solution will be lessened by migration of pore water upward as sediments compact during burial.

These mechanisms of loss of organic matter from bottom sediments can be quantified by *in situ* observations and experiments. Interdisciplinary studies are essential for an understanding of these problems.

Recommendations

1. Sediment traps should be designed and tested for quantitative estimation of particulate input to the bottom.

2. *In situ* studies of the sinking rates of various types of particles are required. The use of radioactive labeled material introduced into natural systems and the development of photographic and optical techniques may be appropriate.

3. The importance of very large debris (e.g., carcasses) must be assessed in terms of the total organic contribution to the bottom.

4. More detailed knowledge of chemical composition of detrital and sedimented organic material and of the extent to which this is available for metabolism by organisms is required. Comparable information on composition and availability, as well as on the vertical distribution in and above the sediment, is required for dissolved organic matter.

5. Knowledge of the flow between interstitial and overlying water is required.

6. A sampling device that can retrieve undisturbed cores with overlying water from the deep sea and maintain *in situ* temperature and pressure is required. This could be included in the development and support of deep-water experimental units for use in *in situ* and retrieval investigation.

7. It is necessary to understand the physiological effect of high pressure and its interaction with temperature on metabolic rates of organisms in order to understand the control of deep-sea metabolism.

8. The potential of direct calorimetry in the study of total benthic community metabolism should be developed. This technique could then be applied to assess the reliability of present measures of anaerobic metabolism.

9. Information is required on the relative metabolism of the various components of the communities (e.g., size groups, taxa, metabolic types, and physiological state) based on both aerobic and anaerobic situations.

10. There is a need for further study of the vertical distribution of organisms within the sediment in relation to physical and chemical gradients.

11. Long-term *in situ* observations on macrofaunal behavior are necessary to understand their effect on the bottom and their metabolic requirements.

12. More emphasis on controlled experiments in nature is required. Experimental enhancement and disturbance of natural benthic systems provide data unavailable by other methods.

Key References

Fenchel, T., 1969, The ecology of marine microbenthos. IV. Structure and function of the benthic ecosystem, *Ophelia 6:* 1-182.

Hessler, R. R., 1974, The structure of deep benthic communities from central ocean waters, in: *The Biology of the Oceanic Pacific*, Oregon State University Press, pp. 79-93.

Hood, D. W., (ed.), 1970, Organic matter in natural waters. *Institute of Marine Science, University of Alaska, Occasional Publication No. 1.*, 625 pp.

Jorgensen, C. B., 1966, *Biology of Suspension Feeding*, Pergamon, Oxford, 357 pp.

Mann, K. H., 1969, The dynamics of aquatic ecosystems, *Advances in Ecological Research 6:* 1-81.

Melchiori, and Hopkins, J., (Ed.), 1972, The role of detritus in the aquatic environment, *Memorie dell'Instituto Italiano di Idrobiologia 29* (Suppl.)

Pamatmat, M. M., and Bhagwat, G., 1973, Anaerobic metabolism in Lake Washington sediments, *Limnology and Oceanography 18:* 611-627.

Parsons, T., and Takahasi, M., 1973, *Biological Oceanographic Processes*, Pergamon Press, Oxford, 186 pp.

Riley, G. A., 1970, Particulate organic matter in sea water, *Advances in Marine Biology 8:* 1-118.

Smayda, T. J., 1970, The suspension and sinking of phytoplankton in the sea, *Oceanography and Marine Biology Annual Review 8:* 353-141.

Sokolova, M. N., 1972, Trophic structure of deep-sea macrobenthos, *Marine Biology 16:* 1-12.

Steele, J. H., (ed.), *1970, Marine Food Chains*, Oliver & Boyd, Edinburgh, 552 pp.

Thiel, H., 1971, Die Bedeutung der Meiofauna in Küstenfernen benthischen Lebensgeneinverschiedener geographischer Regionen, *Verhandlungen der Deutschen Zoologischen Gesellschaft 65:* 37-42.

Vinogradov, M. E., 1955, Vertical migration of zooplankton and their importance for the nutrient of abyssal pelagic fauna, *Trudy Institua Okeanologii-Akademiya Nauk SSSR 13:* 71-76.

Participants

Dr. R. A. BERNER
Department of Geology and
  Geophysics
Yale University
New Haven, Conn. 06520, USA

Dr. M. BEWERS
Bedford Institute of Oceanography
Dartmouth, Nova Scotia, Canada

Dr. R. CHESSELET
Centre des Faibles Radioactivités
Laboratoire Mixte CNRS-CEA (CFR)
91190-Gif-sur-Yvette, France

Dr. R. C. COOKE
Department of Oceanography
Dalhousie University
Halifax, Nova Scotia, Canada

Dr. J. M. DAVIES
Department of Agriculture & Fisheries
  for Scotland
Marine Laboratory
P.O. Box 101
Aberdeen, A89 8DB, Scotland, UK

Dr. A. J. DE GROOT
Instituut voor Bodemvruchtbaarheid
Oosterweg 92
Haren (Groningen), Netherlands

Dr. Y. DESAUBIES
pro tem. Woods Hole Oceanographic
  Institution
Woods Hole, Massachusetts 02543, USA

Dr. M. DE VRIES
Delft Hydraulics Laboratory
PO Box 20
Delft, Netherlands

Dr. P. J. DE WILDE
Nederlands Instituut voor Onderzoek
  der Zee
Postbus 59
Texel, Netherlands

Dr. D. J. DÖRJES
Institut für Meeresgeologie und
  Meeresbiologie "Senckenberg"
D-2940 Wilhelmshaven
Schleusenstr. 39A, Germany

Dr. J. C. DUINKER
Nederlands Instituut voor Onderzoek
  der Zee
Postbus 59
Texel, Netherlands

Dr. K. DYER
Institute of Oceanographic Sciences
Crossway, Taunton
Somerset TA1 2DW, England, UK

Prof. Dr. G. EINSELE
Geologisches Institut
Universität Tübingen
D-74 Tübingen
Sigwartstr. 10, Germany

Dr. S. L. EITTREIM
Lamont-Doherty Geological Observatory
Palisades, New York 10964, USA

Dr. R. D. FLOOD
Woods Hold Oceanographic Institution
Woods Hole, Massachusetts 02543, USA
  and
Department of Earth and Planetary Sciences
Massachusetts Institute of Technology
Cambridge, Mass. 02139, USA

311

Prof. Dr. A. FÜHRBÖTER
Technische Universität Braunschweig
Lehrstuhl für Hydromechanik und
  Küstenwasserbau
D-33 Braunschweig, Germany

Dr. C. H. GIBSON
Department of Applied Mechanics and
  Engineering Sciences
5246 Urey Hall
University of California at San Diego
La Jolla, Box 109
California 92037, USA

Dr. J. GIESKES
Scripps Institution of Oceanography
La Jolla, California 92037, USA

Dr. C. M. GORDON
Ocean Sciences Division
Code 8342
Naval Research Laboratory
Washington, D.C. 20375, USA

Dr. W. J. GOULD
Institute of Oceanographic Sciences
Wormely, Godalming
Surrey GU8 5UB, England, UK

Dr. J. S. GRAY
Department of Zoology
Wellcome Marine Laboratory
Robin Hood's Bay, Yorks., England, UK

Dr. G. GUST
Institut für Meereskunde
Universität Kiel
D-23 Kiel
Düsternbrooker Weg 20, Germany

Dr. B. T. HARGRAVE
Fisheries and Marine Service
Marine Ecology Laboratory
Bedford Institute
Dartmouth, Nova Scotia, Canada

Dr. A. D. HEATHERSHAW
Marine Science Laboratories
Menai Bridge
Gwynedd, North Wales, UK

Dr. R. R. HESSLER
Scripps Institution of Oceanography
PO Box 1529
La Jolla, California 92037, USA

Dr. C. HOLLISTER
Woods Hole Oceanographic Institution
Woods Hole, Massachusetts 02543 USA

Dr. G. H. KELLER
NOAA-Atlantic Oceanographic and
  Meteorological Laboratories
15 Rickenbacker Causeway
Miami, Florida 33149, USA

Dr. P. D. KOMAR
School of Oceanography
Oregon State University
Corvallis, Oregon 97331, USA

Prof. R. B. KRONE
Department of Civil Engineering
University of California
Davis, California 95616, USA

Prof. L. D. KULM
School of Oceanography
Oregon State University
Corvallis, Oregon 97331, USA

Dr. P. LASSERRE
Institut de Biologie Marine
Université de Bordeaux
2 rue du Prof. Jolyet,
33120 Arcachon, France

Dr. A. LERMAN
Department of Geological Sciences
Northwestern University
Evanston, Illinois 60201, USA

Dr. P. F. LONSDALE
Marine Physical Laboratory
Scripps Institution of Oceanography
La Jolla, California 92037, USA

Dr. D. H. LORING
Bedford Institute of Oceanography
Dartmouth, Nova Scotia, Canada

Dr. A. D. McINTYRE
Department of Agriculture and Fisheries
  for Scotland
Marine Laboratory
PO Box 101
Aberdeen AB9 8DB, Scotland, UK

Dr. J. M. MARTIN
Groupe de Géologie Nucléaire
4 Place Jussieu
75230 Paris, Cedex 05, France

Mr. M. W. OWEN
Hydraulics Research Station
Wallingford, Berkshire, England, UK

Dr. M. M. PAMATMAT
Department of Fisheries
Auburn University
Auburn, Alabama 36830, USA

Dr. J. M. PARKS
Center for Marine and Environmental
  Studies
Lehigh University
Bethlehem, Penn. 18015, USA

Dr. N. B. PRICE
University of Edinburgh
Department of Geology
King's Building
West Mains Road
Edinburgh 9, Scotland, UK

Dr. D. REYSS
Centre Océanologique de Bretagne
Boite Postale No. 337.
Cedex 29 273 Brest, France

Dr. D. C. RHOADS
Department of Geology
Yale University
New Haven, Conn. 06520, USA

Prof. A. F. RICHARDS
Marine Geotechnical Laboratory
Lehigh University
Bethlehem, Penn. 18015, USA

Dr. F. L. SAYLES
Woods Hole Oceanographic Institution
Woods Hole, Mass. 02543, USA

Dr. R. C. SEITZ
Science Department
State University of New York
  and
Maritime College
Ft. Schuyler, Bronx
New York 10465, USA

Dr. J. D. SMITH
Department of Oceanography WB-10
University of Washington
Seattle, Washington, 98195, USA

Dr. J. B. SOUTHARD
Department of Earth and Planetary Sciences
Bldg. 54-1018
Massachusetts Institute of Technology
Cambridge, Mass. 02139, USA

Dr. F. N. SPIESS
US Office of Naval Research
223/231 Old Marylebone Road
London NW1 5TH, England, UK

Dr. E. SUESS
Geologisches Institut
Universität Kiel
D-23 Kiel
Olshausenstr. 40–60, Germany

Mr. D. TAYLOR-SMITH
Marine Science Laboratories
Menai Bridge,
Gwynedd, North Wales, UK

Dr. J. M. TEAL
Woods Hole Oceanographic Institution
Woods Hole, Mass. 02543, USA

Dr. P. G. TELEKI
Coastal Engineering Research Center
Kingman Building
Ft. Belvoir, Virginia 22060, USA

Dr. J. H. J. TERWINDT
Rijkswaterstaat—Delta Dienst
PO Box 8
s'Heer Arendskerke, Netherlands

Dr. H. THIEL
Institut für Hydrobiologie und
  Fischereiwissenschaft
Hydrobiologische Abteilung
Universität Hamburg
D-2 Hamburg 50
Palmaille 55, Germany

Prof. Tj. H. VAN ANDEL
School of Oceanography
Oregon State University
Corvallis, Oregon 97331, USA

Dr. G. L. WEATHERLY
Department of Oceanography
Florida State University
Tallahassee, Florida 32306, USA

Prof. J. E. WEBB
Department of Zoology
Westfield College
Kidderpore Avenue
London NW3 7ST, England, UK

Dr. S. WELLERSHAUS
Institut für Meeresforschung
285 Bremerhaven-G
Am Handelshafen 12, Germany

Dr. F. WERNER
Geologisch-Paläontologisches Institut
Universität Kiel
D-23 Kiel
Olshausenstr. 40-60, Germany

Dr. M. WIMBUSH
Nova University Ocean Laboratory
8000 North Ocean Drive
Dania, Florida 33004, USA

Dr. T. WOLFF
Universitetets Zoologiske Museum
Universitatsparken 15
2100 Copenhagen ϕ, Denmark

Dr. R. WOLLAST
Laboratory of Industrial Chemistry
Université Libre de Bruxelles
Av. F. D. Roosevelt, 50
1050 Bruxelles, Belgium

Dr. B. ZEITZSCHEL
Institut für Meereskunde
Universität Kiel
D-23 Kiel
Düsternbrooker Weg 20, Germany

Dr. J. J. ZIJLSTRA
Nederlands Instituut voor Onderzoek
  der Zee
Postbus 59
Texel, Netherlands

Organizing Committee

Dr. I. N. McCAVE (Chairman)
School of Environmental Sciences,
University of East Anglia,
Norwich NOR 88C, England, U.K.

Prof. K. F. BOWDEN
Department of Oceanography
University of Liverpool
PO Box 145
Liverpool L69 3BX, England, UK

Dr. S. E. CALVERT
Institute of Oceanographic Sciences
Wormley
Godalming, Surrey, England, UK

Dr. E. D. GOLDBERG
Scripps Institution of Oceanography
La Jolla, California 92037, USA

Dr. E. L. MILLS
Department of Oceanography
Dalhousie University
Halifax, Nova Scotia, Canada

Dr. H. POSTMA
Nederlands Instituut voor Onderzoek
  der Zee
Postbus 59
Texel, Netherlands

Prof. Dr. E. SEIBOLD
Geologisch-Palaontologisches
Institut der Universität Kiel
Olshausenstr. 40–60
23 Kiel, Germany

SECRETARIAT

Dr. E. G. KOVACH
Deputy Assistant Secretary
General for Scientific Affairs
NATO
1110 Bruxelles, Belgium

Miss E. I. AUSTIN
Scientific Affairs Division
NATO
1110 Bruxelles, Belgium

Miss M. GEEL
Nederlands Instituut voor Onderzoek
  der Zee
Postbus 59
Texel, Netherlands

Index